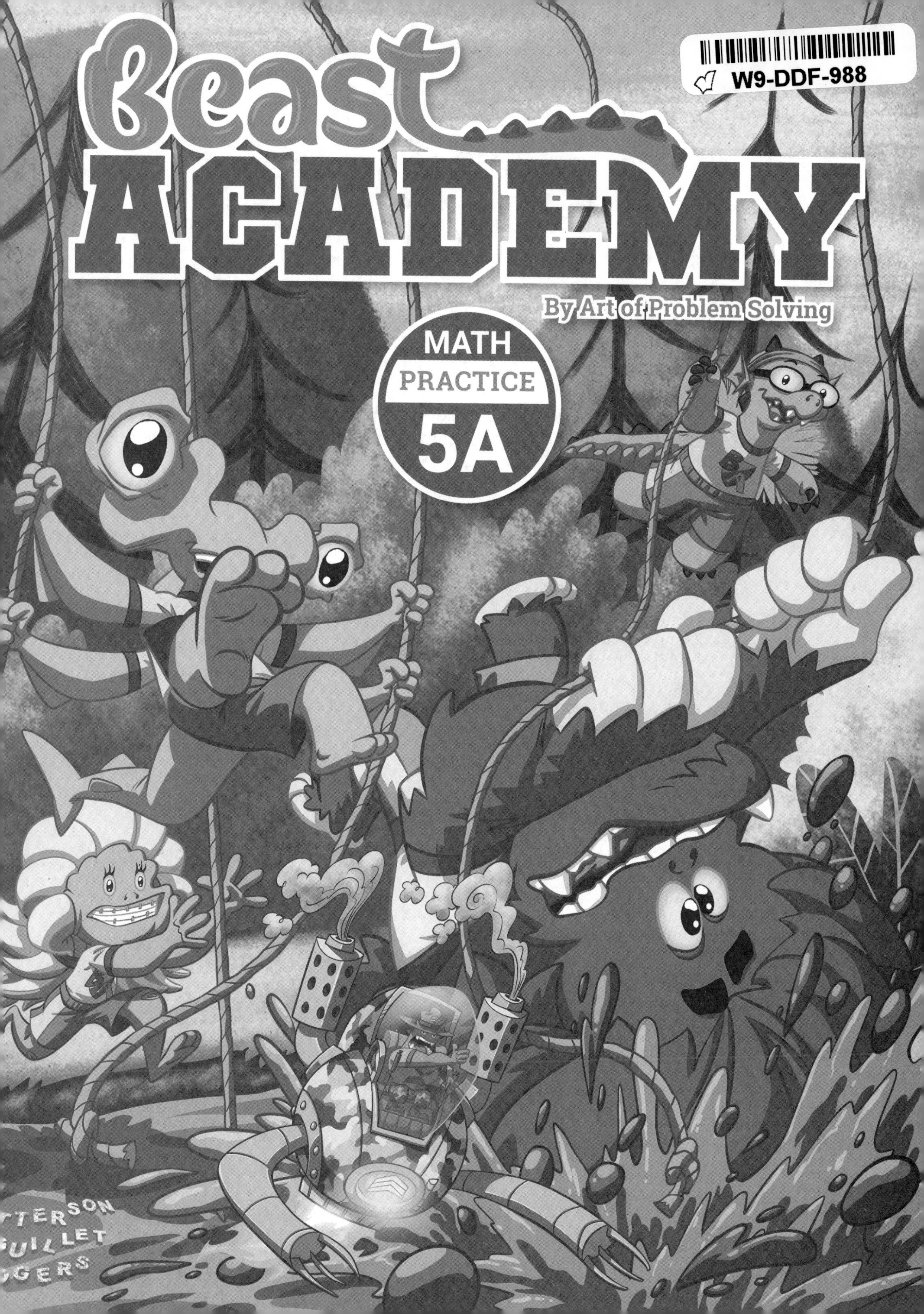

W9-DDF-988
Beast
ACADEMY
By Art of Problem Solving
MATH
PRACTICE
5A

Published by: AoPS Incorporated
10865 Rancho Bernardo Rd Ste 100
San Diego, CA 92127-2102
info@BeastAcademy.com

ISBN: 978-1-934124-61-1

Written by Jason Batterson, Shannon Rogers, and Kyle Guillet
Book Design by Lisa T. Phan
Illustrations by Erich Owen
Grayscales by Greta Selman

Visit the Beast Academy website at www.BeastAcademy.com.
Visit the Art of Problem Solving website at www.artofproblemsolving.com.
Printed in the United States of America.
2017 Printing.

Contents:

This is Practice Book 5A in a four-book series.

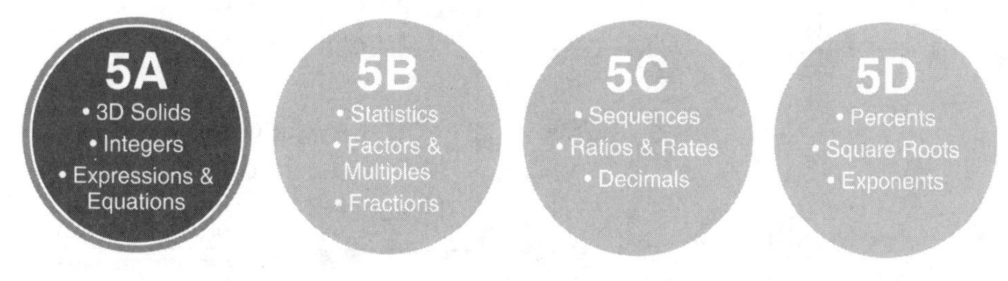

For more resources and information, visit BeastAcademy.com.

INTRODUCTION

How to Use This Book

This is Beast Academy Practice Book 5A.
Each chapter of this Practice book corresponds to a chapter from Beast Academy Guide 5A.
MATH PRACTICE 5A
MATH GUIDE 5A

The first page of each chapter includes a recommended sequence for the Guide and Practice books.
You may also read the entire chapter in the Guide before beginning the Practice chapter.

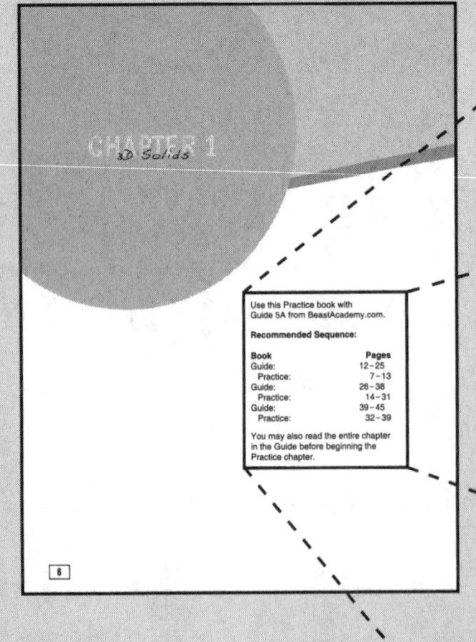

Use this Practice book with Guide 5A from BeastAcademy.com.

Recommended Sequence:

Book	**Pages**
Guide:	12–25
Practice:	7–13
Guide:	26–38
Practice:	14–31
Guide:	39–45
Practice:	32–39

You may also read the entire chapter in the Guide before beginning the Practice chapter.

Some problems in this book are very challenging. These problems are marked with a ★. The hardest problems have two stars!

Every problem marked with a ★ has a ***hint!***

Hints for the starred problems begin on page 102.

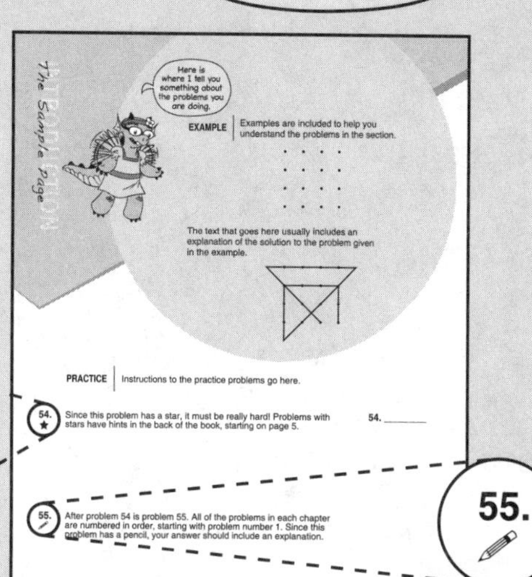

54. ★

55. ✎

Other problems are marked with a ✎. For these problems, you should write an explanation for your answer.

42	Guide Pages: 39-43

Some pages direct you to related pages from the Guide.

None of the problems in this book require the use of a calculator.

Solutions are in the back, starting on page 106.

A complete explanation is given for every problem!

Pretend Solutions 144

54. Each of the solutions in the back of this book includes a complete explanation of how to solve the problem, not just the answer. Sometimes, we will provide a solution that is different than the way you solved the problem. It is useful to read the solution to every problem even if you got the answers correct. You may learn a different way to solve the problem. Or, you may find a mistake that we made! Wouldn't it be fun to write the authors to point out a mistake they made.

55. These are just fake solutions that we wrote so that it looked like there was text here. If we put the real solutions here, it would give away some of the answers to the problems that show up in the book. Then, you wouldn't be able to have fun trying to solve them on your own! Here are some fake diagrams to go with the fake solutions:

56. If you are having trouble with a problem, it can be tempting to just look in the back of the book for the answer. The more you think about a problem, the more you learn by solving it (this is usually true even if you get the wrong answer). Do your best to figure out a problem before looking for the answer. Don't forget to use the hints for the starred problems before reading the solution.

57. Here are some more diagrams to make these fake solutions look like real ones:

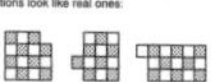

58. Here is some mathy stuff:

$51+52+53=50+50+50+1+2+3=150+6=156$.

So, the answer is **156**. Notice that the answer is bold. This makes it easier to find the answer to each problem quickly.

59. The header below separates sections of problems. This makes it easier to find the solution you are looking for.

Fake Solutions 145

60. Did you notice the number on the right side of the header above? That is the page number where you can find the problems in the Practice book. That way, when you are all finished reading a solution, you can flip back to the page you were working on.

61. You may be asking yourself, "Why did the author go to so much trouble writing text that no one can read without super-human vision?" There's no good answer, really.

62. Sometimes we give two possible ways to solve a problem. Usually, these two ways are separated by the word "or" and some lines like the ones below:

— or —

The second solution would come here. Generally, the quicker or more clever solution comes second.

63. Here is another fake solution with some diagrams. It's from a different book in the Beast Academy series. Can you figure out what is going on just by looking at the diagrams?

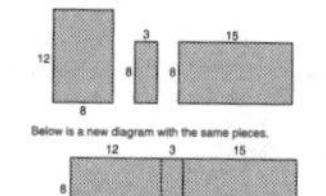

Below is a new diagram with the same pieces.

12 3 15
8

Can you guess what the question that goes with the diagrams above was? Here's a hint: The answer is **240 squares.**

64. There's another header below.

Complete Jibberish 146-148

65. The jibberish below is called Lorem Ipsum. It is complete jibberish used as placeholder text to fill the rest of the space below.

66. Curabitur velit nisl, eleifend nec interdum id, commodo id magna. Nam venenatis neque ac augue ultrices volutpat. Aenean quis diam ut ligula volutpat dapibus.

67. Makes no sense does it? I think some of those words are Latin.

68. Morbi rutrum consequat eros a rutrum. Praesent pellentesque urna ac nibh dictum consequat. Maecenas eget velit magna. Duis volutpat auctor neque sit amet congue. Aliquam erat volutpat. Suspendisse potenti. Pellentesque ut nulla lorem. Morbi consectetur purus volutpat enim vestibulum rhoncus sodales ligula aliquet.

69. Ready to get started on some math problems? Get to it!

Pretend Solutions 111

CHAPTER 1

3D Solids

Use this Practice book with
Guide 5A from BeastAcademy.com.

Recommended Sequence:

Book	**Pages:**
Guide:	12-25
Practice:	7-13
Guide:	26-38
Practice:	14-31
Guide:	39-45
Practice:	32-39

You may also read the entire chapter in the Guide before beginning the Practice chapter.

We know how to compute the perimeter and area of many flat ***two-dimensional*** shapes!
3D SOLIDS

PRACTICE | Compute the perimeter and area of each polygon. Remember to include units where necessary.

1.

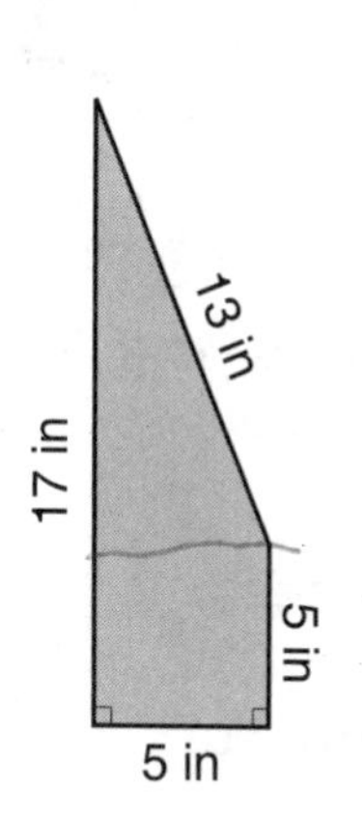

2.

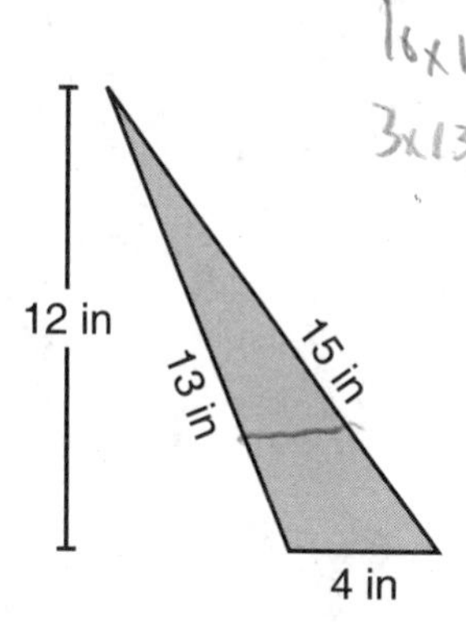

1. Perimeter = ________

Area = ________

2. Perimeter = ________

Area = ________

3. What is the perimeter of a regular pentagon with side length 35 cm?

3. ________

4. What is the area of a right triangle whose leg lengths are 7 mm and 13 mm?

4. ________

5. A square with side length 4 meters is cut out of a 5-by-6-meter rectangle. What is the area of the remaining shape?

5. ________

6. Six equilateral triangles, each with perimeter 16 ft, are attached to create a hexagon. What is the perimeter of the hexagon?

6. ________

"3D" stands for "three-dimensional." 3D objects take up space in three dimensions. We often call these three dimensions ***length***, ***width***, and ***height***.

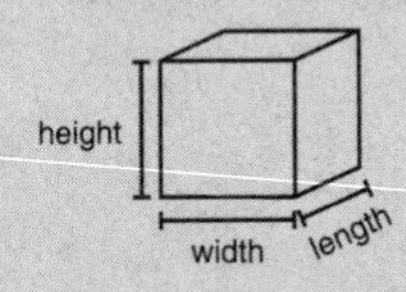

Geometric solids are three-dimensional shapes that take up space.

A flat side of a geometric solid is called a ***face***.→

A line segment where the faces of a solid meet is called an ***edge***.→

A point where edges meet is called a ***vertex***.→

A geometric solid with no curved surfaces whose faces are all polygons is called a ***polyhedron***.

The plural of vertex is ***vertices***.

The plural of polyhedron is ***polyhedra*** or ***polyhedrons***.

When we draw 3D objects on paper, we often use dashed lines to show hidden edges.

PRACTICE | Answer each question below.

7. How many faces, edges, and vertices does this polyhedron have?

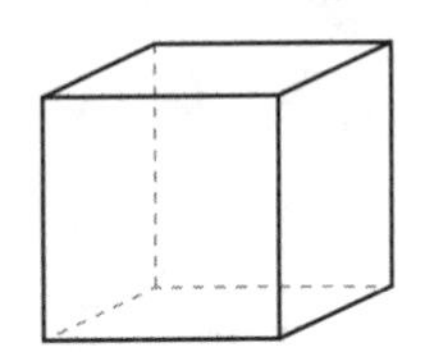

Faces: ________

Edges: ________

Vertices: ________

8. How many faces, edges, and vertices does this polyhedron have?

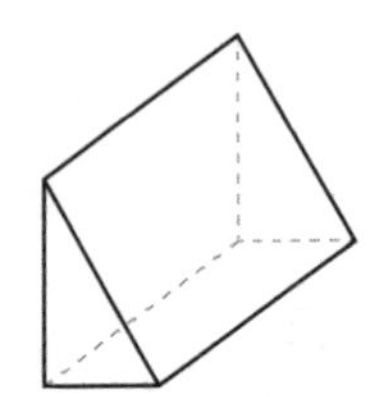

Faces: ________

Edges: ________

Vertices: ________

9. How many faces, edges, and vertices does this polyhedron have?

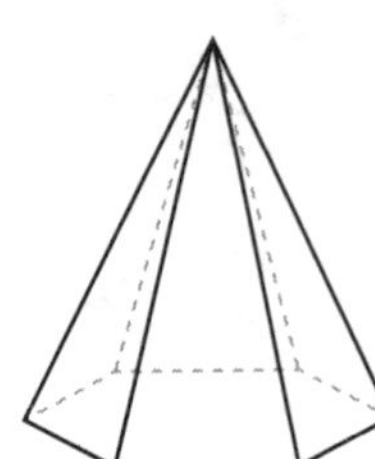

Faces: ________

Edges: ________

Vertices: ________

10. How many faces, edges, and vertices does this polyhedron have?

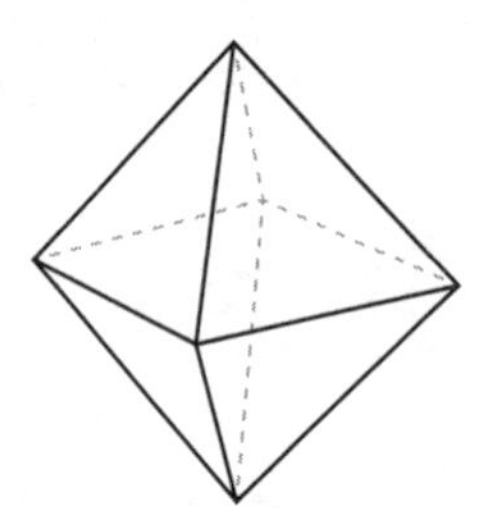

Faces: ________

Edges: ________

Vertices: ________

11. ★ Four equilateral triangles are attached to create a polyhedron with four faces. How many edges and vertices does the polyhedron have?

11. Edges: ________

Vertices: ________

A ***prism*** is a polyhedron with two congruent faces that are parallel.

The congruent faces are called the ***bases*** of the prism.
We name the prism by the shape of its bases.
The bases of a prism are connected by parallelograms.

The ***height*** of a prism is the distance between its bases.

height
height
height

Rectangular prism | Triangular prism | Heptagonal prism

The faces that connect the bases of a prism are its ***lateral faces.***
All of the prisms in this chapter are called ***right*** prisms and have lateral faces that are rectangles.

PRACTICE | Shade the bases of each prism below. Then, draw a line to connect each prism to its name.

12. **13.** **14.** **15.** **16.** **17.**

Triangular prism | Rectangular prism | Pentagonal prism | Hexagonal prism | Octagonal prism | Decagonal prism

18. Write the number of faces, edges, and vertices of each prism in the table below.

Prism	***Faces***	***Edges***	***Vertices***
Triangular Prism			
Rectangular Prism			
Pentagonal Prism			
Hexagonal Prism			
Octagonal Prism			
Decagonal Prism			

19. A nonagon is a polygon with 9 sides. How many faces, edges, and vertices does a nonagonal prism have?

19. Faces: ________

Edges: ________

Vertices: ________

A ***pyramid*** is a polyhedron that has one polygon as a base. All other faces of the pyramid are triangles that meet at a single vertex called the pyramid's ***apex***.

Like a prism, a pyramid is named by the shape of its base.

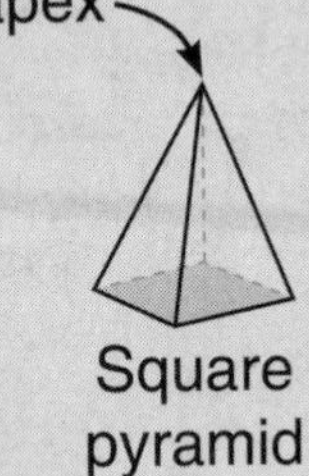

Square pyramid

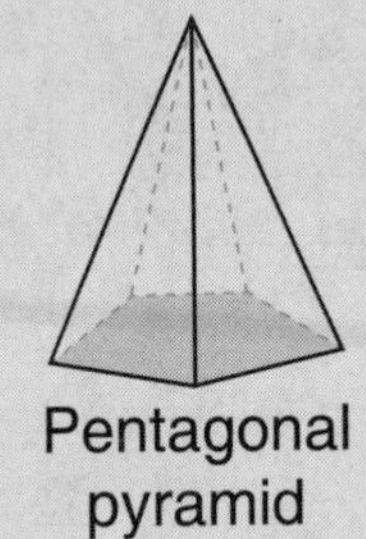
Pentagonal pyramid

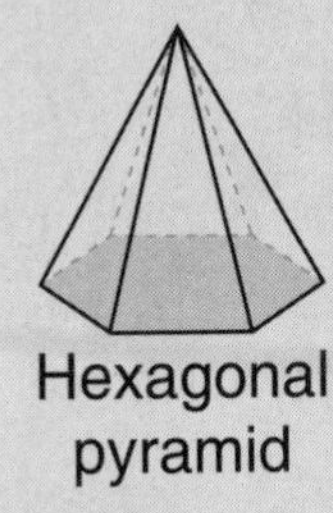
Hexagonal pyramid

PRACTICE | Shade the base of each pyramid below. Then, draw a line to connect each pyramid to its name.

20. 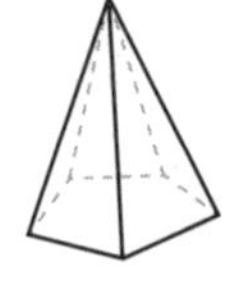

21. 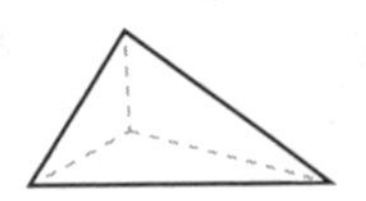

22. 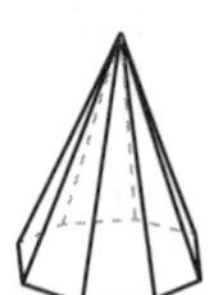

23. 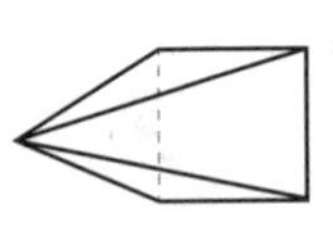

24. 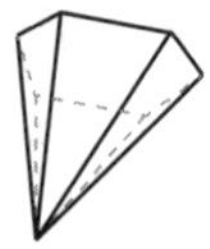

Triangular pyramid | Square pyramid | Pentagonal pyramid | Hexagonal pyramid | Octagonal pyramid

25. Write the number of faces, edges, and vertices of each pyramid in the table below.

Pyramid	*Faces*	*Edges*	*Vertices*
Triangular Pyramid			
Square Pyramid			
Pentagonal Pyramid			
Hexagonal Pyramid			
Octagonal Pyramid			

26. A heptagon is a polygon with 7 sides. How many faces, edges, and vertices does a heptagonal pyramid have?

26. Faces: ________

Edges: ________

Vertices: ________

27. An icosagon is a polygon with 20 sides. How many ***more*** edges does an icosagonal prism have than an icosagonal pyramid?

27. ________

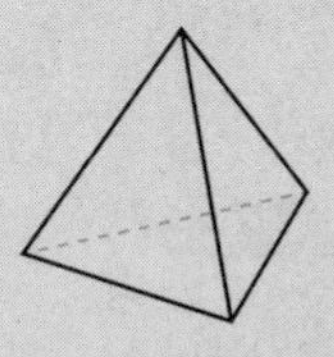

A ***regular tetrahedron*** is a special type of pyramid whose four faces are equilateral triangles.

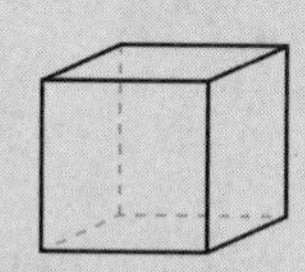

A ***cube*** is a special type of rectangular prism in which all six faces are squares.

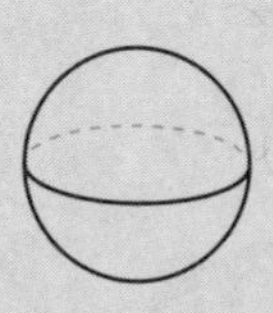

A ***sphere*** is a round 3D object, like a ball.

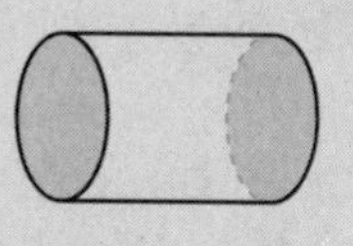

A ***cylinder*** is like a prism, but with circles as its bases.

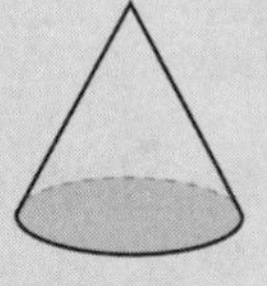

A ***cone*** is like a pyramid, but with a circle as its base.

PRACTICE | Answer each question below.

28. Which solids at the top of this page are polyhedra? Which are not? Explain. (You can review the definition of a polyhedron on page 8.)

29. A solid rubber sphere is cut into two pieces with one straight cut. What is the shape of the new flat sides of both pieces? Does it matter where the sphere is cut?

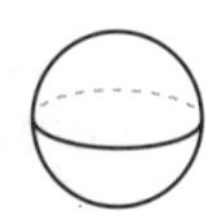

30. ★ Phyllis splits a ***cylinder*** of cake into two pieces with one straight cut. Circle all of the shapes below that could be the flat sides of both new pieces.

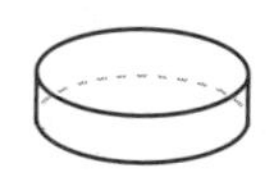

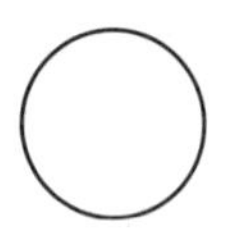 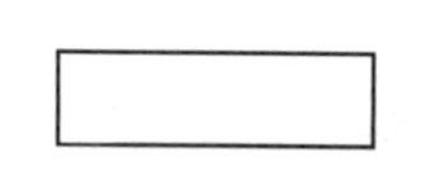 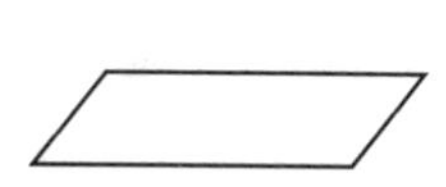 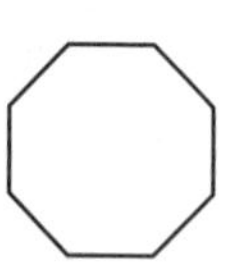 

31. ★ James splits a ***cone*** of cake into two pieces with one straight cut. Circle all of the shapes below that could be the flat sides of both new pieces.

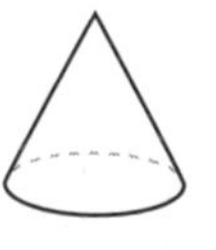

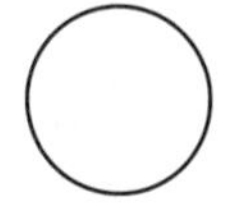 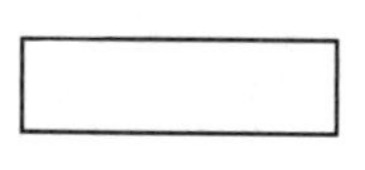 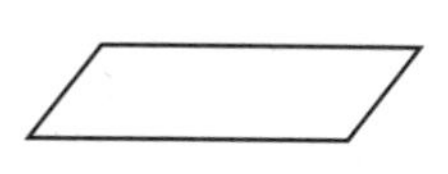  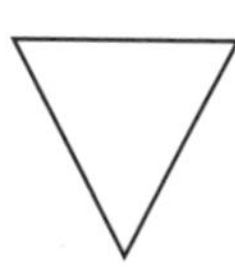

PRACTICE | Answer each question below. Include units where necessary.

32. A prism has exactly 10 faces. What is the shape of its base?

32. ________

33. A prism has exactly 18 edges. What is the shape of its base?

33. ________

34. ★ A prism has exactly 22 vertices. How many faces does it have?

34. ________

35. ★ Ed draws a prism whose bases are congruent polygons with n sides each. Fill in each blank on the right with an expression below to describe the number of faces, edges, and vertices of Ed's prism. An expression may be used more than once.

$n+1$ $\quad$ $2\times n$ $\quad$ $n+2$ $\quad$ $3\times n$ $\quad$ $n+3$

35. Faces: ________

Edges: ________

Vertices: ________

36. Alex draws a prism whose bases are regular heptagons. The height of his prism is 24 inches, and the perimeter of each base is 28 inches. What is the area of one lateral face of the prism?

36. ________

37. A hole in the shape of a rectangular prism is cut out of a cube, as shown below. How many faces, edges, and vertices does the resulting solid have?

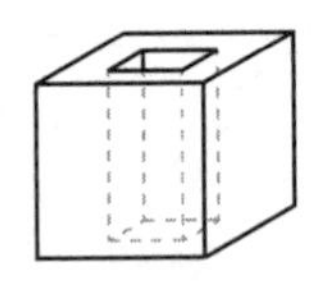

37. Faces: ________

Edges: ________

Vertices: ________

PRACTICE | Answer each question below. Include units where necessary.

38. What is the shape of the base of a pyramid that has exactly 10 faces? **38.** ________

39. What is the shape of the base of a pyramid that has exactly 20 edges? **39.** ________

40. ★ How many faces does a pyramid with exactly 22 vertices have? **40.** ________

41. ★ Al draws a pyramid whose base is a polygon with n sides. Fill in each blank on the right with an expression below to describe the number of faces, edges, and vertices of Al's pyramid. An expression may be used more than once.

41. Faces: ________

Edges: ________

Vertices: ________

$n+1$ $2\times n$ $n+2$ $3\times n$ $n+3$

42. The base of a pyramid is a regular pentagon. Each triangular face of the pyramid is an equilateral triangle with a perimeter of 15 cm. What is the sum of the lengths of all the edges of the pyramid? **42.** ________

43. ★ Phil draws a polyhedron with 12 edges. Can you tell how many faces or vertices the polyhedron has? If so, how many? If not, describe two different polyhedra that each have 12 edges but different numbers of faces or vertices.

Nets

A ***net*** is a 2-dimensional shape that can be folded to make the surface of a 3-dimensional solid.

For example, below is a net for a tetrahedron. We can fold the net along the dashed lines as shown to create a tetrahedron.

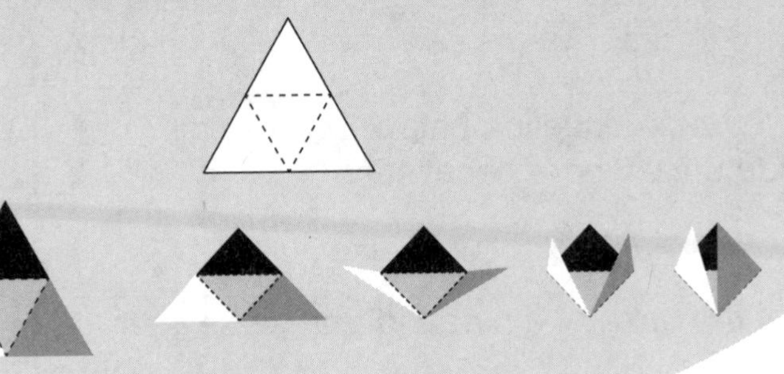

PRACTICE | Draw a line to connect each net below with its corresponding solid.

44. 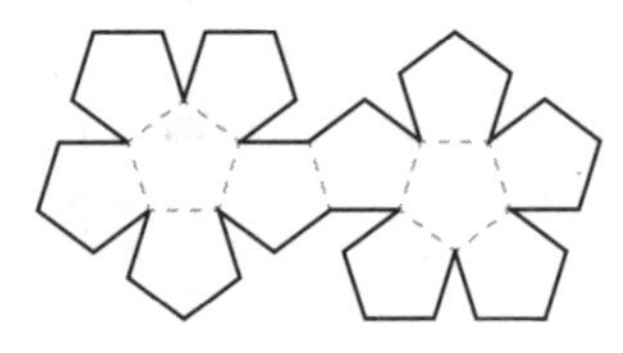

45. 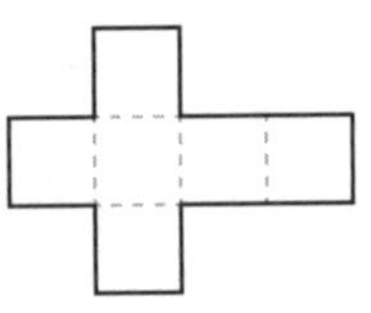

46. 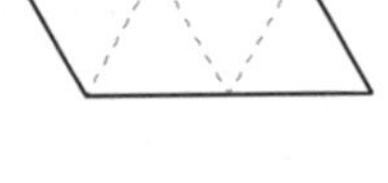

47. 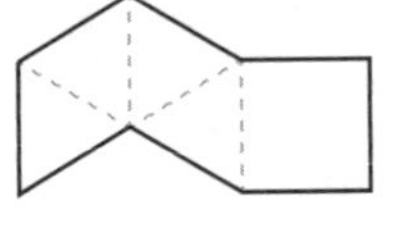

48. 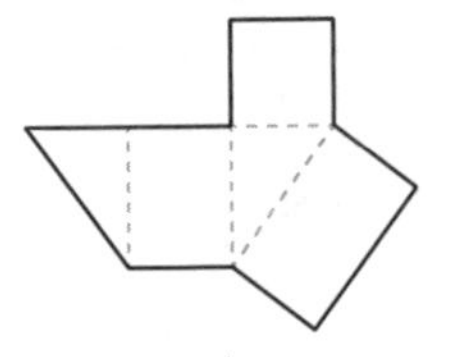

49. 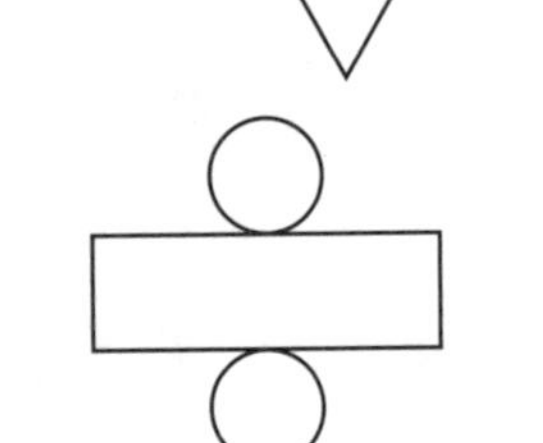

50.

Triangular Prism 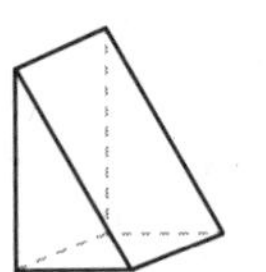

Square Pyramid 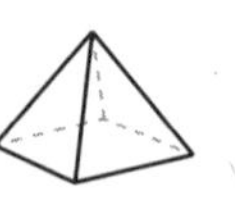

Cube 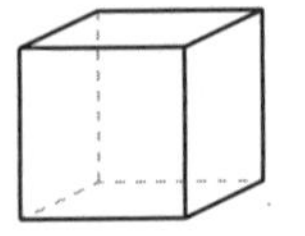

Dodecahedron 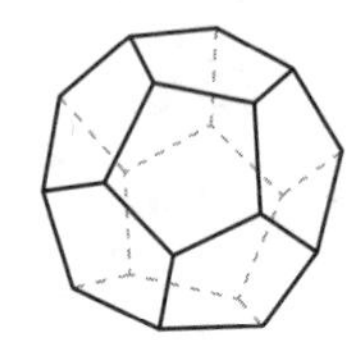

Cylinder 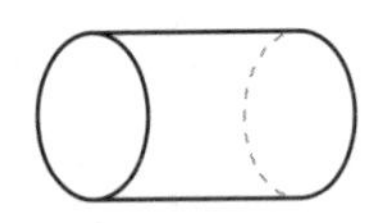

Regular Tetrahedron 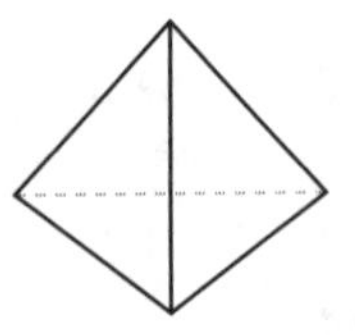

Regular Octahedron 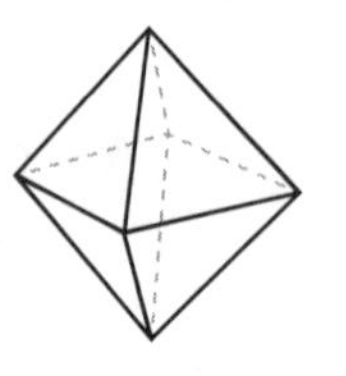

An *icosahedron* is a solid that has 20 faces that are all equilateral triangles.

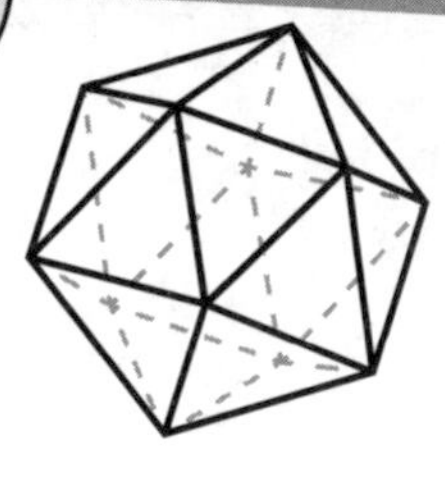

Cut out the shape below along the solid lines. Then, fold along the dashed lines and tape to create your own icosahedron!

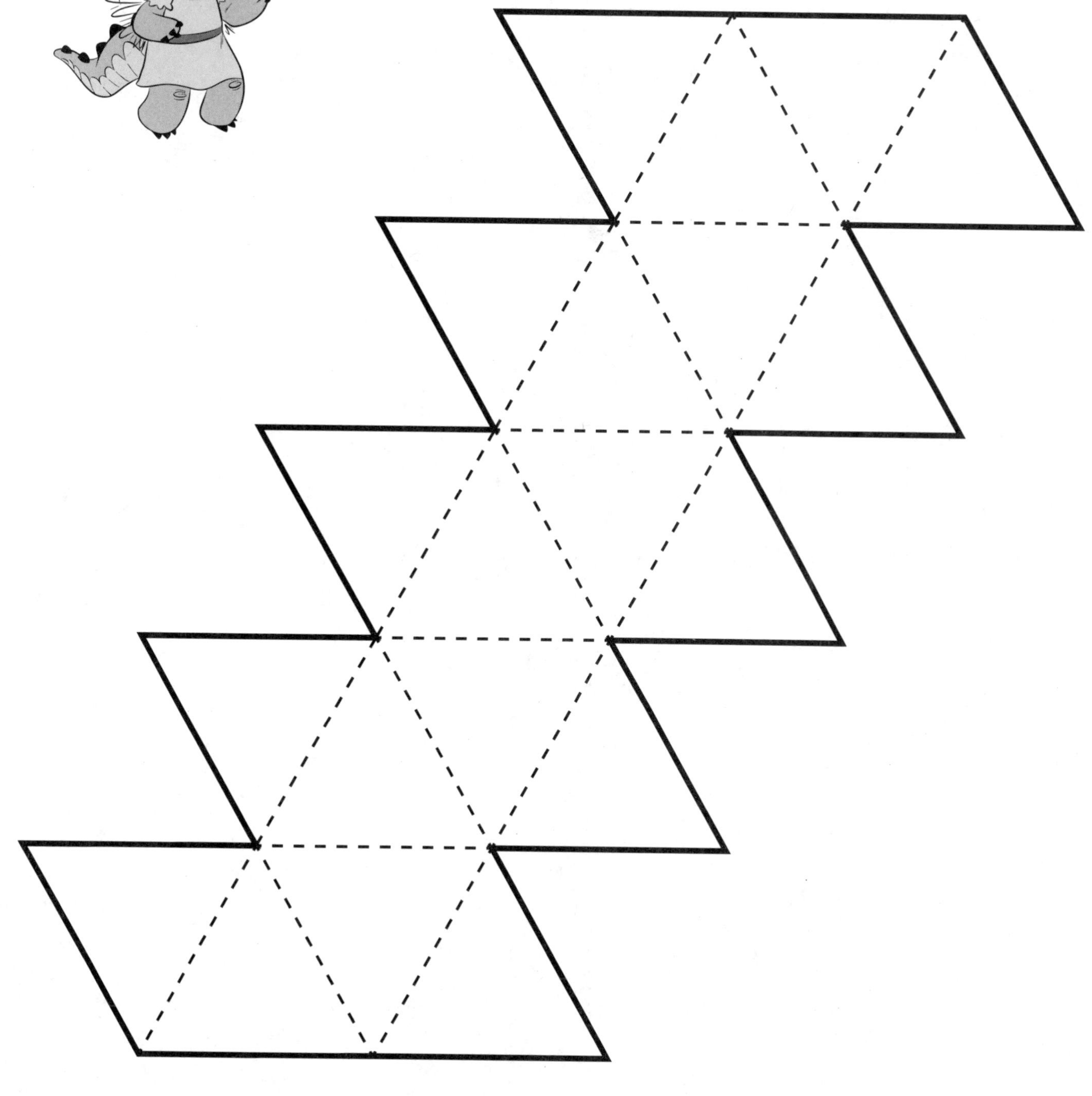

Find more nets to cut out at BeastAcademy.com!

In a **This End Up** puzzle, the goal is to label the faces of a cube net so that when the shaded face is on the bottom, the top face contains a circle, and the four remaining faces have arrows that point to the top.

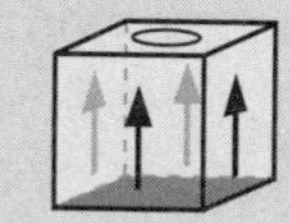

EXAMPLE | Solve the This End Up puzzle to the right.

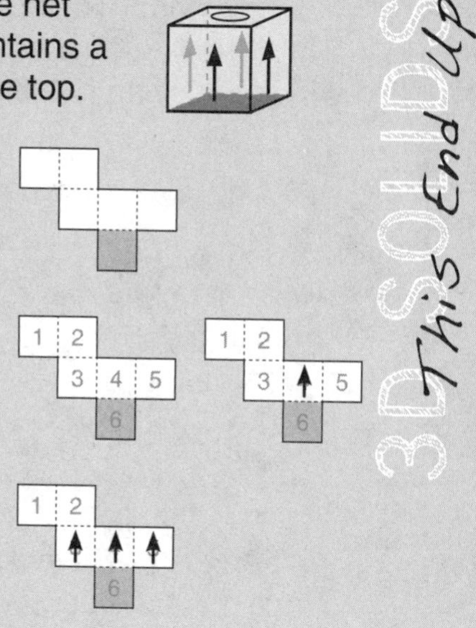

We first note that any arrow that points ***away*** from the bottom of a cube points ***toward*** the top of the cube.

We number the squares of the net as shown. Since face 6 is the bottom of the cube, face 4 must have an arrow that points away from face 6.

The arrow on face 4 does not point toward or away from faces 3 or 5. So, faces 3 and 5 are both sides of the cube (not top or bottom). Therefore, both faces must contain arrows. Those arrows must point the same direction as the arrow on face 4.

The arrow on face 3 points to face 2. Therefore, face 2 is the top of the cube. We place a circle on face 2. Finally, we place an arrow on face 1 that points to face 2.

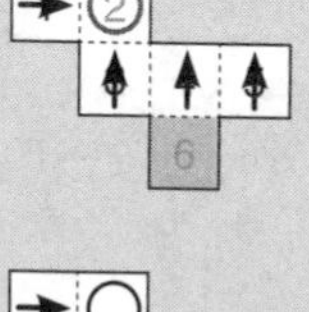

Our final net is shown to the right.

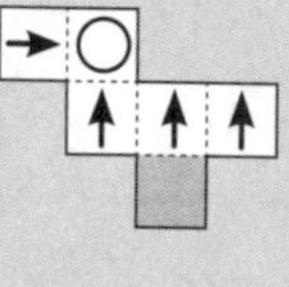

Below, we show how the net can be folded into a cube.

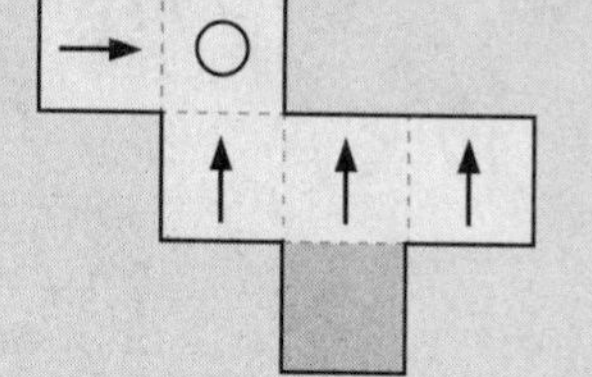
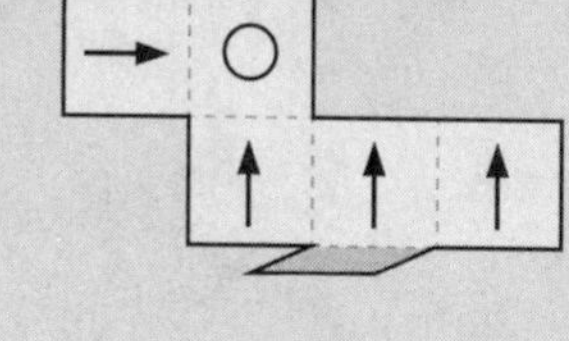
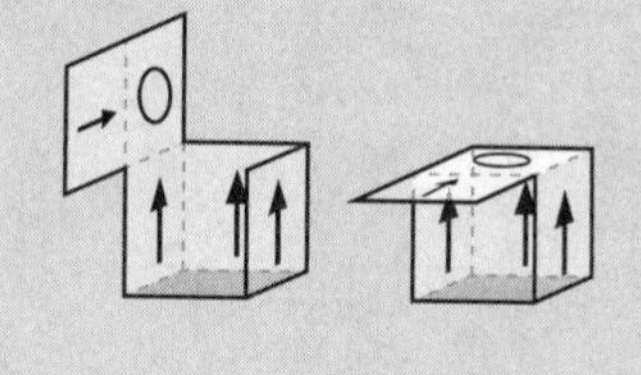
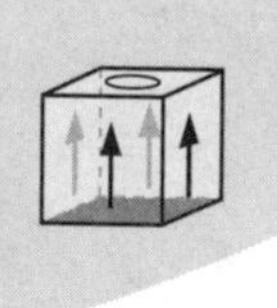

PRACTICE | Solve each This End Up puzzle below. In some puzzles, the bottom is noted with a shaded square. In others, an arrow has been given on one side.

51.

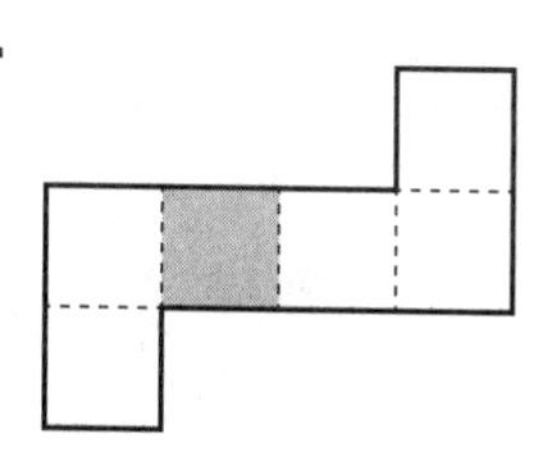

52.

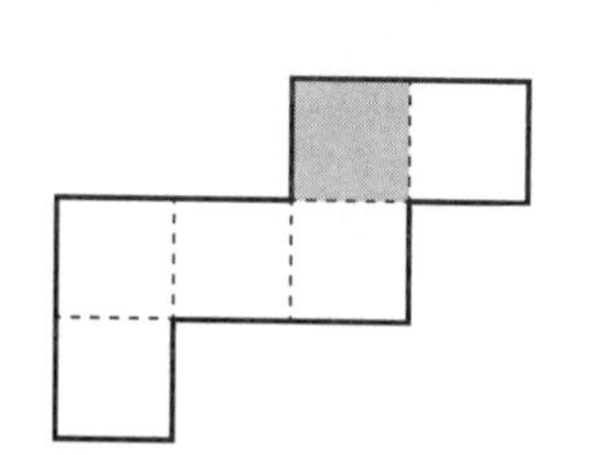

53.

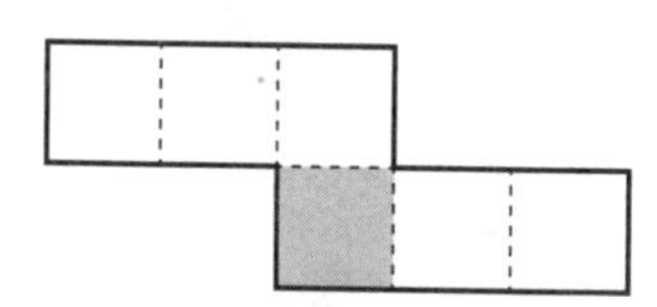

54.

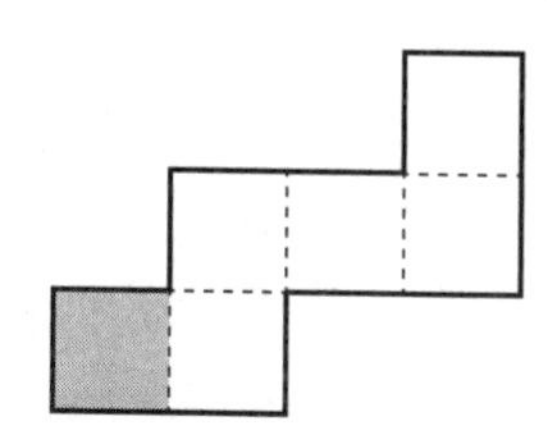

55.

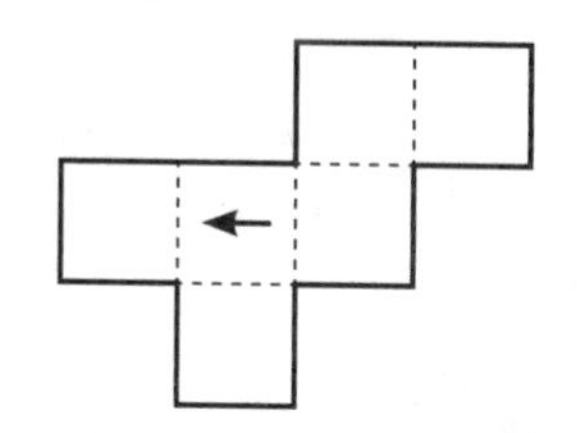

56.

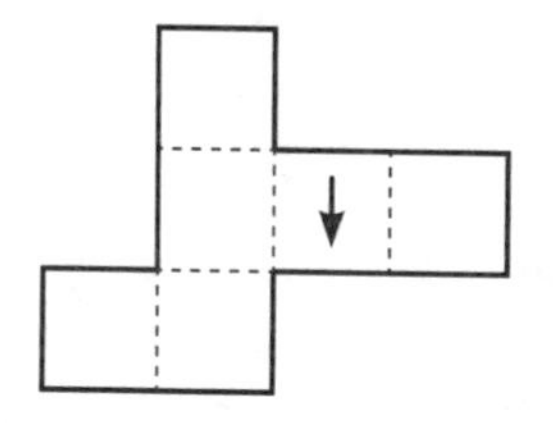

On a standard 6-sided die, the sum of the numbers on opposite faces is always 7.
In a **Die Net** puzzle, the net of a die is given with a few numbers filled in. The goal is to fill in the missing numbers on the net so that the numbers on opposite faces always sum to 7.

EXAMPLE | Solve the Die Net puzzle to the right.

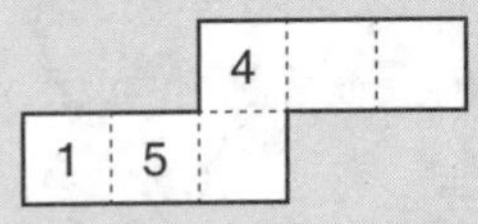

We label the blank faces A, B, and C. Then, we visualize folding the net.
Below is one way to visualize the net being folded into a cube:

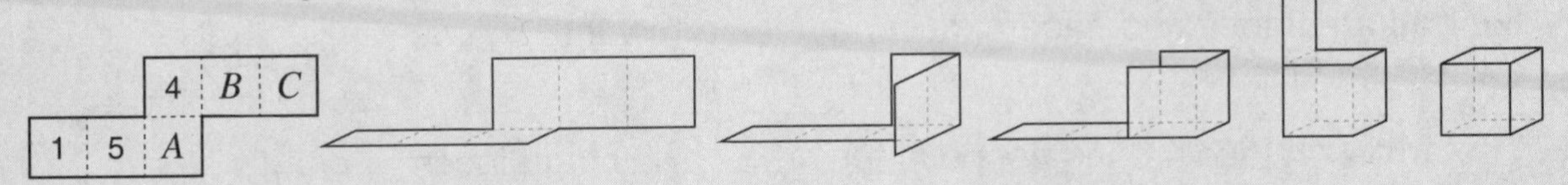

When the net is folded as shown above, face A is the bottom, face 4 is the back, face B is the right side, face C is the front, face 5 is the left side, and face 1 is the top.

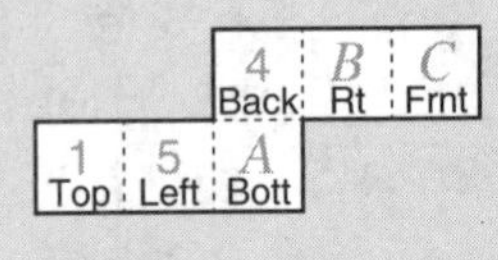

So, face A is opposite face 1, face B is opposite face 5, and face C is opposite face 4. We label the faces as shown:
$A=7-1=\mathbf{6}$, $B=7-5=\mathbf{2}$, and $C=7-4=\mathbf{3}$.

4 **2** **3**
1 5 **6**

If you're having trouble visualizing, you can label the net as we did in **This End Up** to identify pairs of opposite faces. For example, if face 5 is the bottom, we can label the net as shown to see that face B is the top. So, face B is opposite face 5.

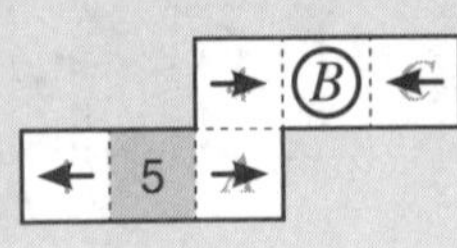

You may have visualized folding the cube net in a different way.
Folding the cube net differently will not change which faces are opposites.

PRACTICE | Solve each Die Net puzzle below.

57.

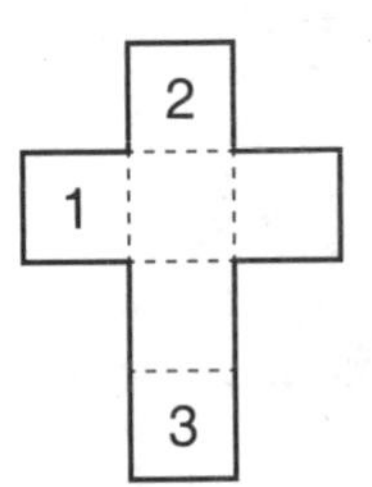

58.

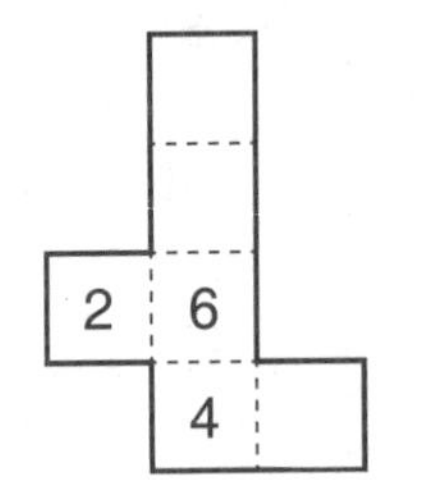

59.

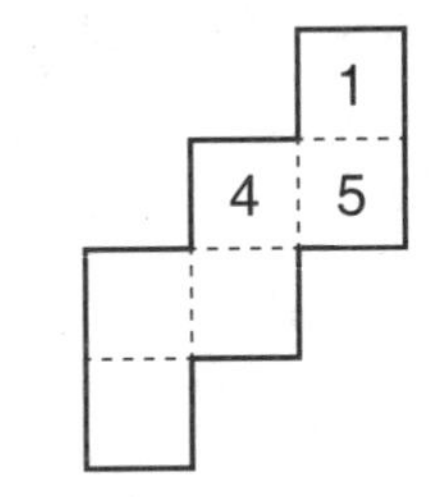

60.

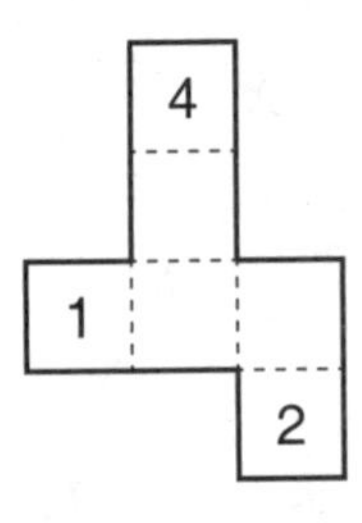

61.

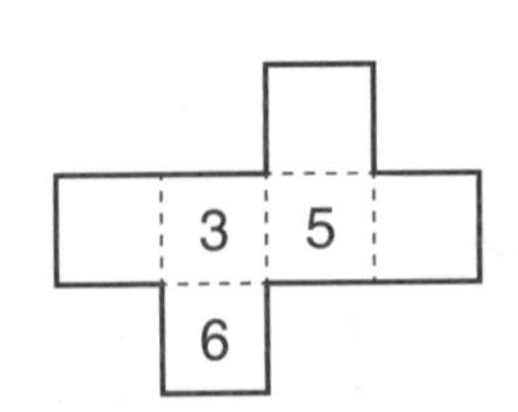

62.

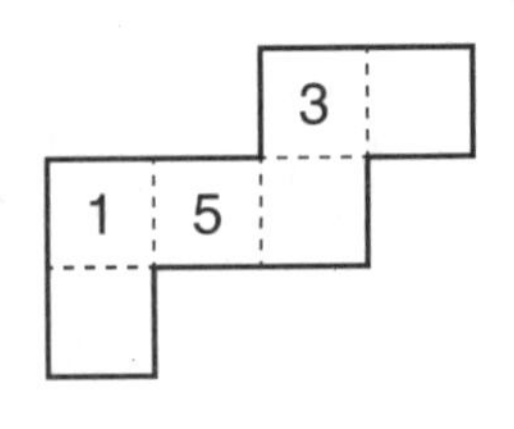

63. The three-square arrangement to the right is part of a die net. What is the sum of the two numbers that could fill the blank square?

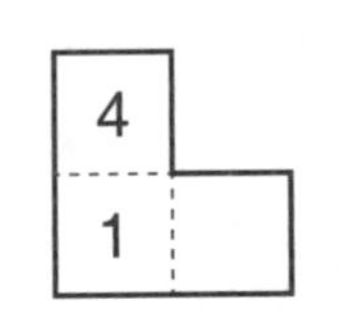

63. ______

A ***heptomino*** is a polygon made by attaching 7 congruent squares along their sides. Some heptominoes can be folded into a cube in which exactly two of the squares overlap.

EXAMPLE | Which two squares overlap when we fold this heptomino into a cube?

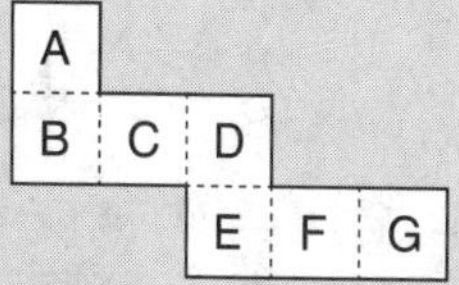

Below is one way to visualize the net being folded into a cube:

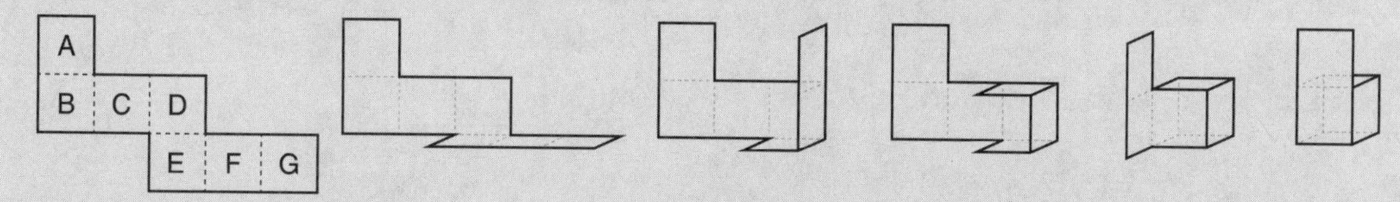

When the net is folded as shown above, face D is the back, face E is the bottom, face F is the right side, face G is the top, face C is the left side, face B is the front, and face A is the top. You may have folded differently, but the pairs of opposite faces will be the same.

Faces A and G are both the top of the cube when face E is the bottom. So, when this heptomino is folded into a cube, **faces A and G** overlap.

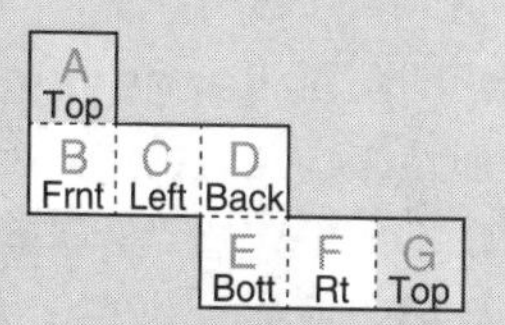

To check, we can use the strategy from **This End Up** to see that when face E is the bottom, faces A and G are both the top and must overlap.

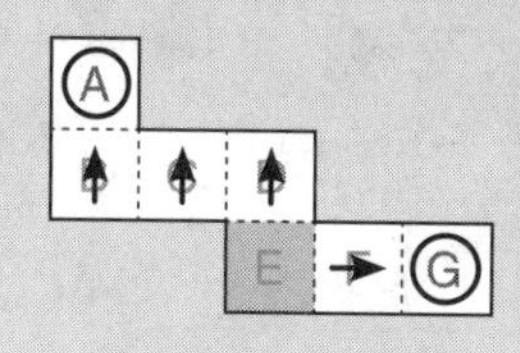

PRACTICE | Circle the letters on the two faces that overlap when we fold each heptomino below into a cube.

64.

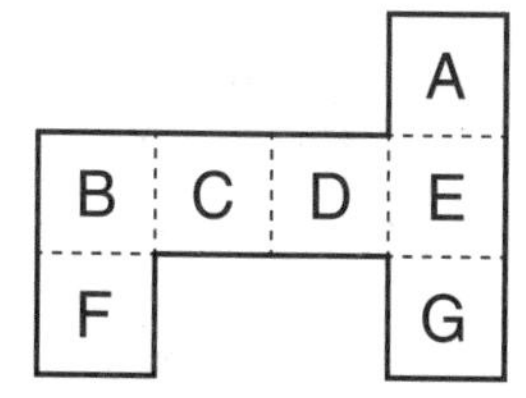

65.

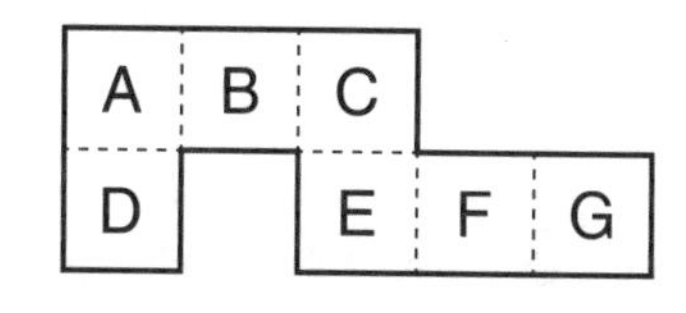

66.

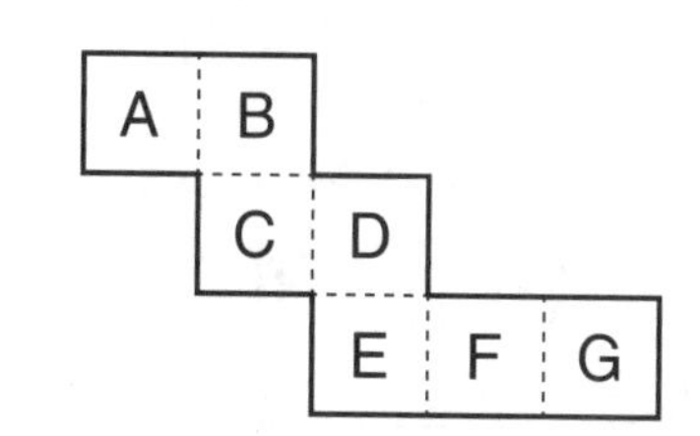

67.

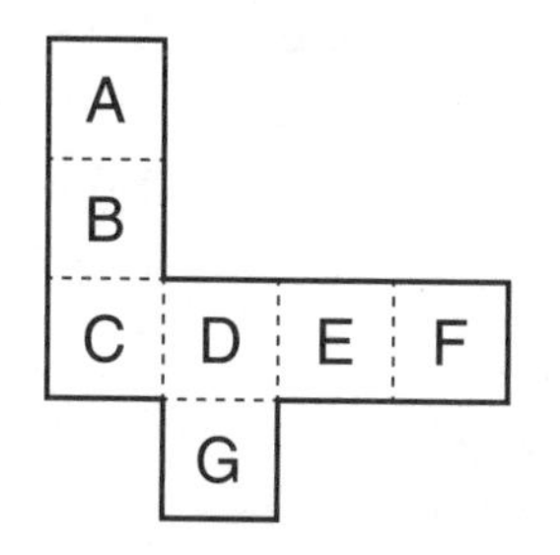

68.

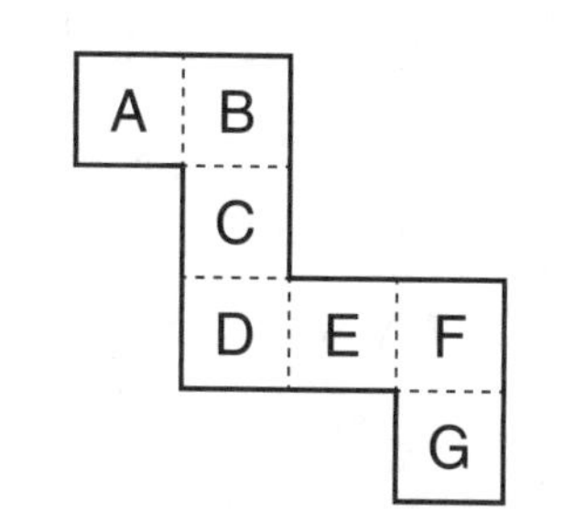

69.

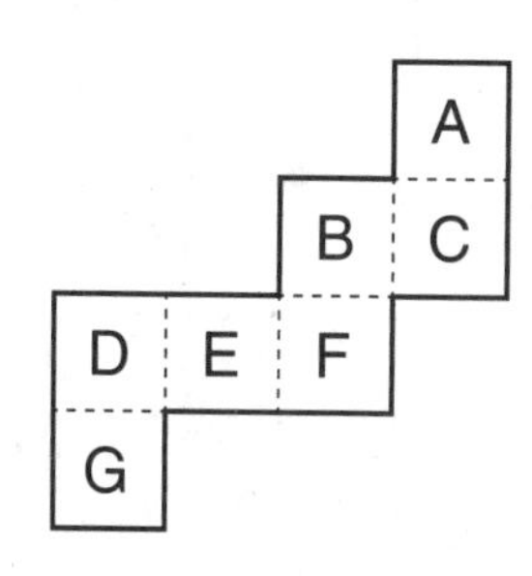

A **Dot Cube** has dots painted on two of its opposite vertices, as shown below.

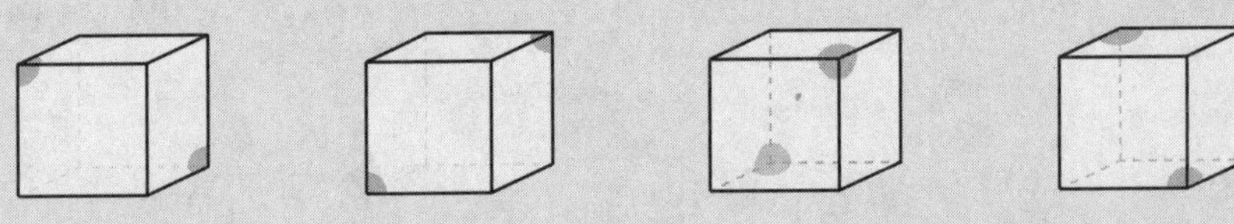

In a **Dot Cube** puzzle, a net of a Dot Cube is shown with some paint missing. The goal is to draw the rest of the paint so that when the net is folded into a cube, two opposite vertices of the cube have dots painted on them.

EXAMPLE | Solve the Dot Cube puzzle to the right.

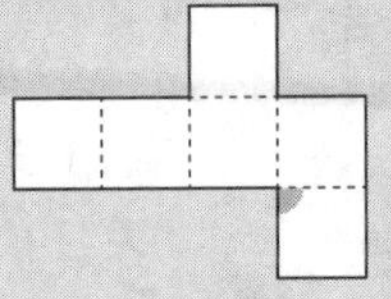

We label each square with a number as shown. One of the painted vertices is the upper-left corner of face 6. Faces 4, 5 and 6 meet at this vertex, so the bottom-right corner of face 4 and the bottom-left corner of face 5 must be painted to complete the spot on this vertex.

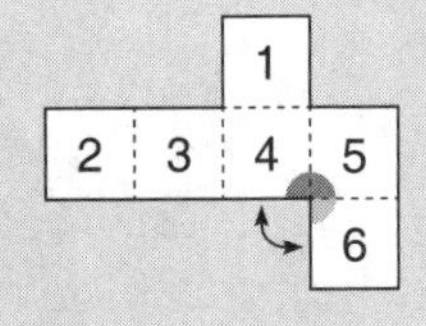

Exactly three faces meet at every vertex of a cube, and each square has some corner painted. If one vertex where three faces meet is painted, then the other painted vertex is where the three ***opposite*** faces meet. So, the other dot must be painted on the vertex where faces 1, 2, and 3 meet.

When we fold this net into a cube, the upper-left corner of face 1 will be attached to the upper-left corner of face 3 and the upper-right corner of face 2. We paint these corners to complete the dot on this vertex.

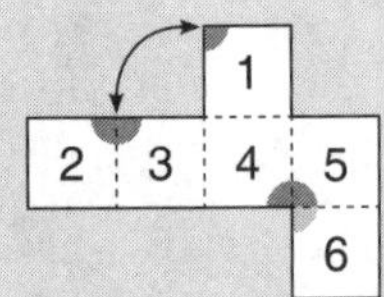

Below, we show one way that the net can be folded into a cube.

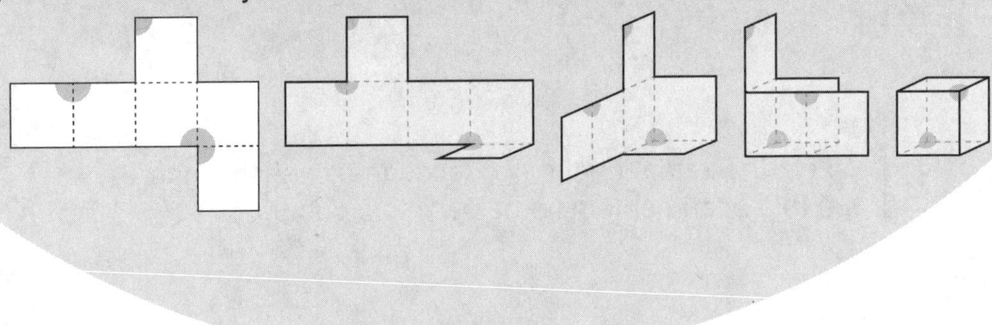

PRACTICE | Solve each Dot Cube puzzle below.

70.

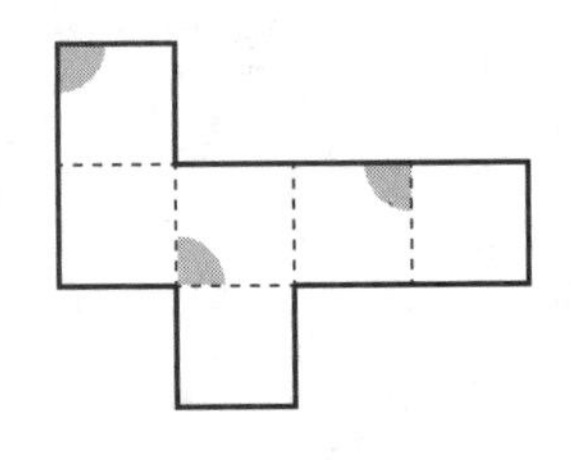

71.

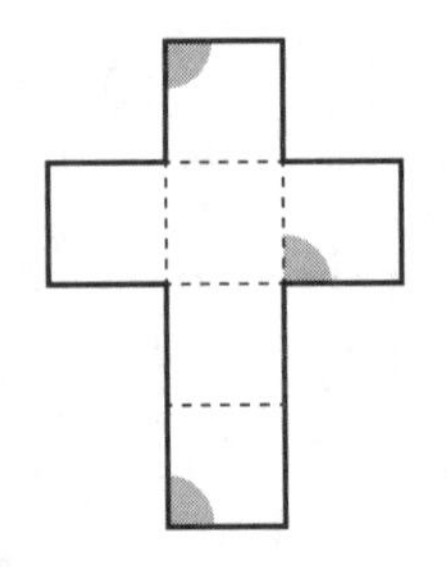

72.

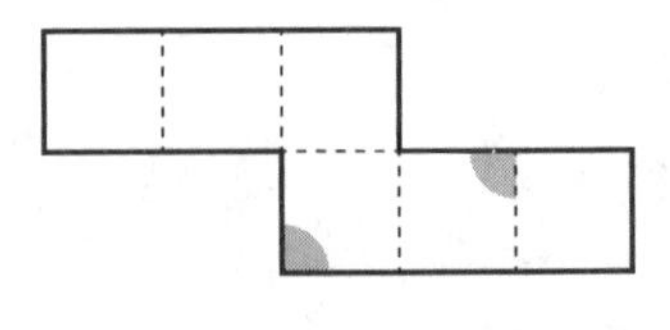

PRACTICE | Solve each Dot Cube puzzle below.

73.

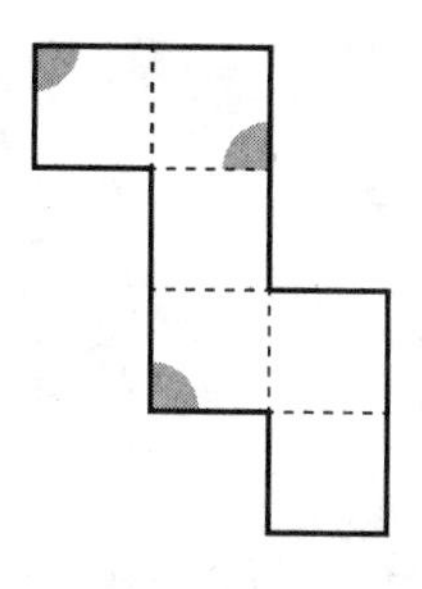

74.

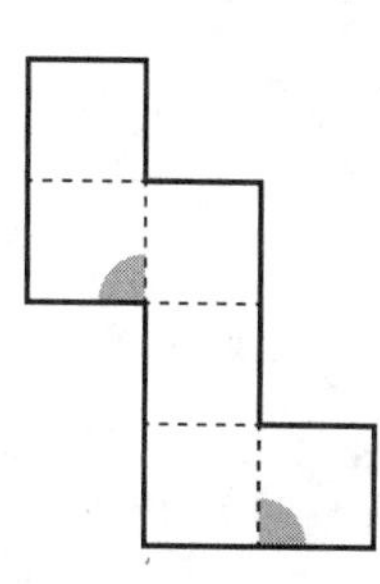

75.

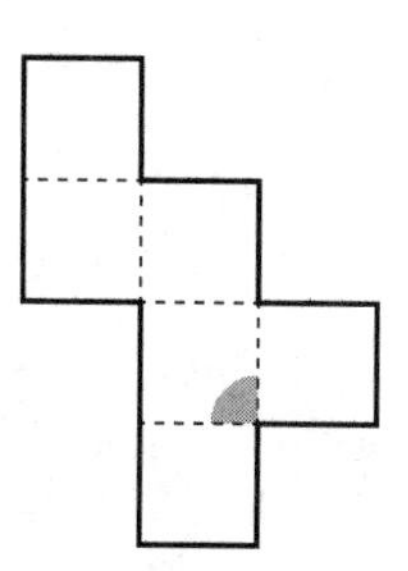

76.

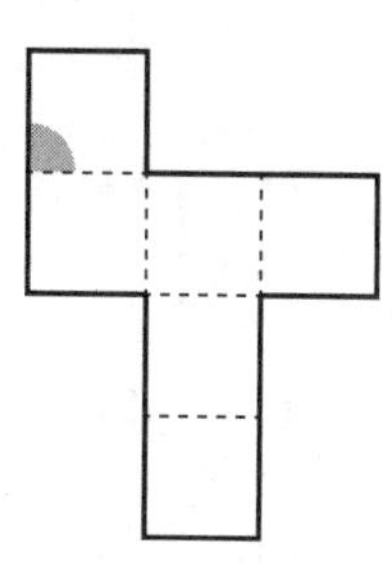

77.

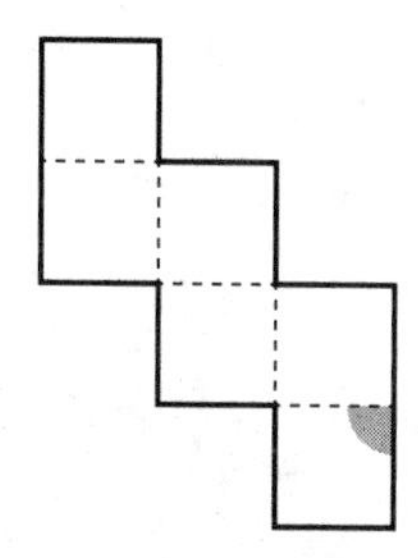

78.

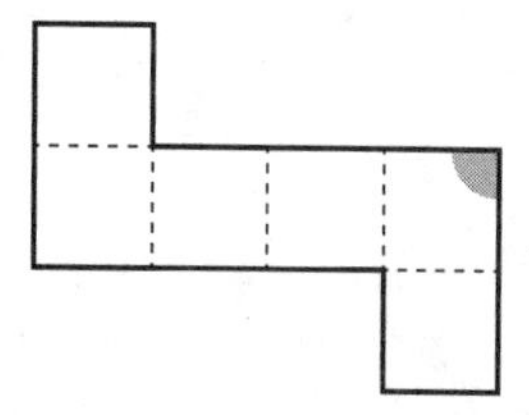

79. The two-square arrangement below is part of a Dot Cube net with one corner already painted. Which additional corner(s) of the two given squares must also be painted? Shade the appropriate corner(s) below.

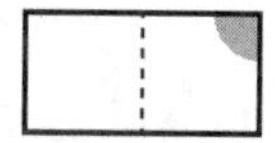

In a **Grid Net** game, players alternate drawing 1-unit squares on a 5-by-6 dot grid.
Player 1 begins by drawing a single 1-unit square anywhere on the grid.
On all other moves, a player must draw a square that is attached to one side of exactly one other square.

The game is complete when six squares have been drawn. If the six-square arrangement (hexomino) makes a cube net, then Player 2 wins. If the arrangement is ***not*** a cube net, then Player 1 wins.

Sample Game: Lizzie (Player 1) vs. Alex (Player 2)

1. Lizzie draws.

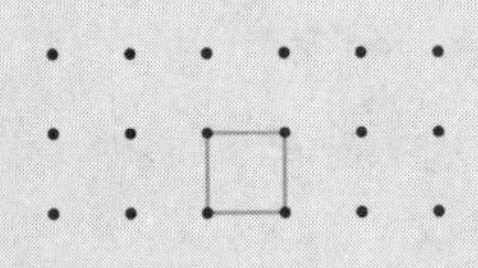

2. Alex draws.

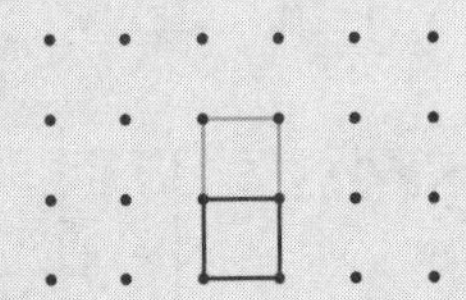

3. Lizzie draws.

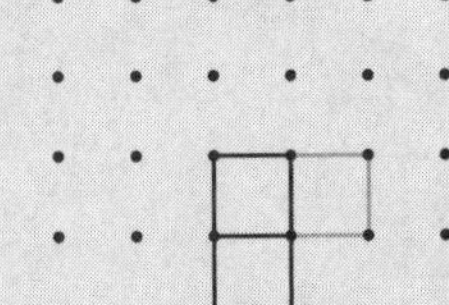

4. Alex cannot draw the square marked with × because this square would be attached to sides of ***two*** squares. He may draw any of the other squares marked with ✓. He draws the square in the lower-left.

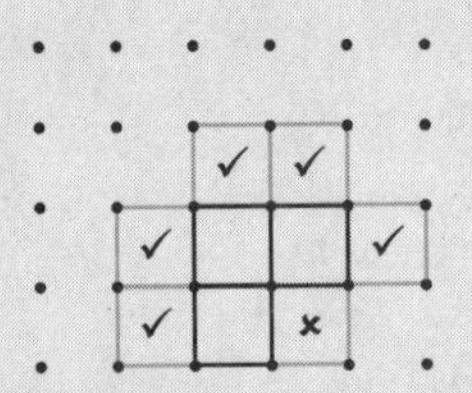

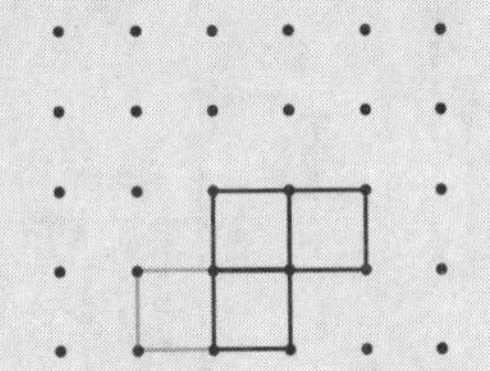

5. Lizzie draws.

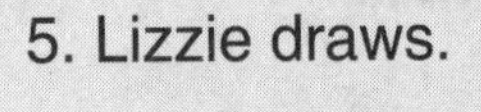

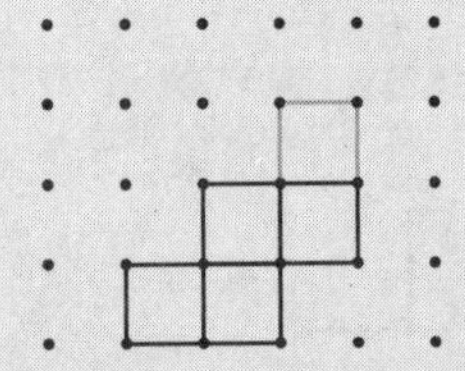

6. Alex draws.

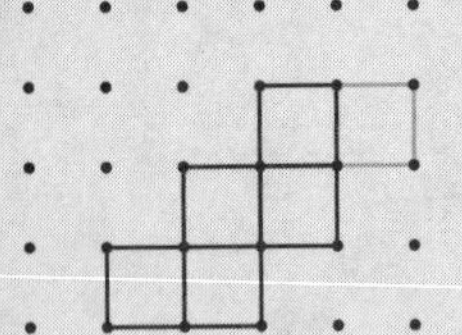

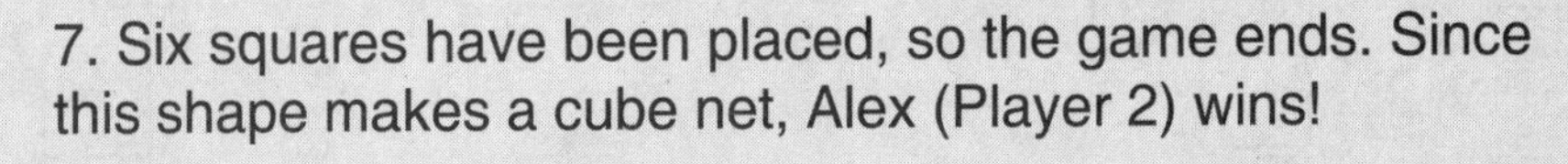

7. Six squares have been placed, so the game ends. Since this shape makes a cube net, Alex (Player 2) wins!

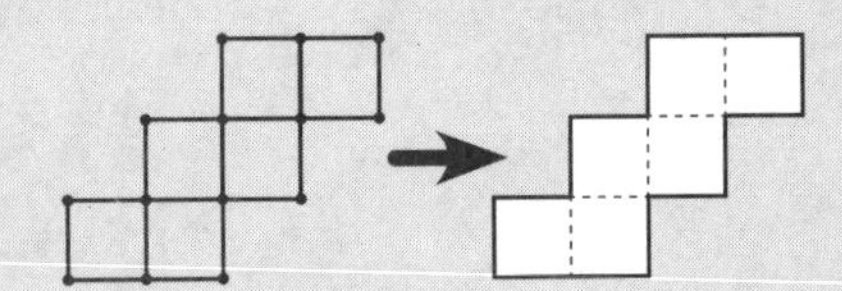

If Alex had instead placed the last square as shown on the right, the result would ***not*** be a cube net. In this hexomino, faces 1 and 4 are both opposite face 6, while face 2 has no opposite.

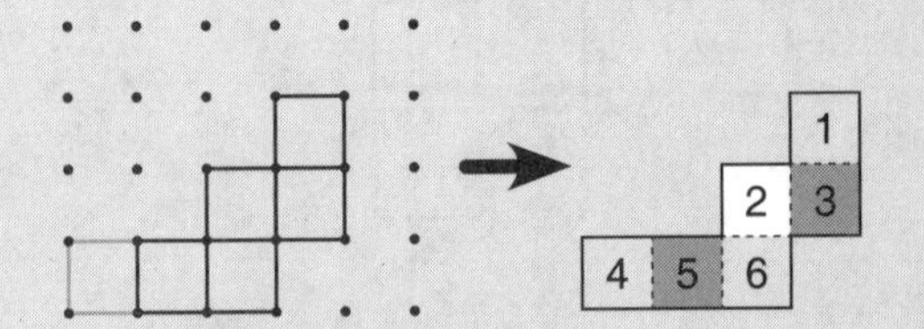

Try a few games with a friend. Print game boards at BeastAcademy.com. Then, explore this game on the following pages!

Only 11 of the 35 different hexominoes are cube nets. (If two hexominoes can be flipped or turned to look identical, we consider them the same.) Try to draw all 11 hexominoes that make a cube net to help you determine the winner of each game! Compare your drawings to the hexominoes page at BeastAcademy.com.

PRACTICE In each game below, it is ***Player 2***'s turn. Find a winning move that creates a cube net in each game below.

80.

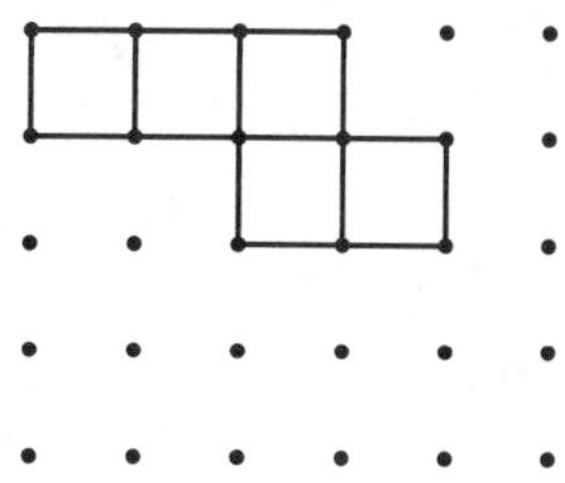

81.

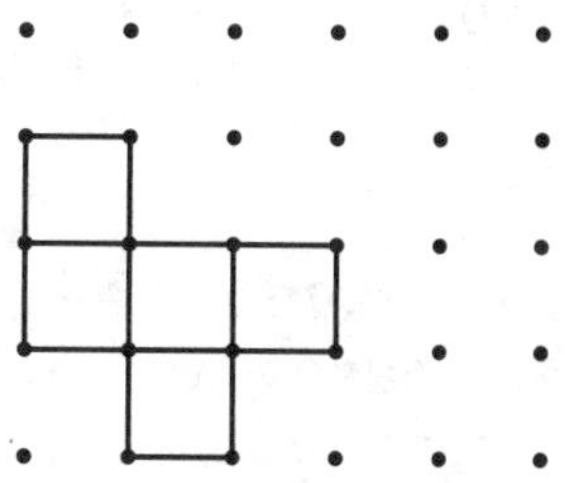

82.

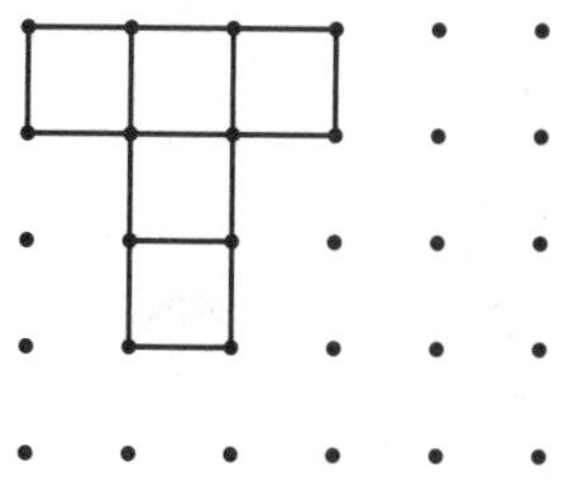

83. Player 2 has **two** winning moves in the game shown to the right. Find both winning moves.

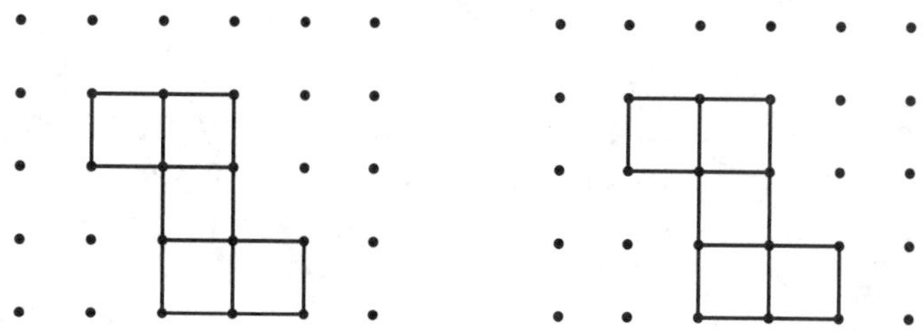

PRACTICE In each game below, it is ***Player 1***'s turn. Find a move for Player 1 that guarantees that Player 2 will not be able to create a cube net with the final move.

84.

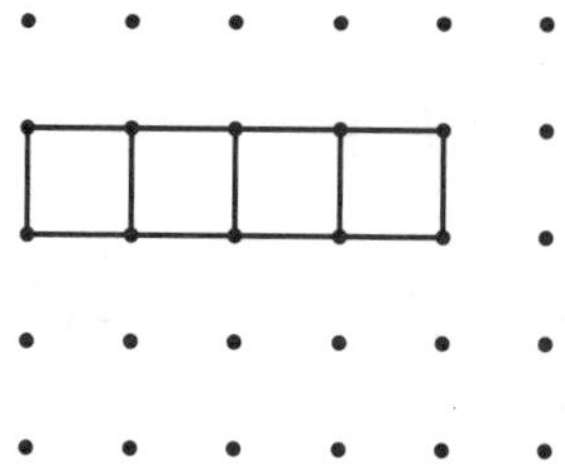

85.

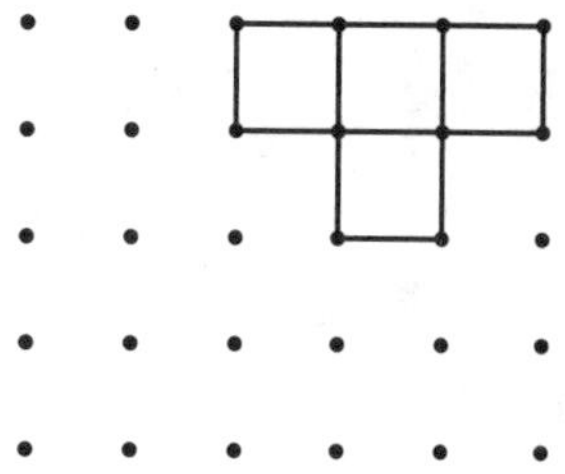

86.

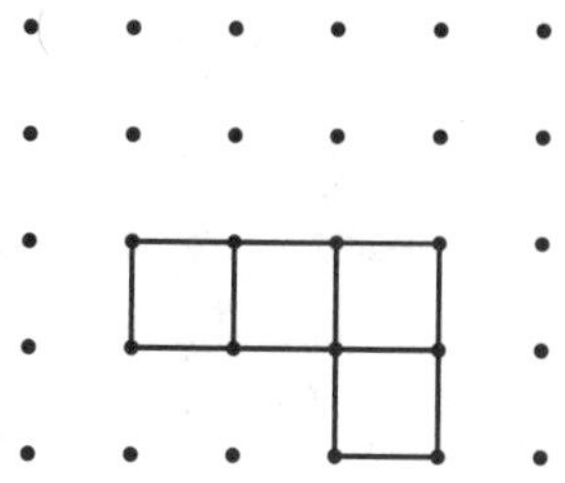

87. Player 1 has **two** winning moves in the game shown to the right. Find both winning moves.

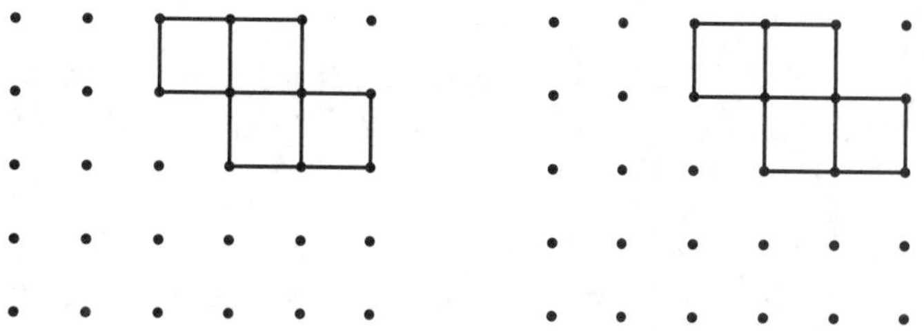

PRACTICE | Answer each question below about the game of Grid Net.

88. It is Player 1's turn to play on the board to the right. Which player is going to win this game? Explain your answer. *(Remember that Player 2 wins if a cube net is formed. Player 1 wins if a cube net is **not** formed.)*

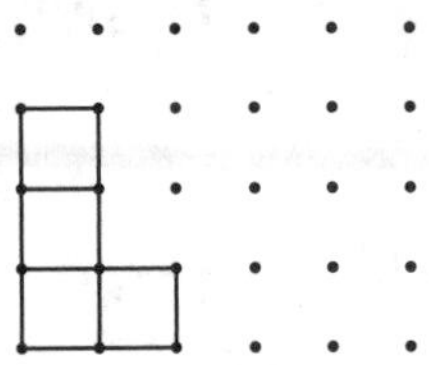

89. It is Player 2's turn to play on the board to the right. Which player is going to win this game? Explain your answer. *(Remember that Player 2 wins if a cube net is formed. Player 1 wins if a cube net is **not** formed.)*

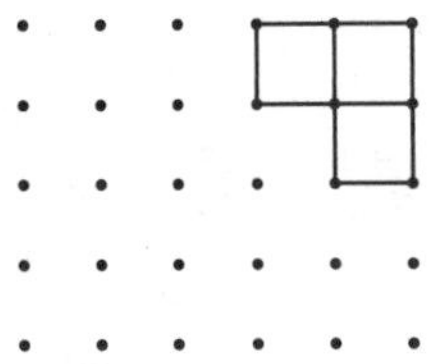

90. ★ In a game of Grid Net played on a 5-by-6 dot grid, Player 1 can always win if he or she plays correctly. Describe Player 1's winning strategy.

91. ★ It is Player 1's turn to play on the ***extended*** board below. If both players make their best moves, who is going to win this game?

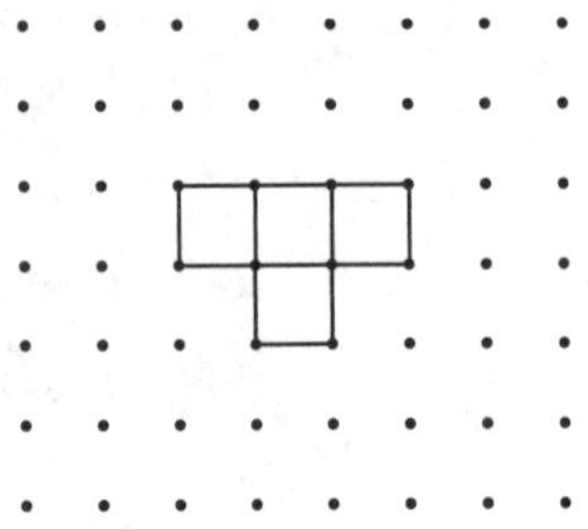

92. On the ***extended*** board below, where can Player 2 add a square to guarantee he or she will be able to create a cube net, regardless of Player 1's next move?

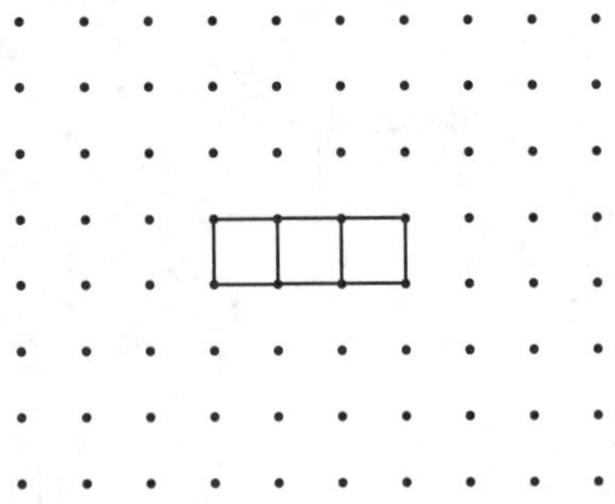

93. ★ In a game of Grid Net played ***on an unlimited dot grid***, Player 2 can always create a cube net if he or she plays correctly. Describe Player 2's winning strategy.

Platonic solids are polyhedra made up of congruent faces of regular polygons, with the same number of faces meeting at each vertex.

There are only five platonic solids. They are listed below.

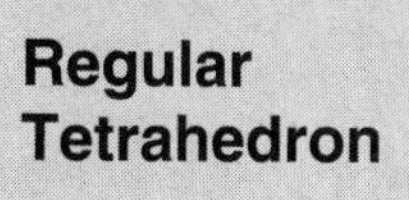

Regular Tetrahedron

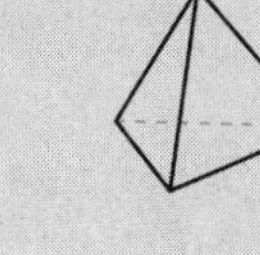

Cube

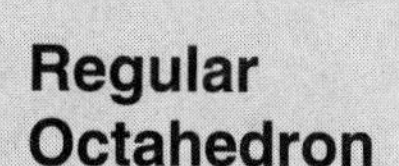

Regular Octahedron

Find nets to cut out and create your own Platonic solids at **BeastAcademy.com!**

Regular Dodecahedron

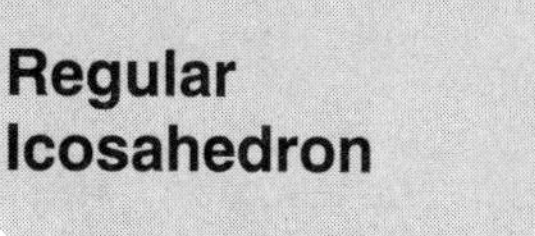

Regular Icosahedron

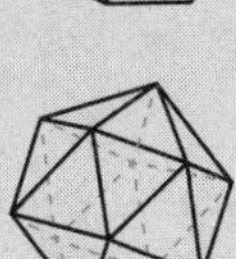

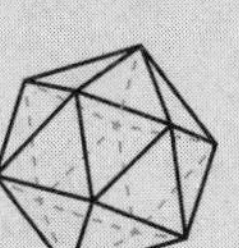

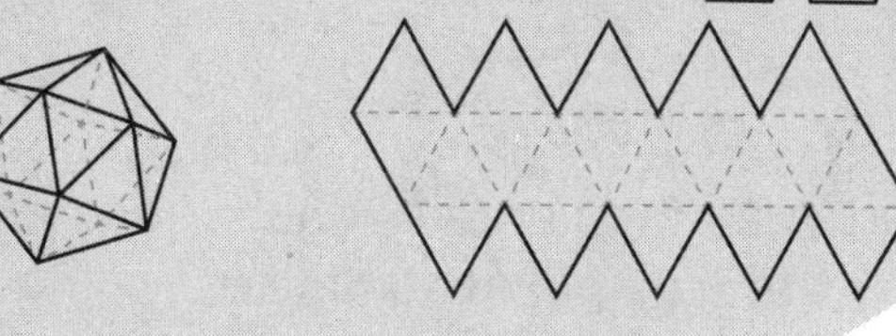

PRACTICE | Fill in the table below for each platonic solid.

94.

Platonic Solid	*Face Shape*	*Number of Faces*	*Number of Edges*	*Number of Vertices*
Regular Tetrahedron	Equilateral triangle			
Cube				
Regular Octahedron				
Reg. Dodecahedron				
Reg. Icosahedron				

95. For each platonic solid above, compute the sum of the numbers of faces and vertices ($f+v$). Compare this sum to the number of edges (e). Describe a relationship between $f+v$ and e that is true for the five platonic solids.

96. ★ Is the relationship above true for all pyramids and prisms? If so, explain why you know this is true. If not, give an example. (Review your work with these solids on pages 9-10.)

EXAMPLE Compute the surface area of the rectangular prism below. Lengths are given in centimeters.

The ***surface area*** of a solid is the total area of all its faces.

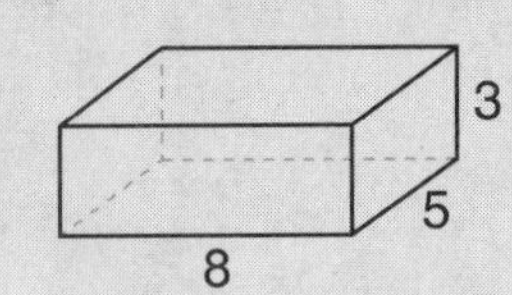

This prism has six rectangular faces. We compute the area of each. Two faces are 3-by-5, two faces are 5-by-8, and two faces are 3-by-8.

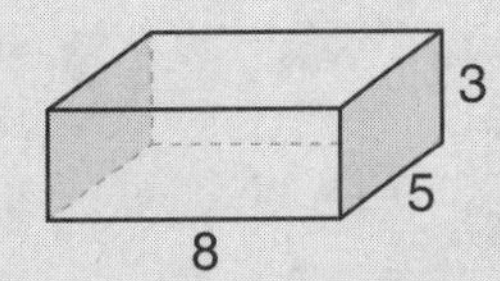

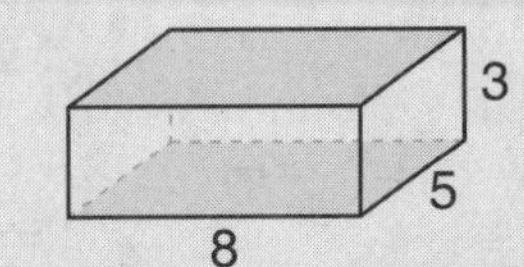

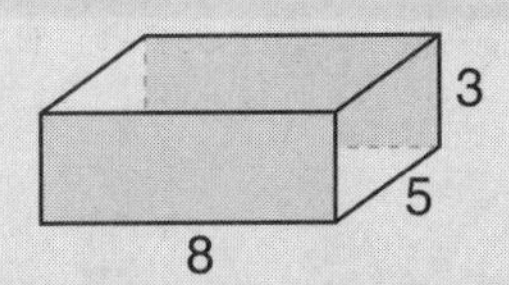

Therefore, the surface area of the prism is

$$\begin{aligned} 2\times(3\times5)+2\times(5\times8)+2\times(3\times8) &= 2\times15+2\times40+2\times24 \\ &= 30+80+48 \\ &= \mathbf{158\ sq\ cm}. \end{aligned}$$

PRACTICE Compute the surface area of each rectangular prism below. All lengths are given in centimeters.

97.

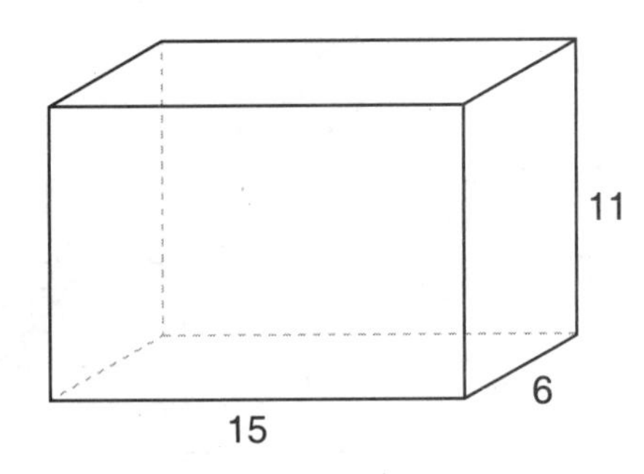

98.

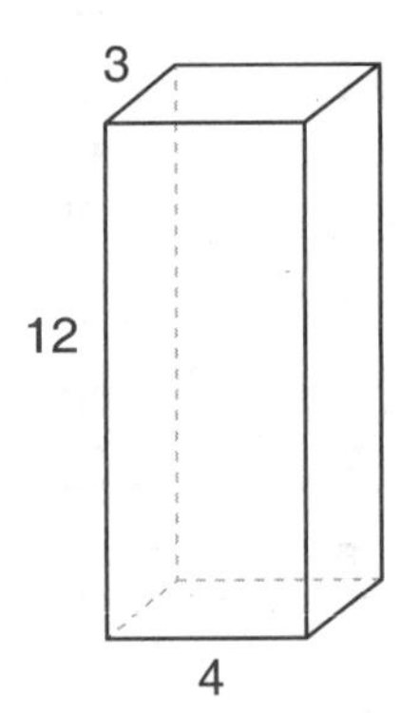

97. __________

98. __________

99.

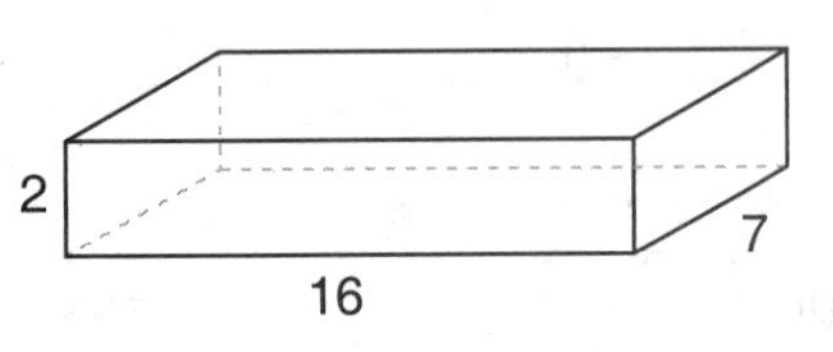

100.

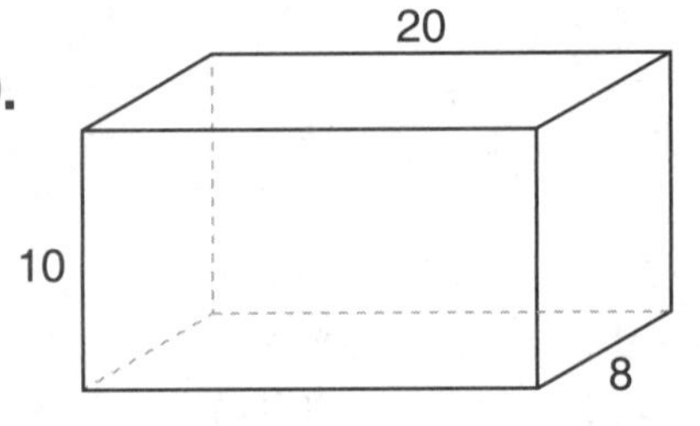

99. __________

100. __________

101. What is the surface area of a cube with edge length 9 ft?

101. __________

102. ★ Write an expression for the surface area of a cube with edge length n cm.

102. __________

EXAMPLE Compute the surface area of the triangular prism to the right. Lengths are given in inches.

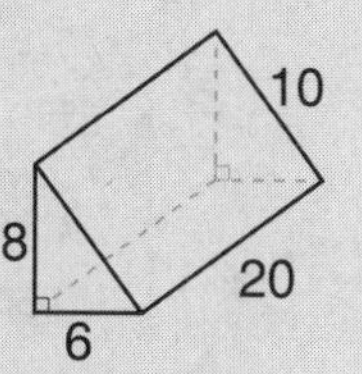

This triangular prism has five faces: two right triangles and three rectangles. We compute the area of each.

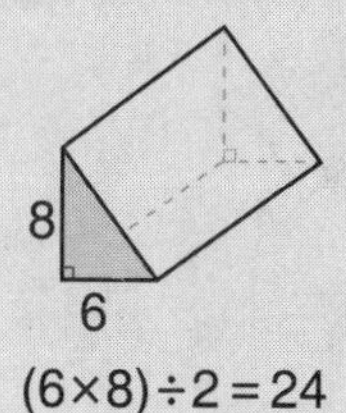

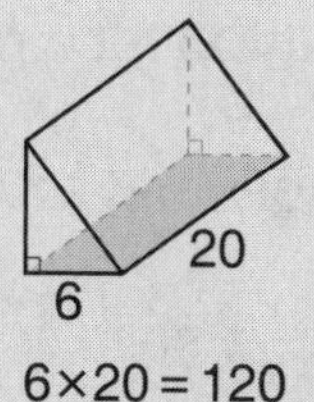

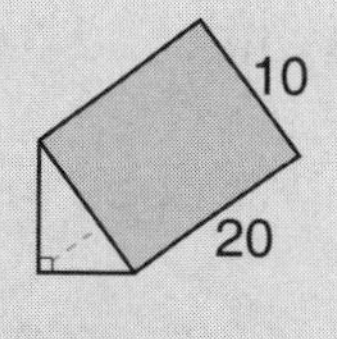

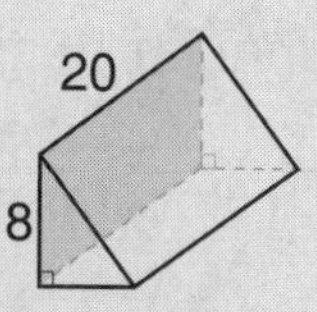

$(6\times8)\div2=24$ $(6\times8)\div2=24$ $6\times20=120$ $10\times20=200$ $8\times20=160$

Therefore, the surface area of the prism is $24+24+120+200+160=$ **528 sq in.**

PRACTICE Compute the surface area of each triangular prism below. All lengths are given in inches.

103.

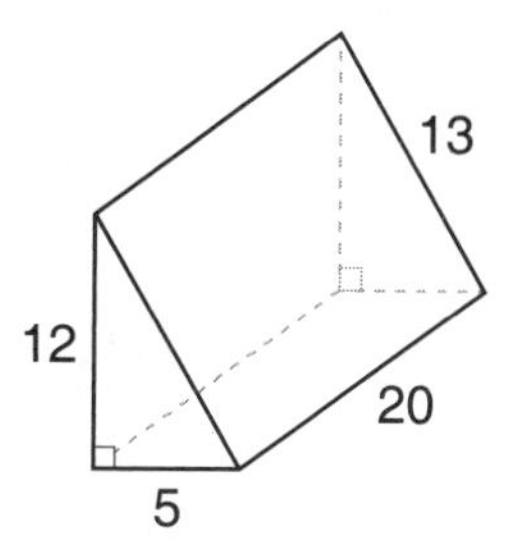

104.

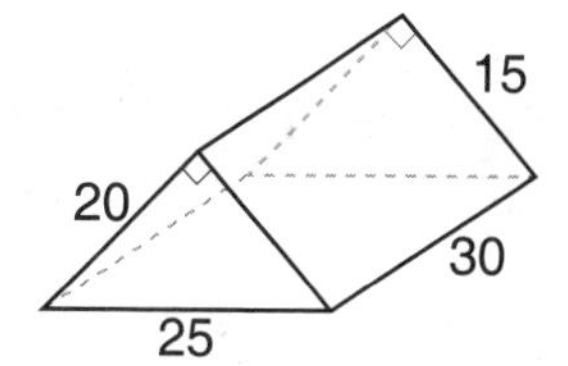

105.

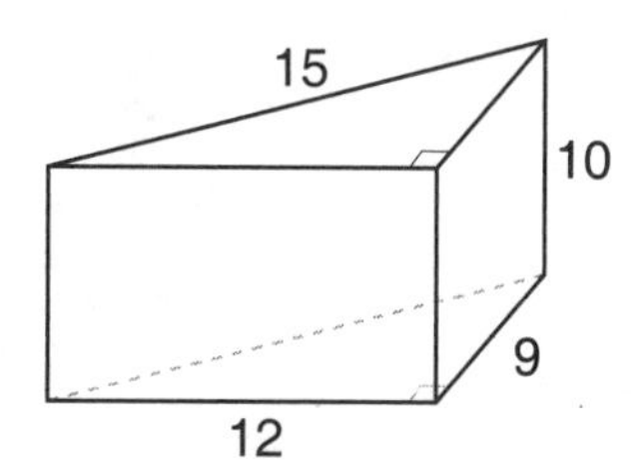

106.

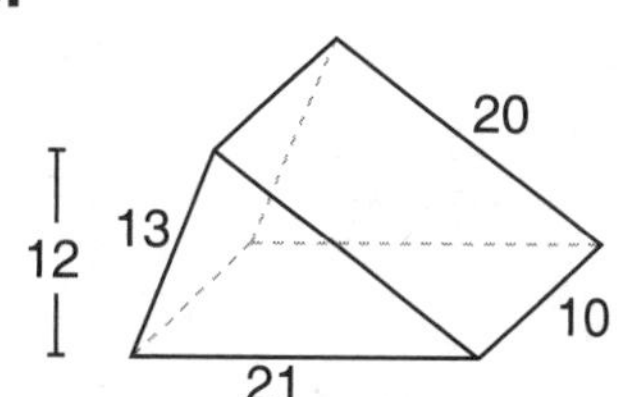

103. ________

104. ________

105. ________

106. ________

107. ★ Jared creates a prism whose base is a right triangle with leg lengths 10 m and 24 m and hypotenuse 26 m. The surface area of the prism is 3,240 sq m. What is the prism's height?

107. ________

108. ★ The base of a triangular prism has a perimeter of 42 cm and an area of 168 sq cm. The height of the prism is 15 cm. What is the surface area of the prism?

108. ________

PRACTICE | Answer each surface area question below. Remember to include units where necessary.

109. A triangular prism is attached to the top face of a cube as shown.

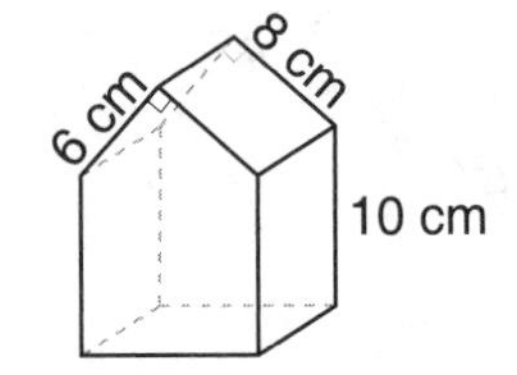

a. How many faces does the solid have? **a.** ________

b. What is the surface area of the solid? **b.** ________

110. All faces of a solid wooden cube are painted red. Then, a hole in the shape of a 3-by-2-by-8-foot rectangular prism is cut out of the cube, as shown.

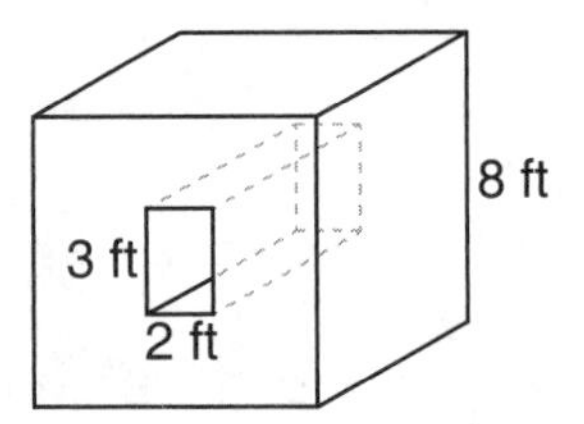

a. How many faces does the new solid have? **a.** ________

b. What was the red surface area of the original cube? **b.** ________

c. After the hole is cut, how many square feet of red surface remain? **c.** ________

d. After the hole is cut, how many square feet of unpainted surface are there? **d.** ________

e. What is the surface area of the new solid? **e.** ________

111. ★ What is the surface area of the rectangular prism with the net shown below?

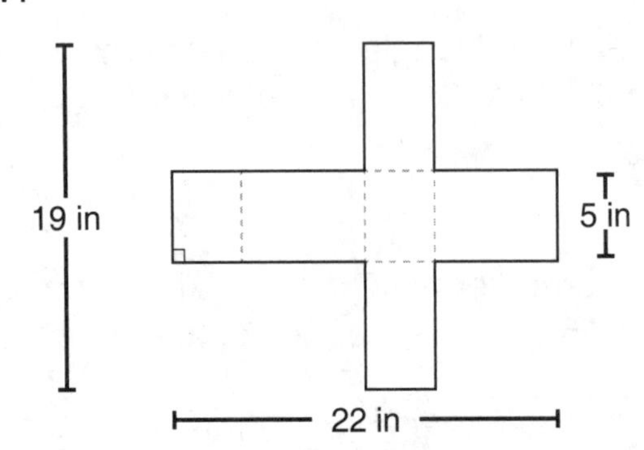

PRACTICE | Answer each surface area question below. Remember to include units where necessary.

112. A green cube with edge length 4 meters is attached to the top of a blue cube with edge length 8 meters, as shown.

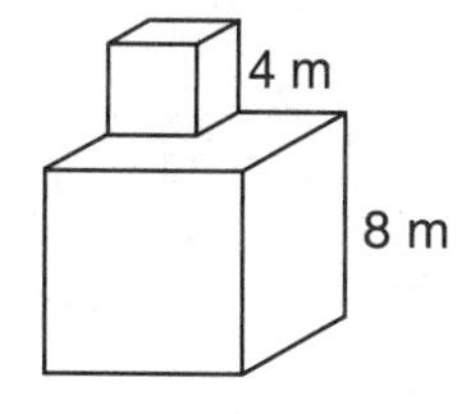

a. How many square meters of the surface of the new solid are green? **a.** ________

b. How many square meters of the surface of the new solid are blue? **b.** ________

c. What is the surface area of the new solid? **c.** ________

113. ★ Alice stacks 7 cubes to make a sculpture. The bottom cube has 9-inch edges. The edge length of every other cube is 1 inch less than the cube directly below it. What is the surface area of the entire sculpture? **113.** ________

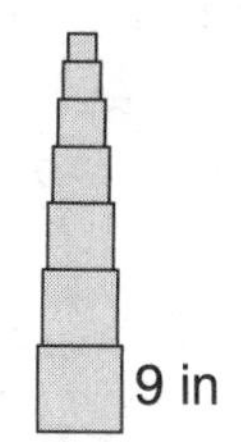

114. ★ The perimeter of a cube net is 56 inches. What is the surface area of the cube it creates? **114.** ________

115. ★ The surface area of a cube is 96 square inches. What is the surface area of a cube whose edges are three times as long? **115.** ________

116. ★ Fill in the blank below with the correct number.

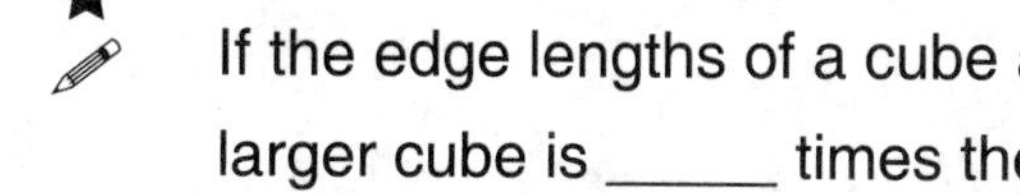

If the edge lengths of a cube are doubled, the surface area of the larger cube is ______ times the surface area of the original cube.

EXAMPLE Winnie stacks 6 unit cubes as shown below.
What is the surface area of the solid they create?

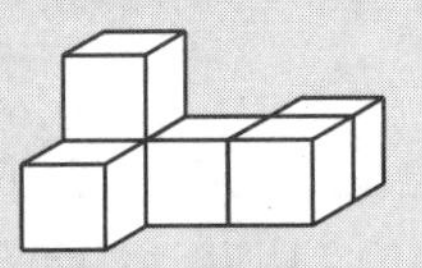

A ***unit cube*** is a cube whose edges are each 1 unit long.

From this view, we see five of the six cubes. The sixth cube must be beneath the raised cube that we see.

We count the visible faces from each of six views.

Front: 4 faces

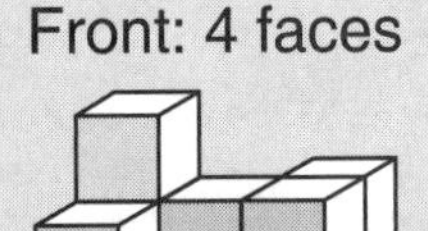

Top: 5 faces

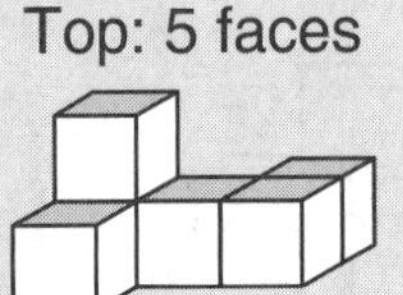

Right: 4 faces

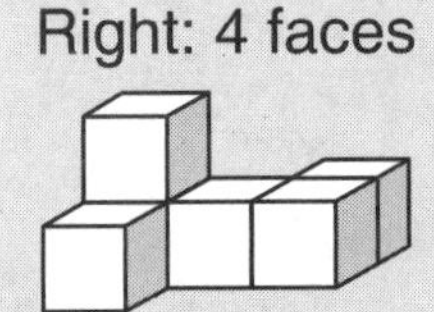

Back: 4 faces

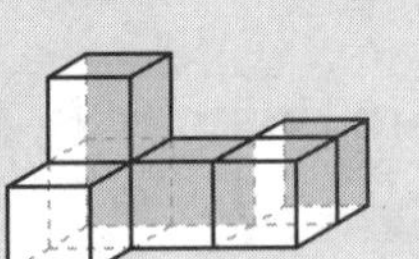

Bottom: 5 faces

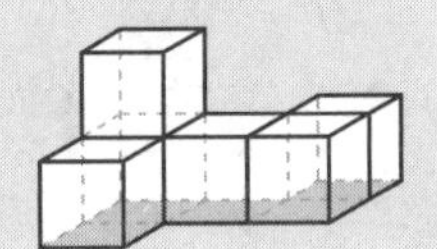

Left: 4 faces

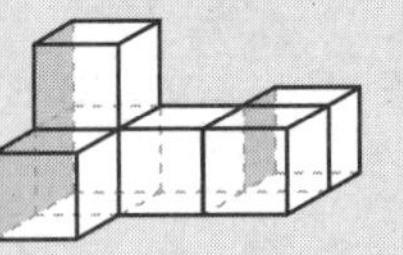

All together, there are $4+4+5+5+4+4=13+13=26$ cube faces visible.
These are unit cubes, so the area of each face is 1 square unit. The surface area of the stack is therefore **26 square units**.

— or —

Before Winnie attached the six cubes, each cube had 6 visible faces for a total of $6\times6=36$ faces. Attaching the cubes hid some of those faces.

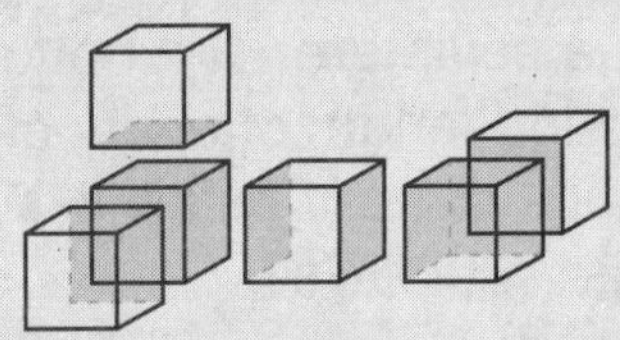

10 faces are hidden, so $36-10=26$ faces are still visible. These are unit cubes, so the area of each face is 1 square unit. The surface area of the stack is therefore **26 square units**.

— or —

We count the number of faces showing for each cube.

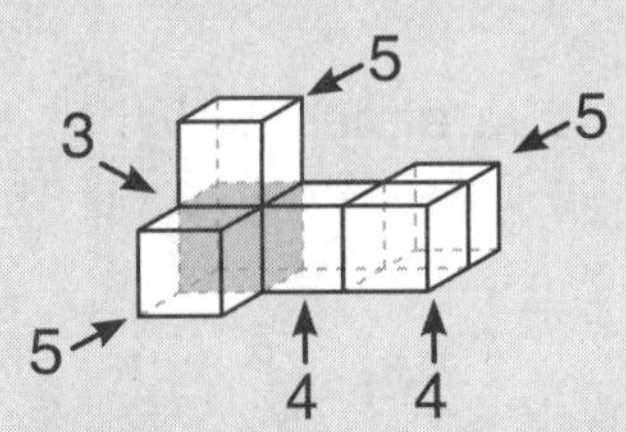

Three cubes have 5 faces showing, two cubes have 4 faces showing, and one cube has 3 faces showing. All together, there are $(5\times3)+(4\times2)+(3\times1)=15+8+3=26$ cube faces visible.
These are unit cubes, so the area of each face is 1 square unit. The surface area of the stack is therefore **26 square units**.

Use any of the methods on the previous page to solve these surface area problems.

PRACTICE Each solid below was created by attaching the specified number of unit cubes. Compute the surface area of each figure.

117. 5 unit cubes

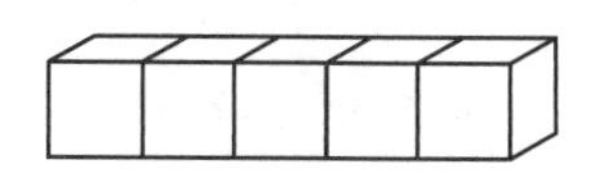

118. 5 unit cubes

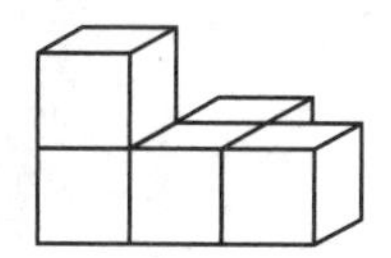

117. ________

118. ________

119. 6 unit cubes

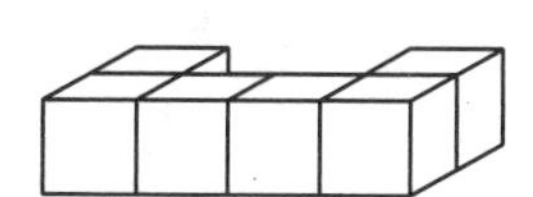

120. 8 unit cubes

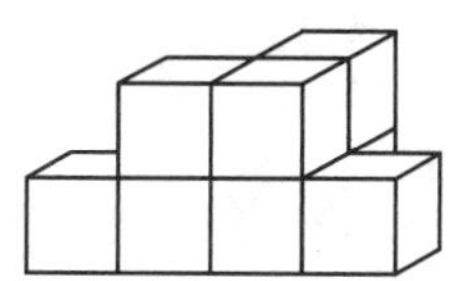

119. ________

120. ________

121. 7 unit cubes

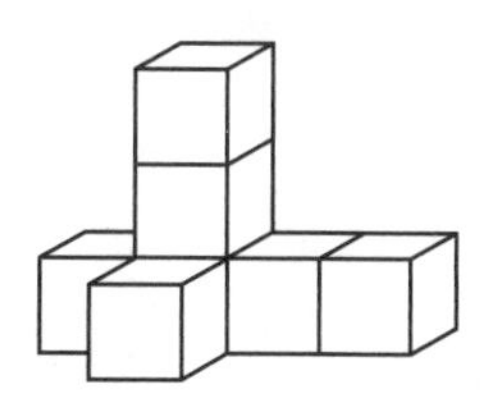

122. ★ 6 unit cubes

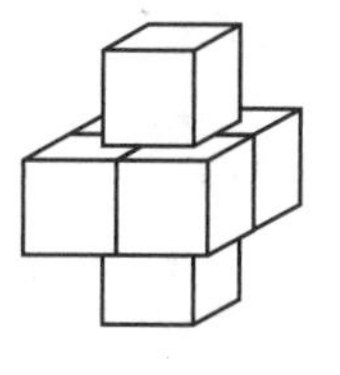

121. ________

122. ________

123. ★ 10 unit cubes

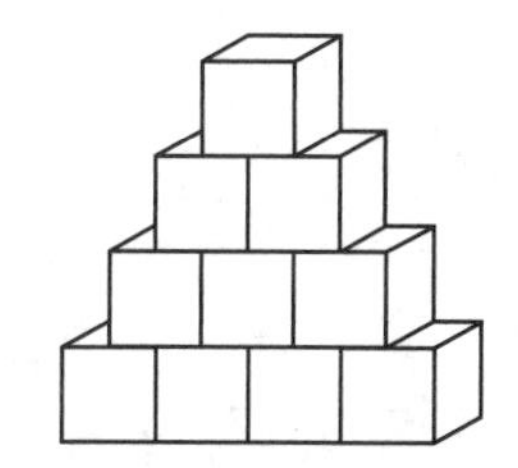

123. ________

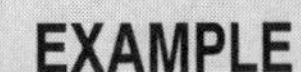

Volume is the amount of space that a three-dimensional object takes up. Volume is given in ***cubic units***, like cubic centimeters, cubic inches, or cubic feet.

EXAMPLE Compute the volume of the rectangular prism to the right. Lengths are given in centimeters.

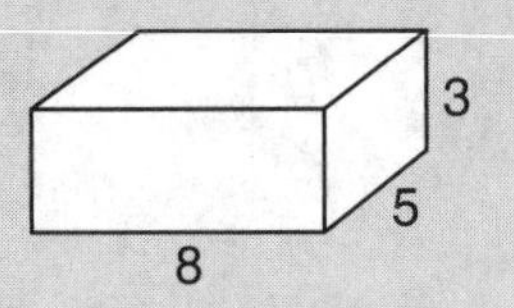

We need to figure out how many cubic centimeters it will take to fill the prism. The base of the prism is an 8-by-5-cm rectangle. We can cover the base using $8\times5=40$ unit cubes.

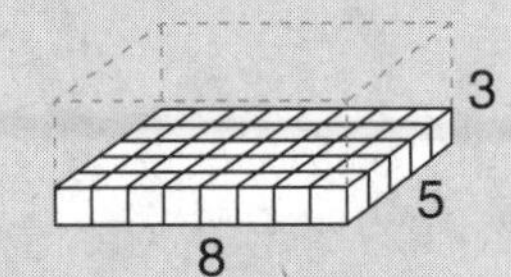

The prism is 3 centimeters high, so we can stack 3 layers of 40 unit cubes each inside the prism.

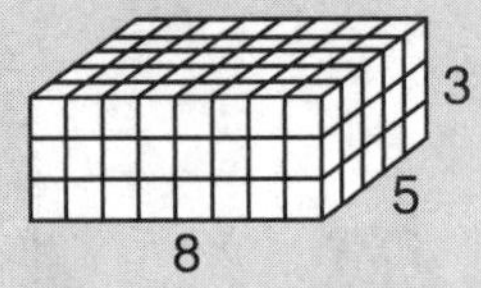

3 layers of 40 cubes is a total of $3\times40=120$ cubes.
So, the volume of the prism is **120 cubic cm**.

PRACTICE Compute the volume of each rectangular prism below. All lengths are given in inches.

124.

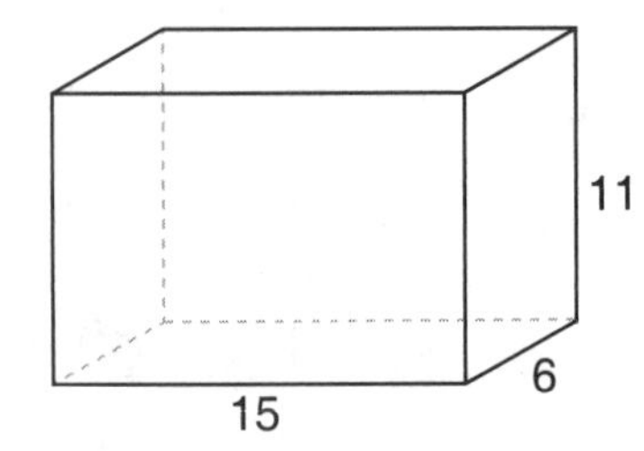

125.

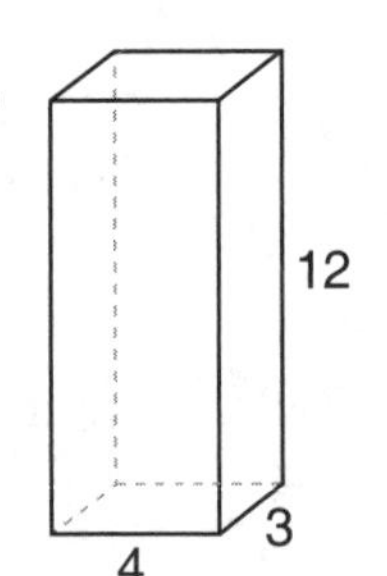

124. __________

125. __________

126.

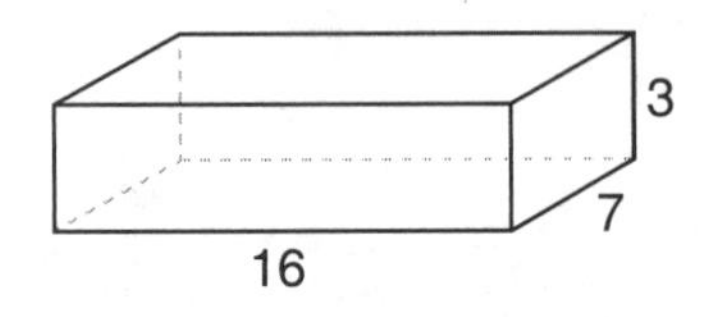

127.

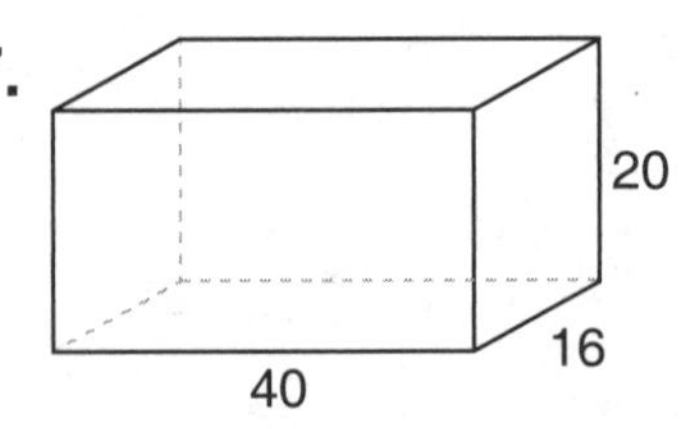

126. __________

127. __________

128. What is the volume of a cube with edge length 5 mm?

128. __________

129. Write an expression for the volume of a cube with edge length n feet.

129. __________

EXAMPLE Compute the volume of the triangular prism to the right. Lengths are given in centimeters.

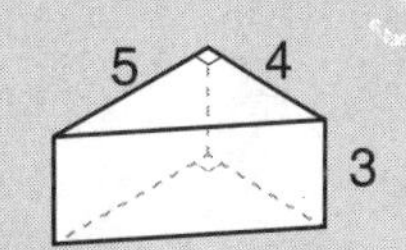

To find the volume of a prism, we start by finding the area of its base. This tells us how many cubes it will take to cover the base.

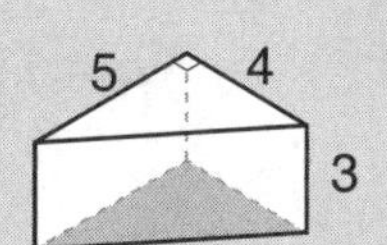

The area of the base is $5\times4\div2=20\div2=10$ sq cm.

Then, the height tells us how many layers of cubes it will take to fill the prism. So, we multiply the area of the base by the height of the prism.

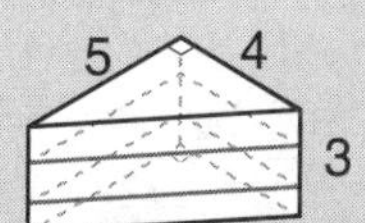

The height of the prism is 3 cm, so its volume is $10\times3=$ **30 cubic cm**.

— or —

We see that cutting a 3-by-4-by-5-cm rectangular prism in half diagonally gives us two copies of this triangular prism.

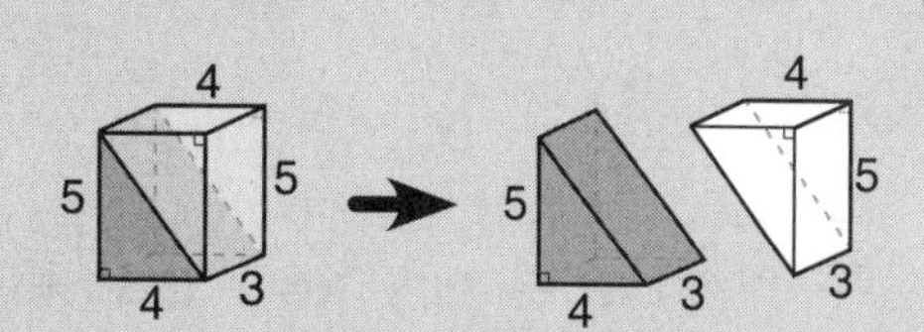

Remember, the bases of this prism are right triangles!

So, the volume of the triangular prism is half the volume of the 3-by-4-by-5-cm rectangular prism: $(3\times4\times5)\div2=60\div2=$ **30 cubic cm**.

PRACTICE Compute the volume of each triangular prism below. All lengths are given in inches.

130.

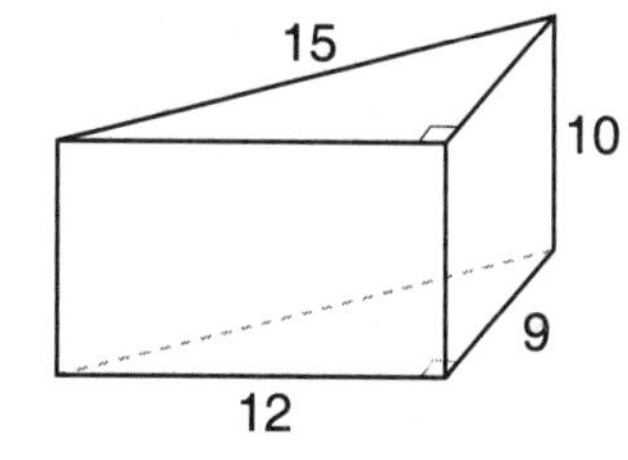

131.

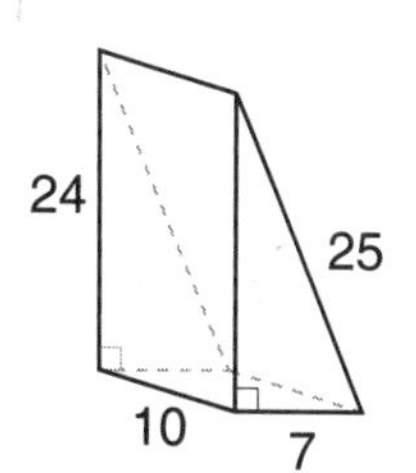

130. ________

131. ________

132.

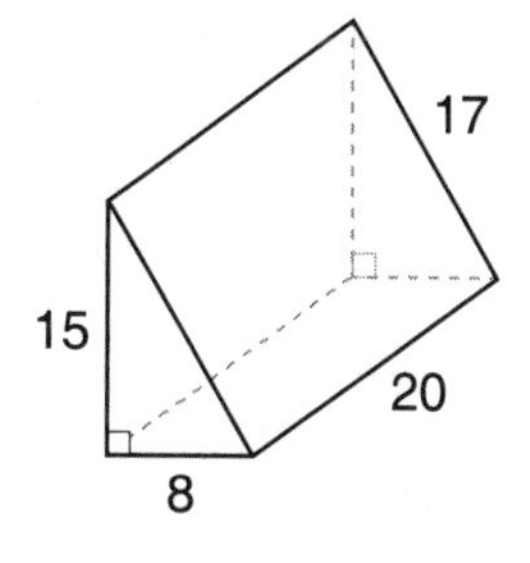

133.

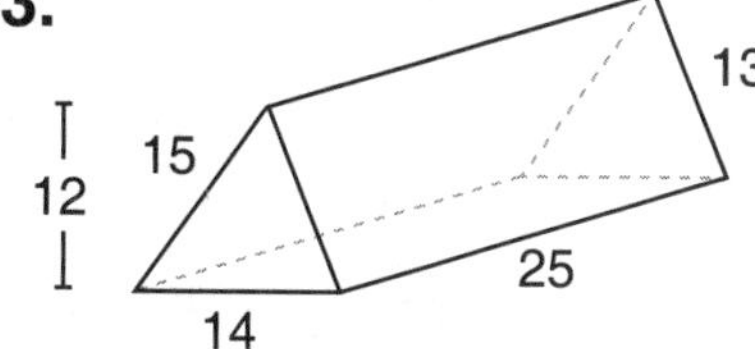

132. ________

133. ________

Volume

No matter the shape of a prism's base, we can calculate the volume of the prism as the product of the area of its base (B) and its height (h):

$$V = B \times h.$$

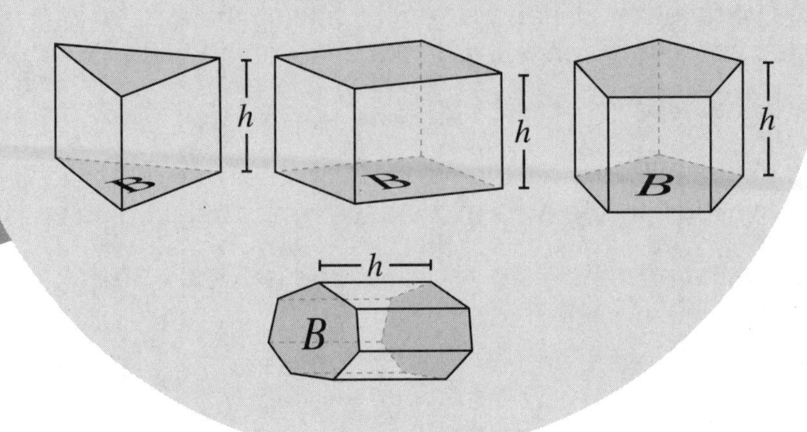

PRACTICE | Answer each question about volume below. Remember to include units where necessary.

134. The area of the base of the decagonal prism below is 50 square feet. Its volume is 600 cubic feet. What is the height of the prism?

134. ________

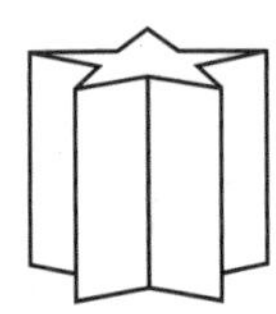

135. The area of each base of the cylinder below is 3 square meters. What is the volume of the cylinder?

135. ________

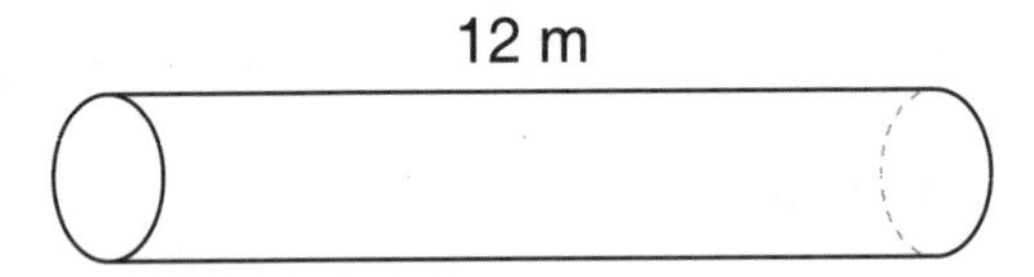

136. The volume of the rectangular prism below is 126 cubic centimeters. What is the value of w?

136. w = ______

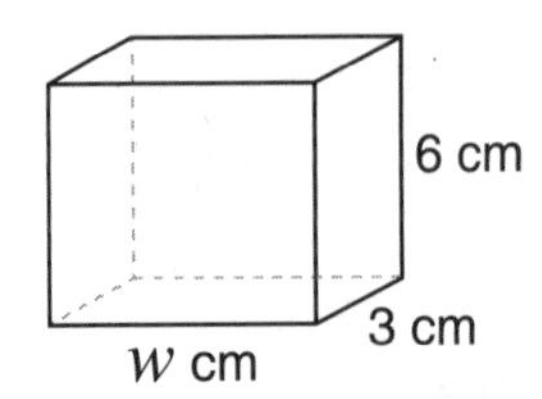

137. ★ There are 3 feet in 1 yard. How many ***cubic feet*** are in 7 ***cubic yards***?

137. ________

PRACTICE | Answer each question about volume below. Remember to include units where necessary.

138. A triangular prism with a height of 10 cm is cut out of a cube with 10-cm edges, as shown below. The base of the triangular prism is a right triangle with sides of length 6 cm, 8 cm, and 10 cm. What is the volume of the remaining solid?

138. 

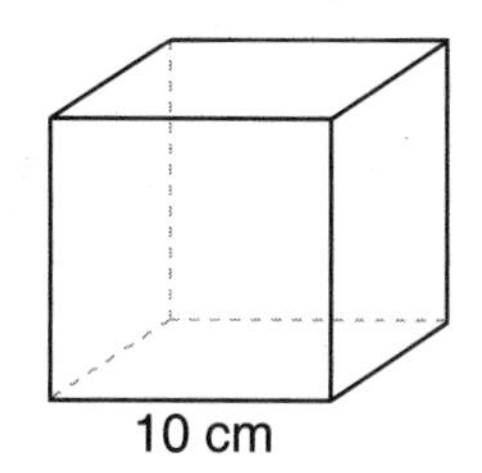

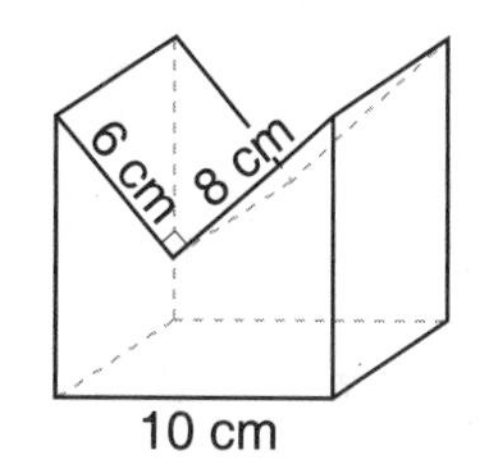

139. A cube with edge length 20 cm is attached to the top of a cube with edge length 40 cm, as shown. What is the volume of the new solid?

139. __________

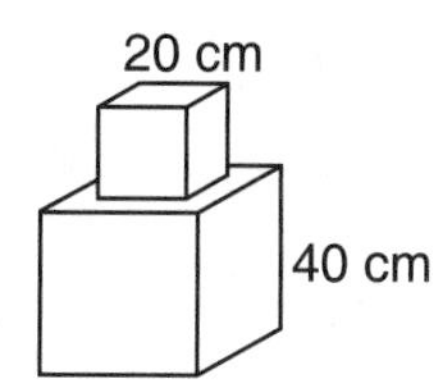

140. What is the volume of a cube whose surface area is 150 square meters?

140. __________

141. ★ Lizzie glues 165 unit cubes together to form a rectangular prism. The perimeter of the base is 28 units. What is the height of the prism?

141. __________

142. ★ The volume of a cube is 1,000 cubic inches. What is the volume of a cube whose edges are three times as long?

142. __________

143. ★ Fill in the blank below with the correct number.

If the edge lengths of a cube are doubled, the volume of the larger cube is ______ times the volume of the original cube.

EXAMPLE

A 3-by-2-by-2-unit wood block is painted on all faces. The block is then cut into unit cubes. How many of these cubes are painted on exactly 2 faces?

The block is cut into $3 \times 2 \times 2 = 12$ unit cubes, as shown below.

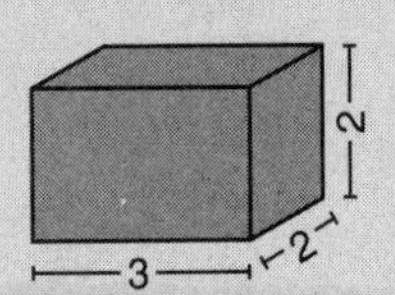

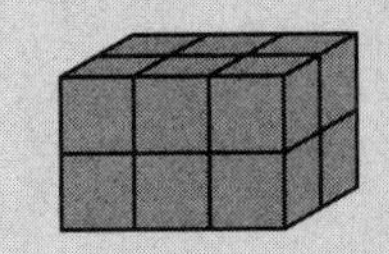
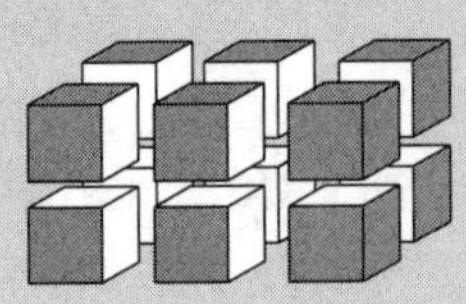

Each of the 8 corner cubes has exactly 3 faces painted.
Each of the **4** other cubes has exactly 2 faces painted.

PRACTICE

Use the information below to answer the questions that follow.

The 4-by-5-by-8-unit rectangular prism to the right is painted on all faces, and then cut into unit cubes.

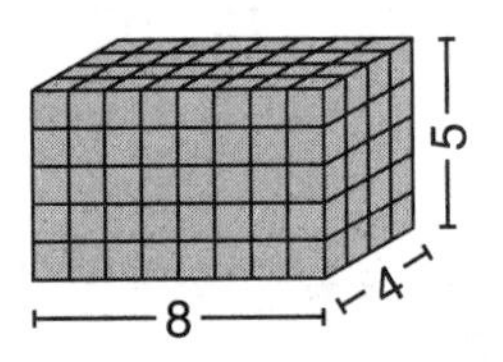

144. How many unit cubes is the block cut into? **144.** ________

145. How many of the unit cubes have exactly three faces painted? **145.** ________

146. How many of the unit cubes have exactly two faces painted? **146.** ________

147. How many of the unit cubes have exactly one face painted? **147.** ________

148. How many of the unit cubes have more than three faces painted? **148.** ________

149. How many of the unit cubes do not have any paint on them? **149.** ________

150. If both 4-by-5 faces of the block were painted red, and the remaining faces were painted green, how many of the unit cubes have at least one face that is painted green? **150.** ________

PRACTICE | Answer each question below.

151. Lizzie assembles 125 wood cubes to make one large cube. She paints the large cube green on all six of its faces, then disassembles it back into smaller cubes. How many of the smaller cubes do not have any paint on them?

151. ________

152. Grogg paints a 5-by-6-by-7-unit block and cuts it into unit cubes. He places all of the unit cubes into a bag, and Alex selects one cube from the bag at random. What is the probability that the cube Alex selects has exactly three faces painted?

152. ________

153. ★ Jim paints a rectangular block on all six faces and then cuts it into exactly 105 unit cubes. Eight of these cubes have exactly three faces painted. How many of these cubes have exactly two faces painted?

153. ________

154. ★ A rectangular block is painted on all faces and then cut into exactly 189 unit cubes. If 35 of the cubes have no paint on them, what is the length of the longest dimension of the block?

154. ________

155. ★★ Cammie paints all the faces of a wood cube orange. She then cuts the painted cube into unit cubes, some of which have no paint on them. The number of unit cubes with 0 painted faces is the same as the number of unit cubes with exactly 1 painted face. How many unit cubes are there all together?

155. ________

Answer each question below. Remember to include units where necessary.

156. The net for a ***square antiprism*** is shown below. It consists of two congruent squares and eight congruent triangles. What is the surface area of the solid with the measurements given by the net below?

156. ________

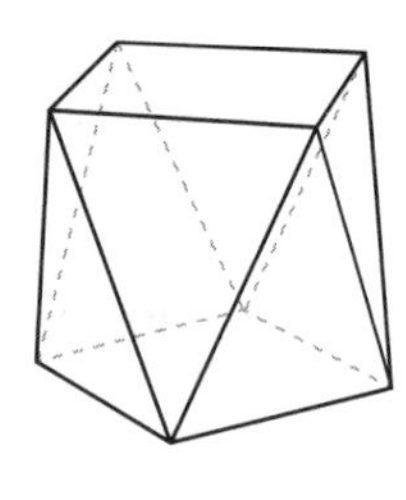

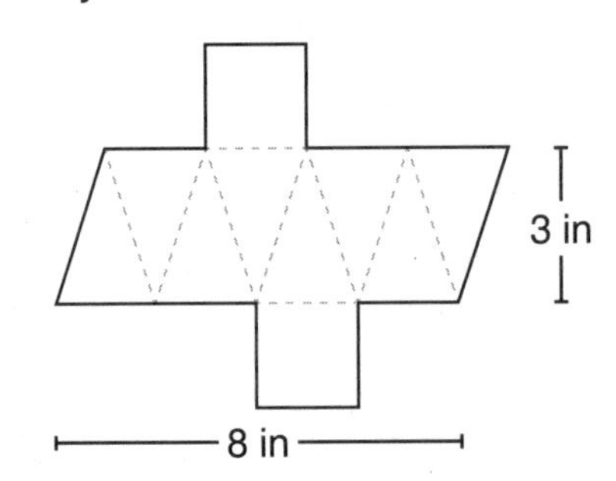

157. How many cubes with 4-inch edges can be placed completely inside of a 3-by-5-by-12-foot prism?

157. ________

158. Phyllis has a small aquarium with a 30-cm square base. She fills the aquarium to a depth of 20 cm. After she puts several rocks at the bottom of the aquarium, the depth of the water in the aquarium is 22 cm. What is the total volume in cubic centimeters of the rocks in the aquarium?

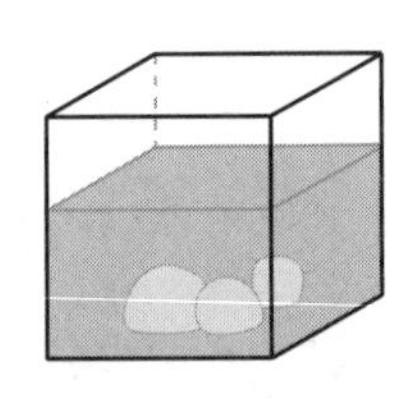

158. ________

159. John uses unit cubes to build a rectangular prism that is 5 by 6 by 7 units. He then removes all eight of the cubes from the corners of the prism. What is the surface area of the resulting solid?

159. ________

160. ★ What is the greatest possible perimeter of a net that can be folded to make the surface of a 2-by-5-by-10-meter rectangular prism?

160. ________

PRACTICE | Answer each question below.
Remember to include units where necessary.

161. ★ A 3-by-4-by-7-foot rectangular prism is cut out of one side of a cube with edge length 10 feet, as shown. What are the surface area and volume of the remaining solid?

161. Surface Area: __________

Volume: __________

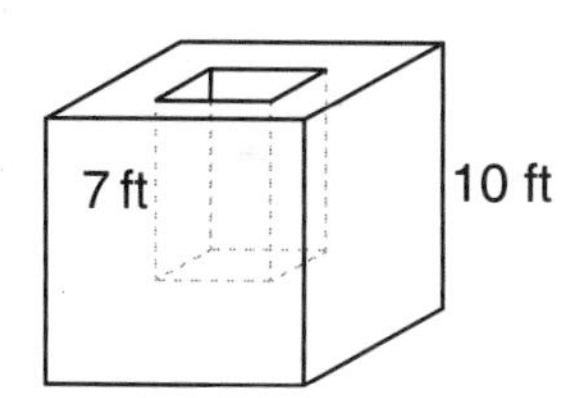

162. ★ Cole uses 27 unit cubes to create a larger cube. Then, Cole removes one unit cube from the center of each face, as well as the unit cube at the center. What are the surface area and volume of the remaining solid?

162. Surface Area: __________

Volume: __________

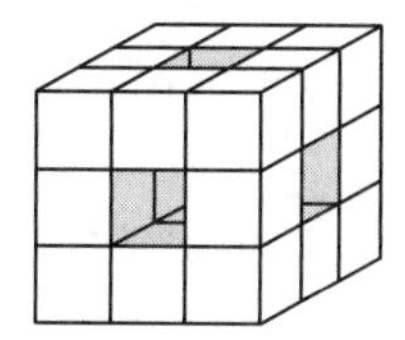

163. ★ Fill in the missing numbers on the octahedron net to the right so that each digit from 1 to 8 appears on exactly one face, and consecutive digits appear on faces that share an edge when the net is folded. For example, "5" must appear on a face that shares edges with faces labeled "4" and "6."

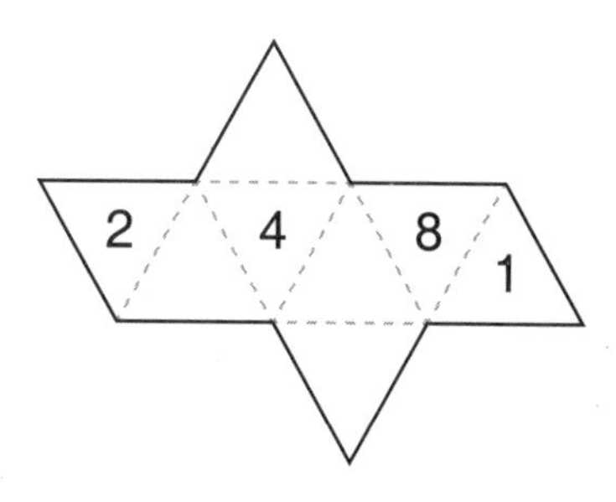

164. ★★ Each cube net below is folded so that the shapes appear on the outside of the cube. Circle the net that creates a cube identical to the cube created by the original net.

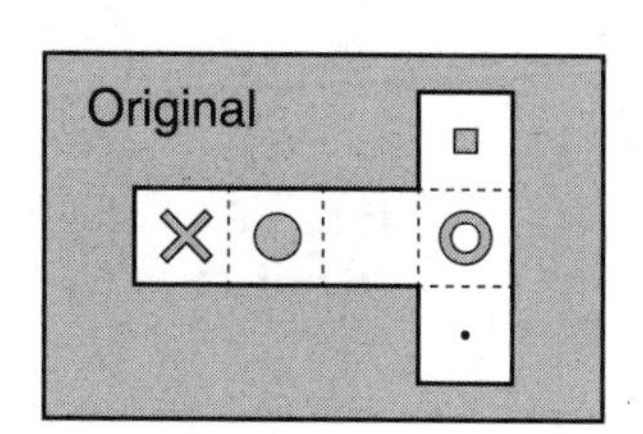

(a) (b) (c)

(d) (e) (f)

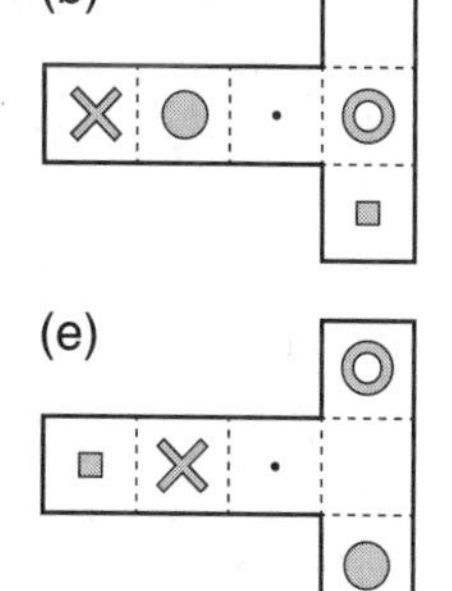

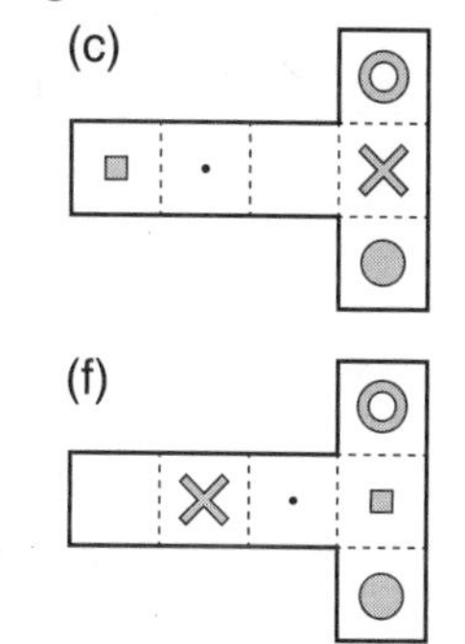

CHAPTER 2
Integers

Use this Practice book with Guide 5A from BeastAcademy.com.

Recommended Sequence:

Book	**Pages:**
Guide:	46-50
Practice:	41-43
Guide:	51-63
Practice:	44-57
Guide:	64-75
Practice:	58-69

You may also read the entire chapter in the Guide before beginning the Practice chapter.

EXAMPLE | Compute -128+56.

Consider the calculation on the number line.

We start at -128 and move 56 units to the right. This brings us 56 units closer to zero. We end up $128-56=72$ units to the left of zero at -72.

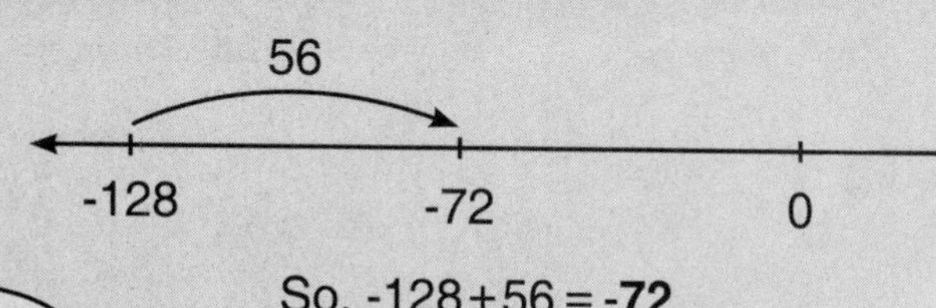

So, -128+56 = **-72**.

When adding integers, the first number tells us where to start on the number line.
The second number tells us which direction to go and how far.
We move right when adding a positive...
...and left when adding a negative.

PRACTICE | Compute the following sums.

1. 7+(-2) = ______

2. 3+(-5) = ______

3. -6+(-8) = ______

4. -9+6 = ______

5. 5+(-6)+(-7) = ______

6. -4+9+(-5) = ______

7. -36+(-42) = ______

8. -94+41 = ______

9. 67+(-28) = ______

10. 120+(-234) = ______

11. -654+789 = ______

12. -75+94+(-26) = ______

EXAMPLE | Compute -37 − (-53).

To subtract a number, we can add its opposite.
So, -37 − (-53) = -37 + 53.

Consider the calculation on the number line. We start at -37 and move 53 units to the right.

We first move 37 units to the right to arrive at 0. Then, we move an additional 53 − 37 = 16 units to the right to arrive at 16.

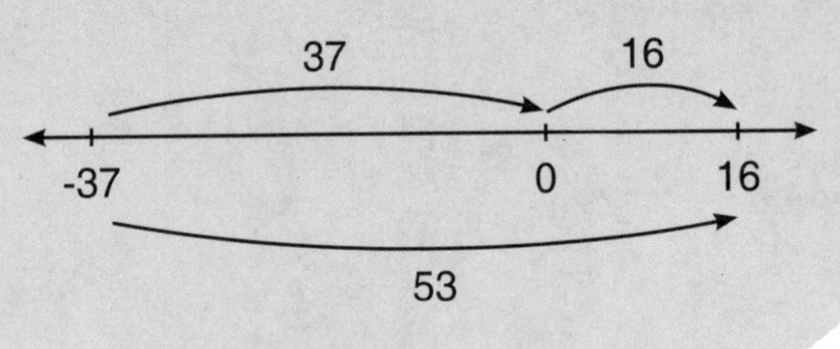

So, -37 − (-53) = **16**.

Sometimes, it's easier to subtract than to add the opposite. For example, 29−13=16, and -34−5=-39.

PRACTICE | Compute the following differences.

13. -7 − 9 = ______

14. 6 − 10 = ______

15. 3 − (-12) = ______

16. -8 − (-8) = ______

17. 8 − (-2) − 12 = ______

18. -9 − 4 − (-5) = ______

19. 34 − (-57) = ______

20. 101 − 117 = ______

21. -58 − 24 = ______

22. 314 − 256 = ______

23. -470 − (-840) = ______

24. -67 − (-24) − 46 = ______

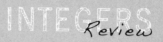

PRACTICE | Answer each question below.

25. Without computing the expressions below, circle those whose results are negative.

-198 − (-299)　　600 + (-803)　　-943 − 500　　-345 + 435　　-639 + 738 − 1,000

26. Grogg picks a positive integer that is less than 1,000. Alex then picks a negative integer that is greater than -1,000. What is the greatest possible difference between Grogg's and Alex's integers?

26. ________

27. Winnie circles three consecutive integers on the number line. Their sum is -21. What is the integer farthest to the left?

27. ________

28. The Polar Yeti lives in environments where the temperature is always within 25 degrees of -13 degrees Fahrenheit. What is the coldest temperature in which the Polar Yeti lives? What is the warmest temperature in which the Polar Yeti lives?

28. Coldest: ________°F

Warmest: ________°F

29. ★ How many ***different*** values can be made by filling in the blanks of the expression ___ − (___ − ___) using each of the numbers -2, 3, and 5 exactly once?

29. ________

30. ★ How many ***different*** sums can be made by adding two different numbers chosen from the forty-one consecutive integers below?

-20, -19, -18, ..., -1, 0, 1, ..., 18, 19, 20

30. ________

EXAMPLE | Compute $5\times(-4)$.

We can think of multiplying a positive integer by another number as repeated addition.

To multiply 5 times ***positive*** 4, we can add 5 copies of 4:

$$5\times4=4+4+4+4+4$$
$$=20.$$

Similarly, to multiply 5 times ***negative*** 4, we can add 5 copies of -4:

$$5\times(-4)=(-4)+(-4)+(-4)+(-4)+(-4)$$
$$=-20.$$

Therefore, we have $5\times(-4)=$ **-20.**

PRACTICE | Compute each of the following products.

31. $3\times(-2)=$ ______

32. $5\times(-3)=$ ______

33. $2\times(-9)=$ ______

34. $6\times(-11)=$ ______

35. $13\times(-4)=$ ______

36. $7\times(-1)=$ ______

37. $150\times(-6)=$ ______

38. $140\times(-20)=$ ______

39. $3\times4\times(-5)=$ ______

40. $7\times8\times(-2)=$ ______

Find more practice problems at BeastAcademy.com!

EXAMPLE | Compute -6×3.

We can use the fact that multiplication is commutative to rewrite -6×3 as $3\times(-6)$.

Then, to multiply $3\times(-6)$ we can add 3 copies of -6:

$$\begin{aligned}3\times(-6) &= (-6)+(-6)+(-6)\\ &= -18.\end{aligned}$$

So, $-6\times3 = \mathbf{-18}$.

When you multiply two numbers that have ***different*** signs, the product is ***negative.***

PRACTICE | Compute each of the following products.

41. -3×2 = ______

42. -5×4 = ______

43. -9×6 = ______

44. -8×12 = ______

45. -15×20 = ______

46. -30×22 = ______

47. -25×24 = ______

48. -4×444 = ______

49. $6\times(-5)\times8$ = ______

50. $-4\times9\times2$ = ______

Find more practice problems at BeastAcademy.com!

EXAMPLE | Look for a pattern in the list of products below to fill in the blanks.

$-3\times3=-9$
$-3\times2=-6$
$-3\times1=-3$
$-3\times0=0$
$-3\times(-1)=$ __
$-3\times(-2)=$ __

As we move down the list, the second factors decrease by 1 and the products increase by 3. We continue the pattern to complete the list of products as shown.

$-3\times3=-9$ (+3)
$-3\times2=-6$ (+3)
$-3\times1=-3$ (+3)
$-3\times0=0$
$-3\times(-1)=$ __
$-3\times(-2)=$ __

→

$-3\times3=-9$
$-3\times2=-6$
$-3\times1=-3$
$-3\times0=0$ (+3)
$-3\times(-1)=$ **3** (+3)
$-3\times(-2)=$ **6**

PRACTICE | Look for a pattern in the list of products below to fill in the blanks.

51.
$-6\times2=$ **-12**
$-6\times1=$ **-6**
$-6\times0=$ **0**
$-6\times(-1)=$ ____
$-6\times(-2)=$ ____
$-6\times(-3)=$ ____

52.
$-4\times2=$ ____
$-4\times1=$ ____
$-4\times0=$ ____
$-4\times(-1)=$ ____
$-4\times(-2)=$ ____
$-4\times(-3)=$ ____

53.
$-7\times2=$ ____
$-7\times1=$ ____
$-7\times0=$ ____
$-7\times(-1)=$ ____
$-7\times(-2)=$ ____
$-7\times(-3)=$ ____
$-7\times(-4)=$ ____

54.
$-9\times2=$ ____
$-9\times1=$ ____
$-9\times0=$ ____
$-9\times(-1)=$ ____
$-9\times(-2)=$ ____
$-9\times(-3)=$ ____
$-9\times(-4)=$ ____

We can use the expression $-5\times(-6+6)$ to show that $-5\times(-6)=30$.

Since $-6+6=0$, and anything times zero is zero, we have

$$-5\times(-6+6)=-5\times0=0.$$

Distributing the -5 gives us:

$$-5\times(-6+6)=0$$
$$(-5\times(-6))+(-5\times6)=0.$$

Two quantities that sum to zero are opposites. $(-5\times(-6))+(-5\times6)=0$, so $-5\times(-6)$ is the ***opposite*** of $5\times(-6)$.

Since $-5\times6=-30$ and the opposite of -30 is 30, we have $-5\times(-6)=\mathbf{30}$.

We can use a similar process to show that the product of any two negatives is always positive. So, we have the following rules for multiplying integers:

$$(+)\times(+)=(+)$$
$$(-)\times(-)=(+)$$
$$(+)\times(-)=(-)$$
$$(-)\times(+)=(-)$$

The product of two numbers with the ***same*** sign is always ***positive.***
The product of two numbers with ***opposite*** signs is always ***negative.***

PRACTICE | Compute each of the following products.

55. $-2\times(-6)=$ ______

56. $-13\times(-1)=$ ______

57. $-4\times7=$ ______

58. $-6\times(-3)=$ ______

59. $7\times(-5)=$ ______

60. $-9\times(-9)=$ ______

61. $-60\times(-8)=$ ______

62. $4\times19=$ ______

63. $-18\times(-5)=$ ______

64. $-130\times(-30)=$ ______

Find more practice problems at BeastAcademy.com!

In a **Block Mountain** puzzle, each block contains an integer. The number in each block is the product of the two numbers below it.

EXAMPLE | Complete the Block Mountain puzzle below.

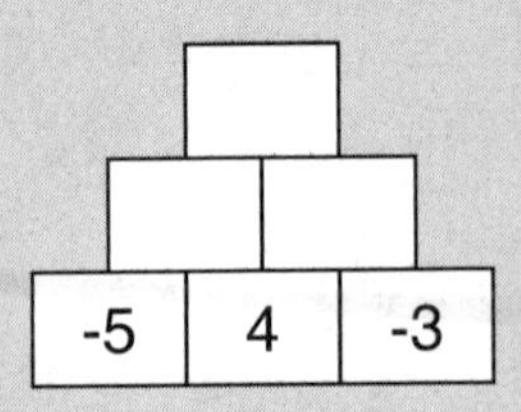

We compute the missing entries as shown below.

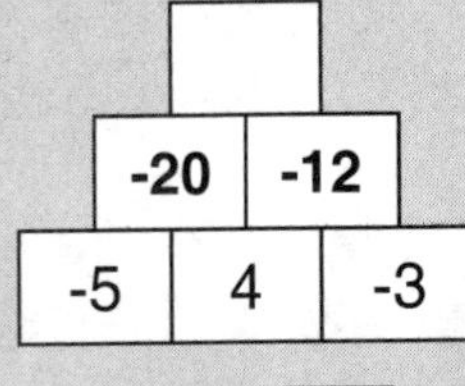

240

-20 | -12

-5 | 4 | -3

$-5\times 4 =$ **-20**.

$4\times(-3) =$ **-12**.

$-20\times(-12) =$ **240**.

PRACTICE | Complete each Block Mountain puzzle below.

65.

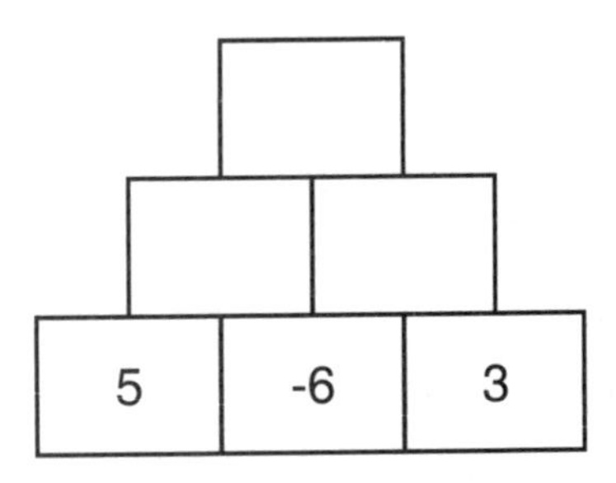

66.

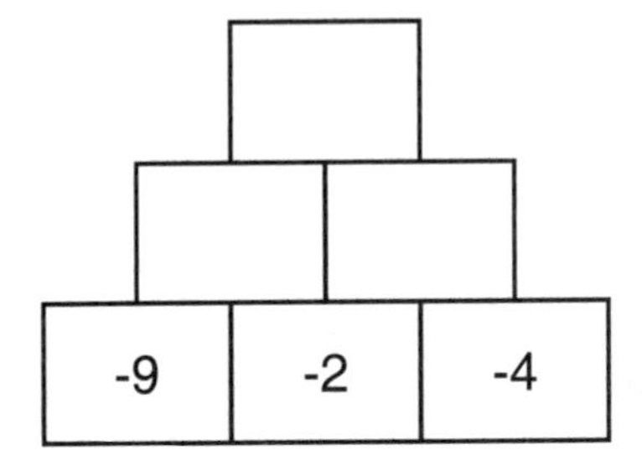

67.

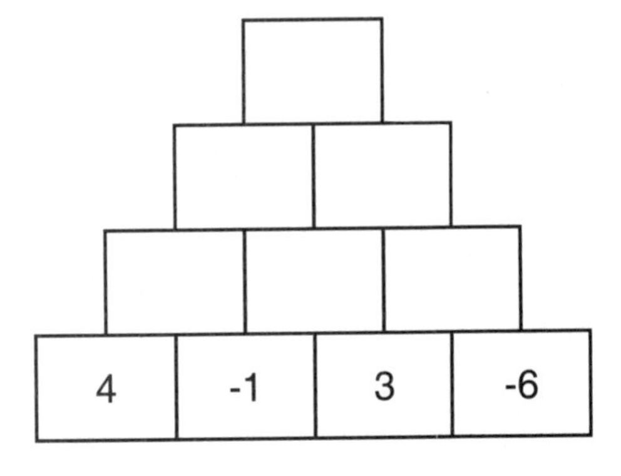

68.

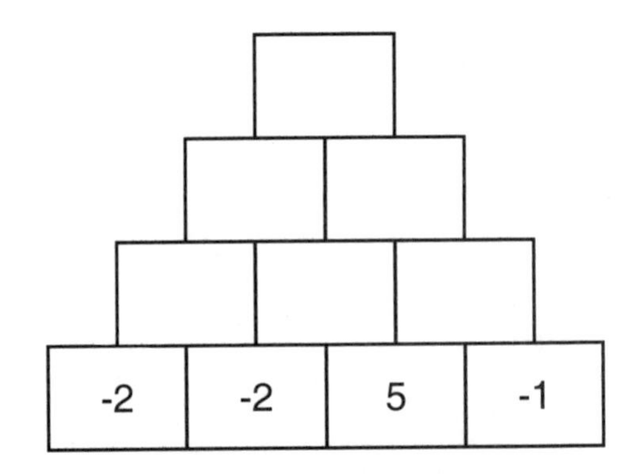

PRACTICE | Complete each Block Mountain puzzle below.

69.

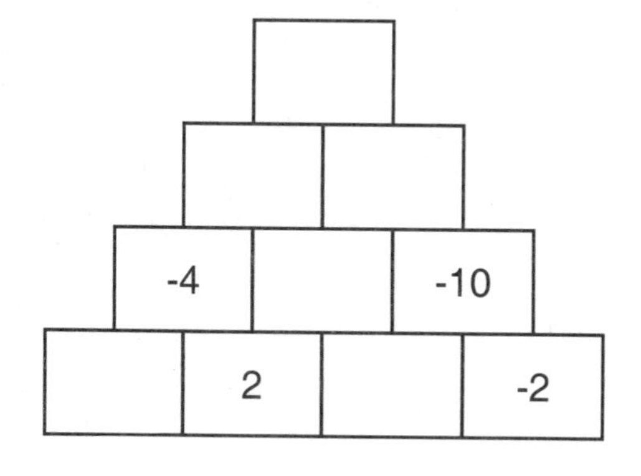

70.

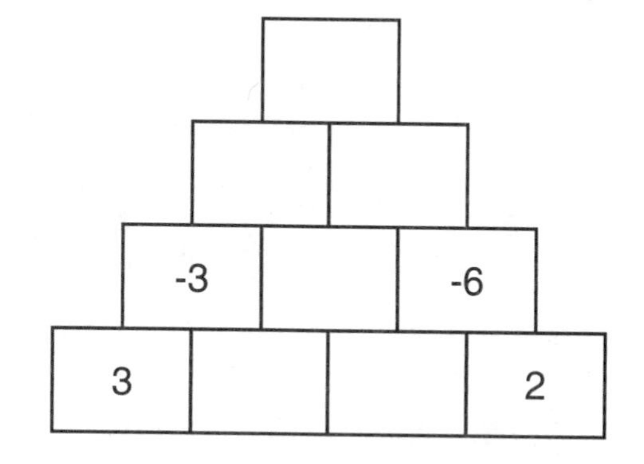

71.

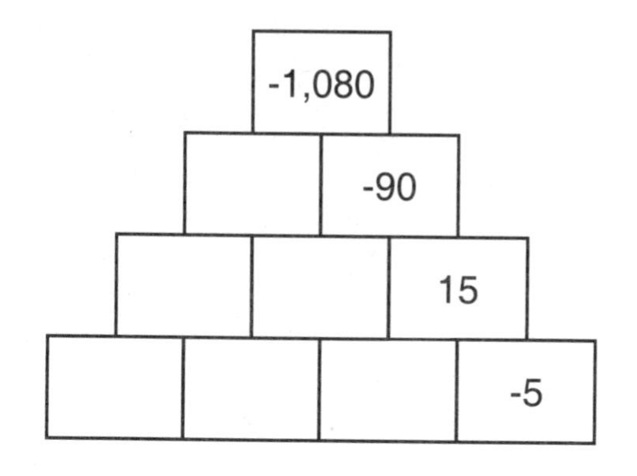

72.

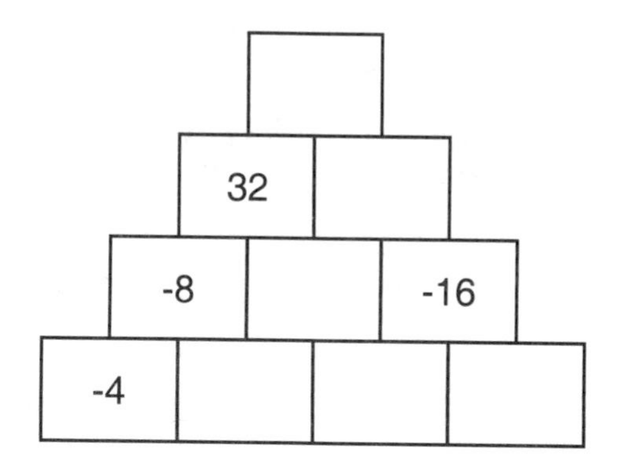

73. ★

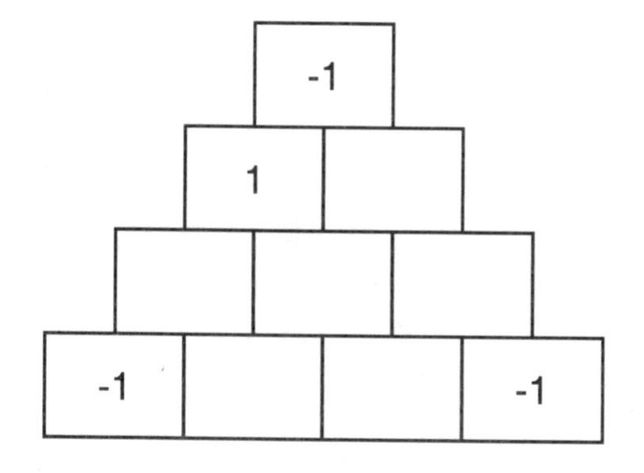

74. ★

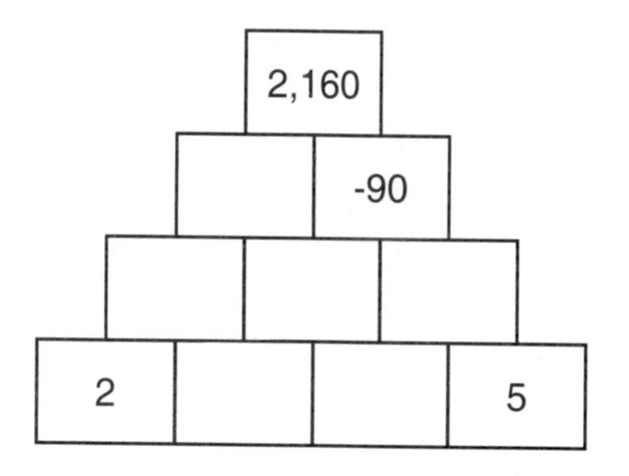

In a **Product Hive** puzzle, the goal is to cross every hexagon in the hive to create a path of ***increasing products***.

Paths may begin at any number, and must always extend to a number in an adjacent hexagon. No hexagons may be crossed more than once.

EXAMPLE | Complete the Product Hive puzzle to the right.

We begin by finding the least possible product in the hive: $-9 \times 3 = -27$. However, every path that begins with -9 and 3 results in a dead end.

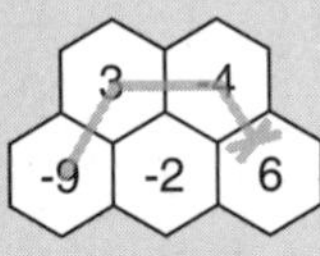

Products:
-27, -12, ~~-24~~

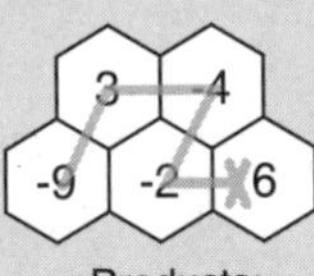

Products:
-27, -12, 8, ~~-12~~

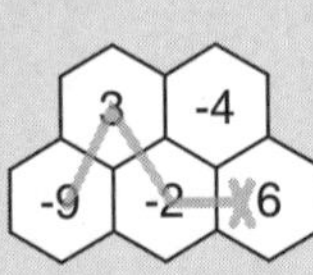

Products:
-27, -6, ~~-12~~

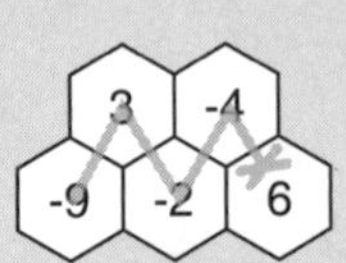

Products:
-27, -6, 8, ~~-24~~

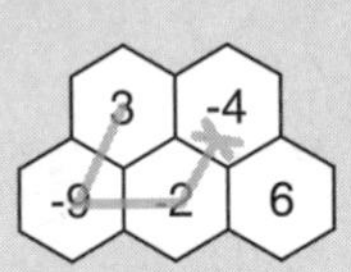

Products:
-27, 18, ~~8~~

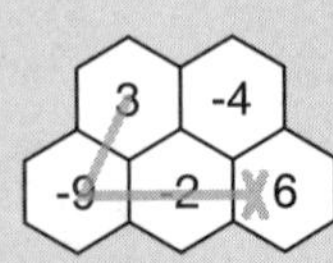

Products:
-27, 18, ~~-12~~

Since -9×3 was the least possible product in the hive, no other path of increasing products can include a move between the -9 and 3 hexagons. So, we draw a "wall" between these two.

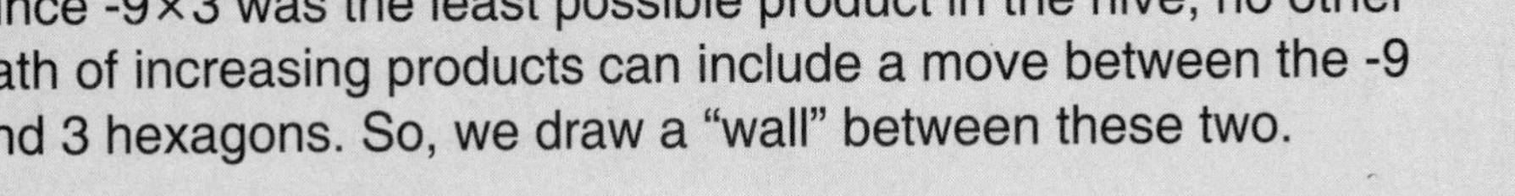

The least remaining product is $-4 \times 6 = -24$. We connect -4 and 6, and look for a path of increasing products. The wall between the -9 and 3 hexagons leaves us with just one possible path that begins with -4 and 6 ***and*** crosses all hexagons.

This is a path of increasing products, as shown to the right. This is the only solution.

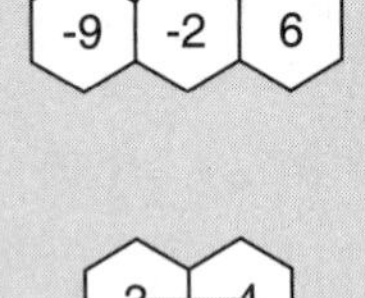

Products:
-24, -12, -6, 18

PRACTICE | Complete each Product Hive puzzle below.

75.

76.

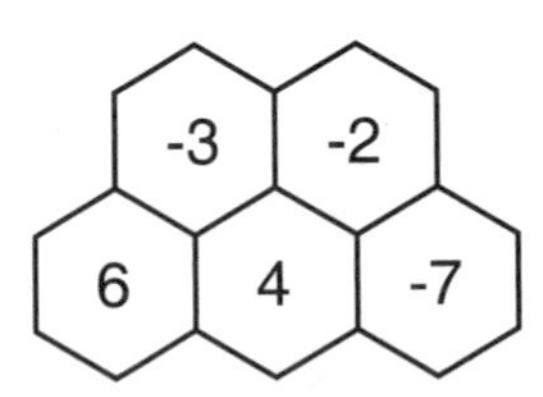

77.

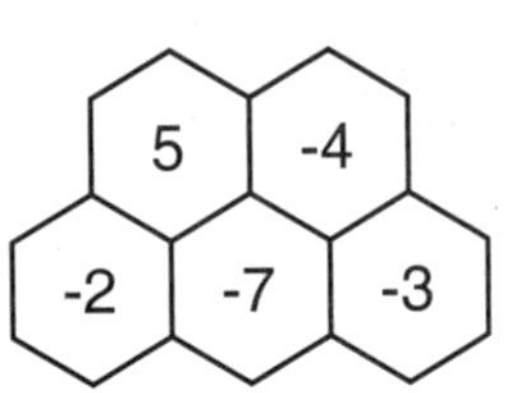

78.

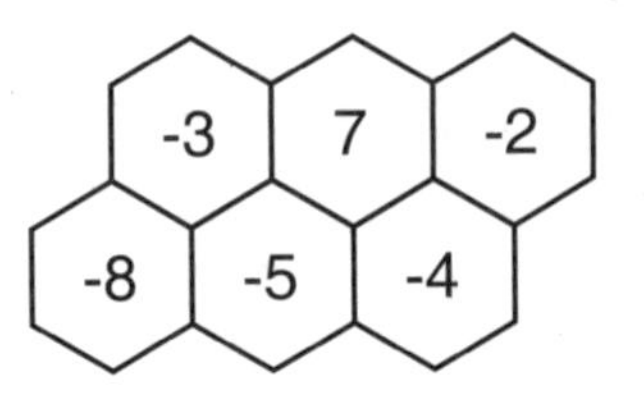

Print more Product Hive puzzles at BeastAcademy.com!

PRACTICE | Complete each Product Hive puzzle below.

79.

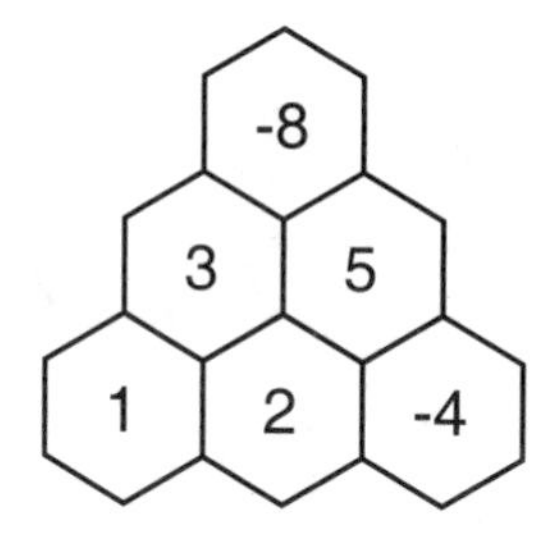

80.

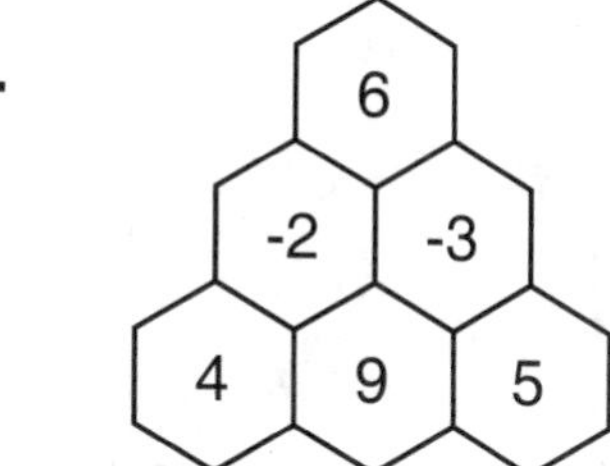

81. ★

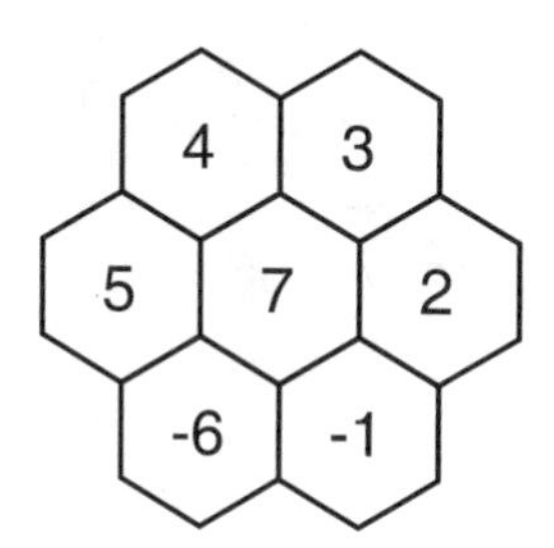

82. ★

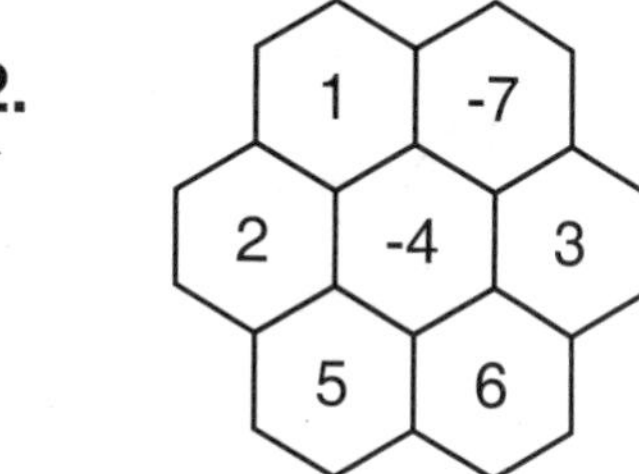

83. ★

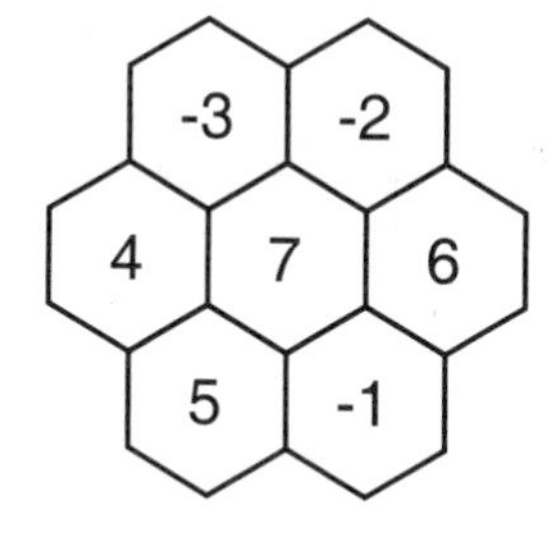

84. ★

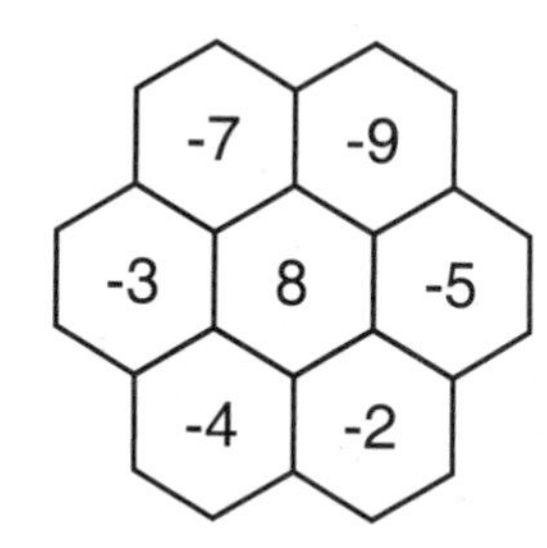

85. ★

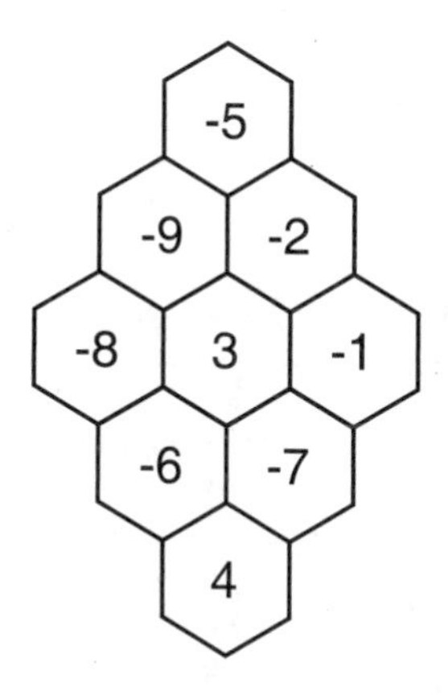

86. ★

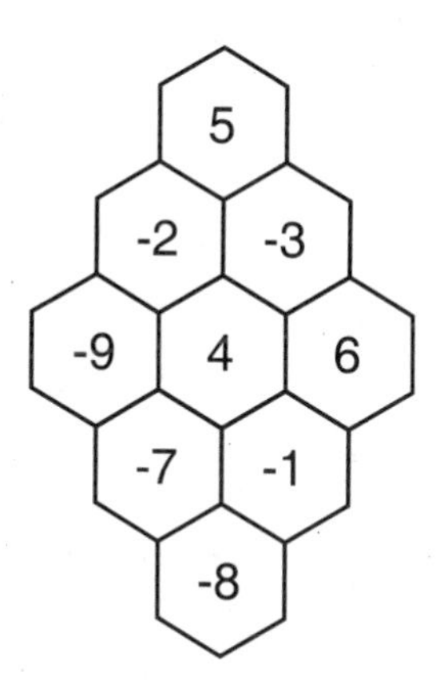

EXAMPLE | Compute (-4)×(-3)×(-2)×(-1).

Working from left to right, we compute the product as shown below:

$$\begin{aligned} & (-4)\times(-3)\times(-2)\times(-1) \\ = &\quad 12 \times (-2)\times(-1) \\ = &\quad -24 \times(-1) \\ = &\quad \mathbf{24}. \end{aligned}$$

– or –

We first multiply the numbers without considering their signs. Then, we find the sign of our answer.

Ignoring the negatives, we have 4×3×2×1 = 24.

Since the number of negatives in the product is even, the final result is positive. So, (-4)×(-3)×(-2)×(-1) = **24**.

An even number of negatives in a product gives a positive result. An odd number of negatives in a product gives a negative result...
...unless one of the integers is zero, which makes the whole product zero.

PRACTICE | Compute each of the following products.

87. 5×(-3)×(-1) = ______

88. 8×(-4)×2 = ______

89. -4×(-5)×(-9) = ______

90. 8×(-6)×4×(-2) = ______

91. -1×(-3)×(-5)×(-7) = ______

92. -100×4×(-5)×(-6) = ______

93. 3×(-2)×(-3)×(-2)×3 = ______

94. 5×(-7)×10×(-7)×5 = ______

95. -6×(-5)×(-4)×(-3)×2×(-1) = ______

96. -7×6×5×(-4)×3×(-2)×(-1) = ______

Find more practice problems at BeastAcademy.com!

PRACTICE | Answer each question below.

97. Compute $(-2)^1 + (-2)^2 + (-2)^3 + (-2)^4$. **97.** ________

98. Compute $(-1)^1 + (-1)^2 + (-1)^3 + \cdots + (-1)^{99} + (-1)^{100}$. **98.** ________

Recall that the three dots (⋯) mean that some numbers are not written in the middle, but the pattern continues.

99. The sum of two integers, a and b, is 76. What is $(-1)^a \times (-1)^b$? **99.** ________

100. Is the following product positive or negative? **100.** ________

$$99 \times (-98) \times 97 \times (-96) \times \cdots \times (-4) \times 3 \times (-2) \times 1.$$

101. Compute the product of all integers from -10 to 10 inclusive. **101.** ________

$$-10 \times (-9) \times (-8) \times (-7) \times \cdots \times 7 \times 8 \times 9 \times 10.$$

102. ★ Winnie multiplies four different integers. The absolute value of each integer is less than 10. Find the least possible product of the four integers. **102.** ________

Sign Wars is a pencil-and-paper game for two players.
The game is played on a 3-by-3 board, as shown on the right.

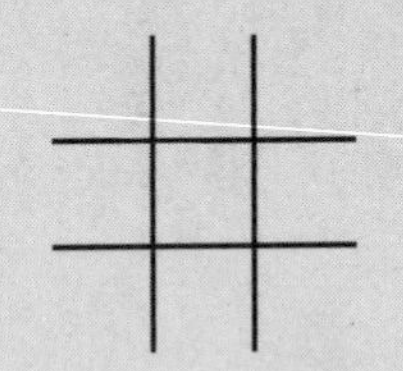

The game begins with the first player placing a 1 in an empty square. Then, the second player places a -1 in any remaining square. Play continues with the first player placing 1's and the second player placing -1's until the board has been filled.

At the end of the game, the product of each column and each row is computed. If a product is positive, the player who placed 1's gets a point. If a product is negative, the player who placed -1's gets a point. The player with the most points wins.

An example of a completed game is shown below. In this game, the player who placed -1's wins by a score of 4 to 2.

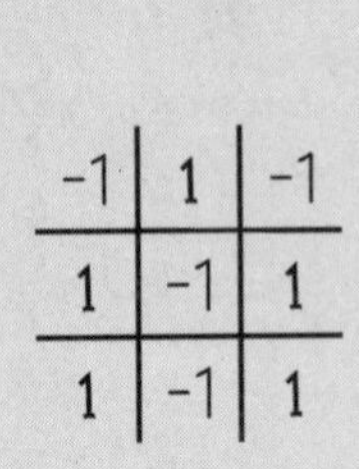

(−) (+) (−)

-1	1	-1	(+)
1	-1	1	(−)
1	-1	1	(−)

Scores
Player 1 (+): 2
Player 2 (−): 4

Player 2 wins!

Find a partner and play!

PRACTICE

In the games below, Grogg is placing 1's and Alex is placing -1's. It is Alex's turn to play. Find the only move Alex can make that guarantees he can win, ***regardless of Grogg's remaining moves.***

103.

-1	1	-1
	1	1
-1	1	

104.

-1	1	1
1		-1
	1	-1

105.

1		-1
	1	-1
-1	1	1

106. ★

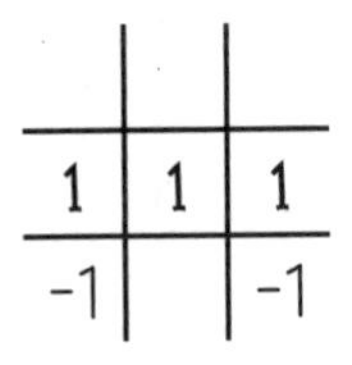

107. ★

-1	1	-1
1		1

108. ★

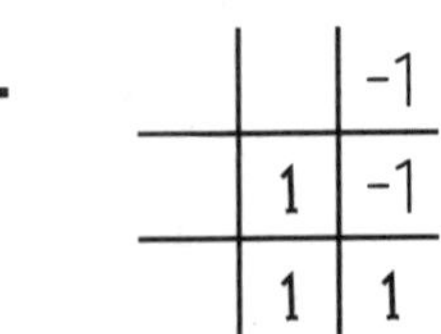

PRACTICE In the games below, Grogg is placing 1's and Alex is placing -1's. It is Grogg's turn to play. Find the only move Grogg can make that guarantees he can win, ***regardless of Alex's remaining moves.***

109.

<table><tr><td></td><td>-1</td><td></td></tr><tr><td></td><td>-1</td><td>1</td></tr><tr><td>-1</td><td>1</td><td>1</td></tr></table>

110.

<table><tr><td>-1</td><td>1</td><td>-1</td></tr><tr><td></td><td>1</td><td>1</td></tr><tr><td></td><td>-1</td><td></td></tr></table>

111.

<table><tr><td></td><td>-1</td><td>1</td></tr><tr><td>1</td><td>1</td><td></td></tr><tr><td></td><td>-1</td><td>-1</td></tr></table>

112. ★

<table><tr><td>1</td><td></td><td>-1</td></tr><tr><td>1</td><td>-1</td><td></td></tr><tr><td></td><td></td><td></td></tr></table>

113. ★

<table><tr><td>-1</td><td>-1</td><td></td></tr><tr><td></td><td></td><td>1</td></tr><tr><td></td><td></td><td>1</td></tr></table>

114. ★

<table><tr><td></td><td></td><td>-1</td></tr><tr><td>1</td><td>1</td><td></td></tr><tr><td>-1</td><td></td><td></td></tr></table>

PRACTICE Answer the following questions about Sign Wars.

115. ★ After playing several games of Sign Wars, Alex says, "If I get three -1's in a row, column, or diagonal, I am guaranteed to win." Is he correct?

116. ★ ✎ Is it possible for the player placing -1's to win with 6 points? If yes, provide an example. If no, explain why not.

117. ★★ ✎ Is it possible to tie in a game of Sign Wars? If yes, provide an example. If no, explain why not.

In an **Integer Blobs** puzzle, we circle "blobs" of two or more integers whose product is a given target number.

Every square in a blob must share at least one edge with another square in the blob. Blobs may not overlap, and every number in the grid must be in a blob.

EXAMPLE | Solve the Integer Blobs puzzle on the right.

Target: ***-24***

-8	3	2
2	-2	-1
6	-3	-4

The 6 in the bottom-left corner must be grouped with the 2 above it or the -3 to its right. Since the prime factorization of $24 = 2^3 \times 3$ has only one 3, we cannot include both the 6 and the -3 in the same blob. So, we group the 6 with the 2 above it.

Then, $6 \times 2 \times (-2) = -24$, so we complete the blob as shown.

-8	3	2
2	-2	-1
6	-3	-4

Next, since $-8 \times 3 = -24$, we make a blob with the -8 in the top-left corner and the 3 to its right.

The remaining integers have product -24, so we complete the puzzle as shown.

-8	3	2
2	-2	-1
6	-3	-4

PRACTICE | Solve each Integer Blob puzzle below.

118. Target: ***-20***

5	-4	2
4	-5	-10
-2	-2	-5

119. Target: ***36***

-2	-3	-3
-2	-9	-4
6	-2	-3

120. Target: ***-225***

5	-5	-3
9	45	-1
-5	25	-3

121. Target: ***126***

3	2	9
-6	-7	7
-2	-9	7

Prime factorizations can help us determine which numbers must be grouped into blobs...
...and which numbers cannot be grouped together.

PRACTICE | Solve each Integer Blob puzzle below.

122. Target: ***-198***

2	-11	9
11	-3	-6
3	2	33

123. Target: ***-84***

-2	-3	2
3	-14	-7
2	-7	12

124. Target: ***450***

25	-5	-9	-6
-2	3	10	-75
-3	30	-1	3
-15	-1	25	-6

125. Target: ***-1,950***

-3	25	-5	26
65	-2	-15	-5
2	-13	-26	15
-5	3	-39	50

126. ★ Target: ***-660***

-5	2	11	-2
-3	-2	2	-11
-3	-5	6	5
22	-30	2	11

127. ★ Target: ***1,155***

21	-33	-35	-3
11	5	-11	-7
5	7	-11	-5
-3	-55	3	-7

EXAMPLE Fill in the missing number in the equation below.

$$24 \div (-8) = \square$$

To divide $24 \div (-8)$, we find the number that can be multiplied by -8 to get 24.

We can use the multiplication fact $\boxed{-3} \times (-8) = 24$ to see that $24 \div (-8) = \boxed{\mathbf{-3}}$.

We can use multiplication to solve division problems!

PRACTICE Connect each division problem on the left with the corresponding multiplication problem on the right that can be used to solve it. Then, fill in the missing blanks in both equations.

128. $24 \div (-3) = \square$ $\square \times 9 = -36$

129. $24 \div (-2) = \square$ $\square \times (-5) = 45$

130. $-24 \div (-12) = \square$ $\square \times (-3) = -33$

131. $-33 \div (-3) = \square$ $\square \times (-2) = 24$

132. $-36 \div 9 = \square$ $\square \times (-11) = -33$

133. $-33 \div (-11) = \square$ $\square \times (-3) = 24$

134. $-45 \div 5 = \square$ $\square \times 5 = -45$

135. $45 \div (-5) = \square$ $\square \times (-12) = -24$

We can use the relationship between multiplication and division to show that the sign rules for dividing integers are the same as they are for multiplying integers.

$(+)\times(+)=\mathbf{(+)}$, so $\mathbf{(+)}\div(+)=(+)$.
$(-)\times(-)=\mathbf{(+)}$, so $\mathbf{(+)}\div(-)=(-)$.
$(+)\times(-)=\mathbf{(-)}$, so $\mathbf{(-)}\div(-)=(+)$.
$(-)\times(+)=\mathbf{(-)}$, so $\mathbf{(-)}\div(+)=(-)$.

EXAMPLE | Compute $-60\div(-5)$.

$60\div5=12$, and two numbers with the same sign have a positive quotient. Therefore, $-60\div(-5)=\mathbf{12}$.

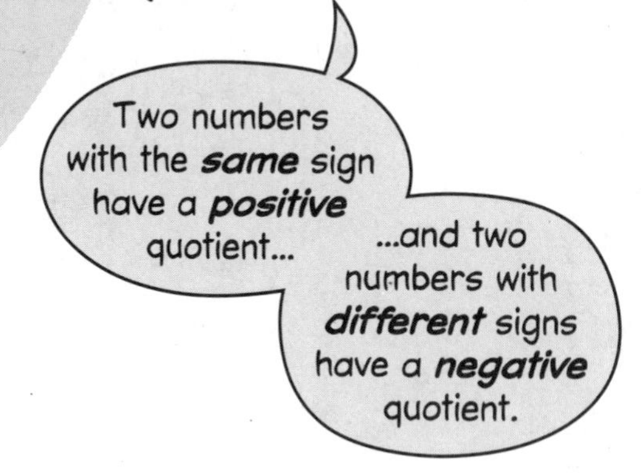

Two numbers with the ***same*** sign have a ***positive*** quotient...
...and two numbers with ***different*** signs have a ***negative*** quotient.

PRACTICE | Compute each of the following quotients.

136. $27\div(-9)=$ _____

137. $-24\div8=$ _____

138. $-48\div6=$ _____

139. $-32\div(-8)=$ _____

140. $45\div(-3)=$ _____

141. $90\div(-5)=$ _____

142. $-420\div(-70)=$ _____

143. $-9{,}600\div12=$ _____

144. $(-21\times10)\div15=$ _____

145. $(-120\div5)\div(-3)=$ _____

PRACTICE | Answer each question below.

146. Grogg computes the quotient $84\div(-2)$, then divides the result by -3. What is his result?

146. _______

147. Lizzie divides -24 by 3, then adds 9. Winnie adds -24 to 9, then divides by 3. How much greater is Lizzie's result than Winnie's result?

147. _______

EXAMPLE Fill in the missing entries in the Cross-Number puzzle below to make all the equations true.

	÷	-12	=	-4
÷	■	÷	■	÷
-8	÷	2	=	
=	■	=	■	=
	÷		=	

We fill in the missing entries in order as shown below.

48 ÷ (-12) = -4.

-8 ÷ 2 = -4.

-12 ÷ 2 = -6.

48	÷	-12	=	-4
÷	■	÷	■	÷
-8	÷	2	=	**-4**
=	■	=	■	=
	÷	**-6**	=	

48 ÷ (-8) = -6.

-4 ÷ (-4) = 1.

48	÷	-12	=	-4
÷	■	÷	■	÷
-8	÷	2	=	**-4**
=	■	=	■	=
-6	÷	**-6**	=	**1**

PRACTICE Fill in the missing entries in the Cross-Number puzzles below to make all the equations true.

148.

-80	÷	-4	=	
÷	■	÷	■	÷
10	÷	-2	=	
=	■	=	■	=
	÷		=	

149.

-15	×		=	
÷	■	÷	■	÷
5	×		=	-25
=	■	=	■	=
	×	-2	=	

You can find more of these puzzles at BeastAcademy.com!

PRACTICE Fill in the missing entries in the Cross-Number puzzles below to make all the equations true.

150.

-144	÷	6	=	
÷		÷		÷
-8	÷		=	-4
=		=		=
	÷		=	

151.

	÷		=	-60
÷		÷		÷
-25	÷		=	-5
=		=		=
-48	÷		=	

152.

-21	×		=	-84
÷		÷		÷
	×		=	
=		=		=
-7	×	-2	=	

153.

55	÷		=	
×		×		×
	÷	-4	=	5
=		=		=
	÷		=	25

154.

	×	5	=	375
÷		×		÷
15	÷		=	-15
=		=		=
	×		=	

155.

-24	÷		=	-3
×		÷		×
-21	×		=	
=		=		=
	÷		=	-252

In a **Zig-Zag** puzzle, there are two kinds of numbers: numbers above the zig-zag line (***top numbers***), and numbers below the line (***bottom numbers***).

Each bottom number is either multiplied or divided by the top number to its right according to the following rule:

> If the bottom number is divisible by the top number to its right, then divide. Otherwise, multiply.

The result of this calculation becomes the next bottom number to the right.

EXAMPLE | Fill in the missing entries to complete the Zig-Zag puzzle below.

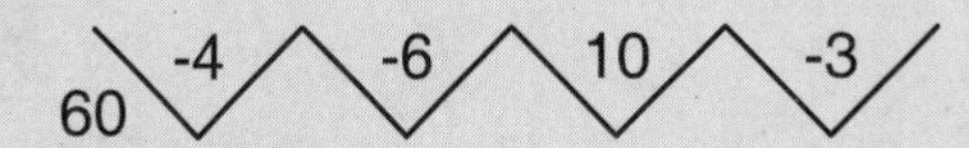

We begin with the 60 at the far left. Since -4 divides 60 evenly, we compute 60÷(-4) and write the quotient as the next bottom number.

Top: -4, -6, 10, -3
Bottom: 60, **-15**

$60 \div (-4) = -15$

Next, since -6 does not divide -15 evenly, we multiply.

Top: -4, -6, 10, -3
Bottom: 60, **-15**, **90**

$-15 \times (-6) = 90$

We repeat this process, and complete the rest of the puzzle as shown:

Top: -4, -6, 10, -3
Bottom: 60, **-15**, **90**, **9**, **-3**

$90 \div 10 = 9$
$9 \div (-3) = -3$

PRACTICE | Fill in the missing entries to complete the Zig-Zag puzzles below.

156. Top: 5, 4, -4, -6; Bottom: -16

157. Top: -2, -3, -4, -5; Bottom: 240

158. Top: 11, -4, 25, -3; Bottom: 44

159. Top: 16, -5, -20, 16; Bottom: -320

PRACTICE | Fill in the missing entries to complete the Zig-Zag puzzles below.

Some of these Zig-Zags are missing top numbers!

And don't forget about the bottom numbers on the left and right ends of the Zig-Zag!

160.

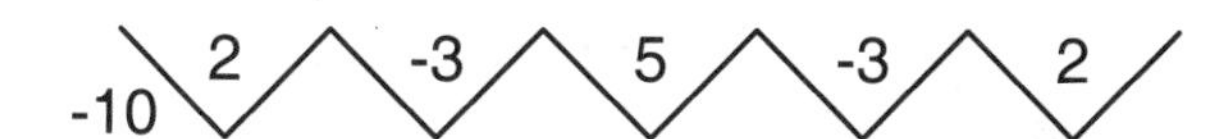

-10, 2, -3, 5, -3, 2

161.

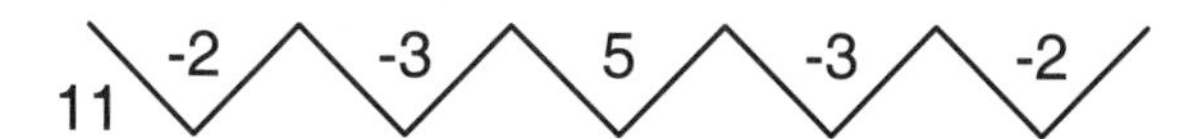

11, -2, -3, 5, -3, -2

162.

Top: 6, 2, 8, 12, -21

Bottom: -5, -15, 3, 14, -1

163.

Top: -9, -2, 7, -4

Bottom: -7, -21, 180, 15, -120

164.

Top: -6, -4, -7

Bottom: 35, -5, -10, 30, 120, 28

165. ★

Top: 3, -3, -4, -9, -6

Bottom: 27, -2, -10, -4

166.

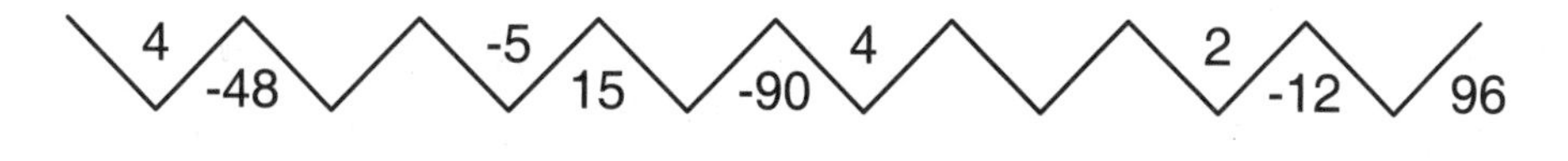

Top: 4, -5, 4, 2

Bottom: -48, 15, -90, -12, 96

EXAMPLE | Compute -(5+7).

Writing a negative in front of an expression means to take its opposite. Since $5+7=12$, we have $-(5+7)=\mathbf{-12}$.

— or —

Taking the opposite of an expression is the same as multiplying by -1. So, we have:

$$\begin{aligned}-(5+7)&=-1\times(5+7)\\&=-1\times 12\\&=\mathbf{-12}.\end{aligned}$$

PRACTICE | Evaluate the expressions below.

167. $-(6+4)=$ ______

168. $-(4-8)=$ ______

169. $-(-4+9-12)=$ ______

170. $-(199+299-300)=$ ______

171. $-(-8-7-13)=$ ______

172. $-(-23-34-45)=$ ______

PRACTICE | Answer each question below.

173. What is the value of $-x$ when $x=3$? **173.** ________

174. What is the value of $-x$ when $x=-5$? **174.** ________

175. What is the value of $-x+y$ when $x=-4$ and $y=-2$? **175.** ________

176. What is the value of $-(x+y)$ when $x=-4$ and $y=-2$? **176.** ________

177. If $-n=13$, what is the value of n? **177.** ________

178. If $-m+4=10$, what is the value of m? **178.** ________

EXAMPLE | Compute -9^2.

The expression -9^2 means "the opposite of 9^2," or $-(9^2)$.

Since $9^2 = 81$, we have $-9^2 = \mathbf{-81}$.

Note that $(-9)^2 = (-9) \times (-9) = 81$.

PRACTICE | Evaluate the expressions below.

179. -2^4 = ______

180. $(-2)^6$ = ______

181. -3^3 = ______

182. $(-3)^5$ = ______

183. $-(-6)^2$ = ______

184. $-(-5^2)$ = ______

PRACTICE | Answer each question below.

185. Circle the expressions below that are equal to -16.

-4^2 $\quad -(4^2)$ $\quad (-4^2)$ $\quad (-4)^2$ $\quad -(-4^2)$

186. Circle the expressions below that are equal to -125.

-5^3 $\quad -(5^3)$ $\quad (-5^3)$ $\quad (-5)^3$ $\quad -(-5^3)$

PRACTICE | Answer each question below.

187. Evaluate each expression below when $a = 3$.

$a^2 =$ ______ $(-a)^2 =$ ______ $-a^2 =$ ______

188. Evaluate each expression below when $a = -3$.

$a^2 =$ ______ $(-a)^2 =$ ______ $-a^2 =$ ______

189. Evaluate each expression below when $b = 6$.

$b^3 =$ ______ $(-b)^3 =$ ______ $-b^3 =$ ______

190. Evaluate each expression below when $b = -6$.

$b^3 =$ ______ $(-b)^3 =$ ______ $-b^3 =$ ______

191. Circle the expressions below that are ***positive*** for all nonzero values of x. If none of the expressions meet the criteria, circle "none of these".

x^2 $-x^2$ $-(x^2)$ $(-x)^2$ $-(-x)^2$ none of these

192. Circle the expressions below that are ***negative*** for all nonzero values of y. If none of the expressions meet the criteria, circle "none of these".

y^3 $-y^3$ $-(y^3)$ $(-y)^3$ $-(-y)^3$ none of these

193. Is the quantity $-a \times a$ always negative for nonzero values of a? Explain why or why not.

EXAMPLE Evaluate the expression $-7^{11} \div (-7)^{10}$.

We start by simplifying. Since 10 is even, we know that $(-7)^{10}$ is positive. So, $(-7)^{10}$ is equal to 7^{10}, and the expression simplifies to

$$-7^{11} \div (-7)^{10} = -7^{11} \div 7^{10}.$$

Next, we look for the number we multiply 7^{10} by to get -7^{11}.
$7^{10} \times \boxed{7} = 7^{11}$, so $7^{10} \times \boxed{-7} = -7^{11}$. Therefore, $-7^{11} \div (-7)^{10} = -7^{11} \div 7^{10} = \mathbf{-7}$.

— or —

We begin by ignoring the signs of the numbers we divide:

$$7^{10} \times \boxed{7} = 7^{11}, \text{ so } 7^{11} \div 7^{10} = 7.$$

Then, we consider the sign of the quotient.
Since 10 is even, $(-7)^{10}$ is positive, while -7^{11} is negative.
Since -7^{11} and $(-7)^{10}$ have opposite signs, their quotient is negative.

So, $-7^{11} \div (-7)^{10} = \mathbf{-7}$.

We check: $(-7)^{10} \times \boxed{-7} = (-7)^{11} = -7^{11}$ ✓

PRACTICE Evaluate the expressions below. You may write your answers using exponents.

194. $12^{15} \div 12^{14} =$ ______

195. $(-8)^{24} \div 8^{23} =$ ______

196. $25^{45} \div (-25^{44}) =$ ______

197. $-19^{30} \div (-19)^{30} =$ ______

198. $(-2)^{100} \div (-2) =$ ______

199. $-(-3)^{36} \div (-3)^{35} =$ ______

200. $(-10 \times 10^{29}) \div 10^{30} =$ ______

201. ★ $(-4)^{30} \div 2^{59} =$ ______

202. ★ $\left(-99^{10} \div (-99)^{10}\right) + \left((-99)^{9} \div (-99^{9})\right) =$ ______

PRACTICE | Answer each question below.

203. Grogg computes the sum

$$(-1)+(-3)+(-5)+\cdots+(-95)+(-97)+(-99).$$

Lizzie computes the sum

$$2+4+6+\cdots+96+98+100.$$

Winnie adds Grogg's sum and Lizzie's sum. What does she get?

203. ________

204. How many different pairs of integers have a product of -36?
Pairs with the same numbers in a different order are considered the same. For example, (2 and -2) is the same pair as (-2 and 2).

204. 

205. ★ Angelica writes down three nonzero numbers. Explain why at least one pair of her numbers has a positive product.

PRACTICE | Answer each question below.

206. Alex takes a nonzero integer, raises it to an even power, then raises the result to an odd power. Grogg starts with the same integer as Alex, raises it to an odd power, then raises the result to an even power. Are the signs of Alex's and Grogg's final numbers the same, different, or do we need more information to tell?

207. Beginning with a list of positive integers, Belinda flips a coin once for each number. If the coin lands heads, the number remains positive. If the coin lands tails, she replaces the number with its opposite.

a. ★ Belinda starts with a list of ***three*** positive numbers and flips a coin to determine the sign of each. What is the probability that the product of the three resulting numbers is negative?

a. ________

b. ★ Belinda starts with a list of ***four*** positive numbers and flips a coin to determine the sign of each. What is the probability that the product of the four resulting numbers is negative?

b. ________

c. ★★ Belinda starts with a list of ***one thousand*** positive numbers and flips a coin to determine the sign of each. What is the probability that the product of the thousand resulting numbers is negative?

c. ________

CHAPTER 3
Expressions & Equations

Use this Practice book with Guide 5A from BeastAcademy.com.

Recommended Sequence:

Book	**Pages:**
Guide:	76-89
Practice:	71-84
Guide:	90-109
Practice:	85-101

You may also read the entire chapter in the Guide before beginning the Practice chapter.

There are many ways to write multiplication. Besides using the × symbol, which can be confused for the variable x, we can also use a dot.

For example, $2 \cdot 5$ means $2 \times 5 = 10$.

We can also indicate multiplication by writing numbers next to each other, using parentheses to separate the numbers.

For example, $-4(5)$ means $-4 \times 5 = -20$.

We could also write $-4(5)$ as $(-4)5$ or $(-4)(5)$.

Using a dot or parentheses to write multiplication helps avoid confusing the × symbol for the variable x!

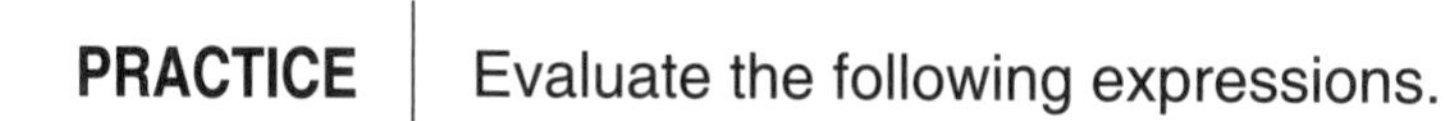

PRACTICE | Evaluate the following expressions.

1. $3 \cdot 7 =$ _______

2. $-4 \cdot 11 =$ _______

3. $5(9) =$ _______

4. $(18)(3) =$ _______

5. $-7 \cdot (-11) =$ _______

6. $2(3)(4) =$ _______

7. $6(8+9) =$ _______

8. $(4-6) \cdot 13 =$ _______

9. $(5 \cdot 6)+(7 \cdot 8) =$ _______

10. $(6-7)(7-8)(8-9) =$ _______

To show the product of a number and a variable, we write them next to each other, with the number to the left of the variable.

For example, $3n$ means $3 \cdot n$, and $3ab$ means $3 \cdot a \cdot b$.

We read $3n$ as "three n," and $3ab$ as "three a b."

EXAMPLE | Compute $6xy$ for $x=2$ and $y=5$.

$6xy$ means $6 \cdot x \cdot y$. So, when $x=2$ and $y=5$, we have

$$6xy = 6 \cdot 2 \cdot 5 = \mathbf{60}.$$

We don't need parentheses when writing the product of a number and a variable.

PRACTICE | Evaluate the expressions below for $a=3$, $b=-4$, and $c=5$.

11. $10a =$ ______

12. $5b =$ ______

13. $-7c =$ ______

14. $ac =$ ______

15. $2ab =$ ______

16. $-6abc =$ ______

17. $2(3b) =$ ______

18. $4(2ac) =$ ______

PRACTICE | Evaluate the expressions below for $x=8$ and $y=6$.

19. $2x+y =$ ______

20. $5y-2x =$ ______

21. $6-2xy =$ ______

22. $5xy+4x+3y+2 =$ ______

Complicated division expressions are often written using a fraction bar.

For example, $(6+9) \div (7-2)$ can be written as $\frac{6+9}{7-2}$.

We evaluate the numerator and denominator of a fraction first, then divide. So, we have:

$$\frac{6+9}{7-2} = \frac{15}{5} = \mathbf{3}.$$

Math beasts almost never use the ÷ symbol for division!
With a fraction bar, it's like the numerator and denominator are inside invisible parentheses!

PRACTICE | Evaluate each expression below.

23. $\frac{4+14}{3} =$ ________

24. $\frac{24}{15-23} =$ ________

25. $\frac{27+8}{3+2} =$ ________

26. $\frac{20}{4(3)} =$ ________

27. $\frac{6(-8)}{4(-6)} =$ ________

28. $\frac{3(4+5)}{(2-11)} =$ ________

PRACTICE | Evaluate the expressions below for $a=4$, $b=6$, and $c=24$.

29. $\frac{ab}{c} =$ ________

30. $\frac{b+c}{a+5} =$ ________

31. $\frac{c-a-b}{2a-6} =$ ________

PRACTICE | Evaluate the expressions below for $r=-2$, $s=3$, and $t=-7$.

32. $\frac{s+t}{r} =$ ________

33. $\frac{12r+s}{-t} =$ ________

34. $\frac{-16r}{3s-t} =$ ________

EXAMPLE Circle the expression below that is equivalent to $18+24\div(6-3)$. Then, evaluate the circled expression.

$$\frac{18+24}{6-3} \qquad 18+\frac{24}{6-3} \qquad 18+\frac{24}{6}-3 \qquad \frac{18+24}{6}-3$$

We can rewrite the division in $18+24\div(6-3)$ as a fraction. Since division comes before addition in the order of operations, only 24 is divided by the grouped quantity $(6-3)$. So, we have numerator 24 and denominator $6-3$.

Therefore, $18+24\div(6-3)$ is equivalent to $\mathbf{18+\frac{24}{6-3}}$.

To evaluate, we compute the denominator of the fraction first, then divide, then add:

$$\begin{aligned} 18+\frac{24}{6-3} &= 18+\frac{24}{3} \\ &= 18+8 \\ &= \mathbf{26}. \end{aligned}$$

PRACTICE Connect each expression on the left with its equivalent expression on the right. Then, evaluate the matched expression on the right.

35. $(30-20)\div(5-3)$ $\qquad$ $30-\frac{20}{5}-3=$ ________

36. $(30-20)\div5-3$ $\qquad$ $30-\frac{20}{5-3}=$ ________

37. $30-20\div(5-3)$ $\qquad$ $\frac{30-20}{5-3}=$ ________

38. $30-20\div5-3$ $\qquad$ $\frac{30-20}{5}-3=$ ________

Remember, when evaluating expressions, we apply the following **order of operations**:

1. Grouped expressions (numerators, denominators, and expressions inside parentheses or absolute value bars)
2. Exponents
3. Multiplication and division (working from left to right)
4. Addition and subtraction (working from left to right)

PRACTICE | Evaluate each expression below.

39. $\frac{6+3}{3}+2=$ ________

40. $5-\frac{8}{6(4)}=$ ________

41. $3\cdot\frac{7+9}{2}=$ ________

42. $\frac{3\cdot 7+9}{2}=$ ________

43. $\frac{-3(4)}{(6-4)^2}=$ ________

44. $\frac{20^2}{2}+\frac{20}{2^2}+\left(\frac{20}{2}\right)^2=$ ________

45. $17-2\left(\frac{1+11}{2\cdot 3}\right)=$ ________

46. $\frac{6^2}{7+5}\cdot\frac{7-6-5}{2}=$ ________

47. $\frac{8(7-3)^2}{-(3-7)^3}=$ ________

48. $\left(\frac{5+7+9}{2^5-5^2}\right)^3=$ ________

A ***term*** is a number, a variable, or a product of numbers and variables. Terms with the same variables are called ***like terms***. For example, $3x$ and $6x$ are like terms, and $-2y$ and y are like terms. However, $5x$ and $5y$ are not like terms.

Numbers without variables, such as 4 and -7, are also like terms.

In a **Like Terms Link** puzzle, each pair of like terms is connected by a path, as shown in the solved example below.

		$5x$	$2n$	-6
	$8n$			
		$3x$	d	
$-d$				
3				

		$5x$	$2n$	-6
	$8n$			
		$3x$	d	
$-d$				
3				

Paths may not travel diagonally, cross another path, or pass *through* a square that contains a term.

Like terms must also have the same exponents.
For example, $4a^2$ and $3a^2$ are ***like terms***, but $4a^2$ and $3a$ are not.

PRACTICE | Solve each Like Terms Link puzzle below. We recommend using a pencil.

Print more Like Terms Link puzzles at BeastAcademy.com.

49.

$12a$			
	$5c$	$6b$	$2a$
	b		$6c$

50.

			$7x$
	-8	$4y$	
		$-3x$	
$8y$			4

51.

	$13t$	$-5r$		
	$-4s$			
$-4r$			$2t$	$-5s$

52.

			$11u$	
	11		$2v$	$-2u$
		w		
			2	
		$2w$	$-9v$	

EXAMPLE | Simplify the expression $4a+3a$.

We can combine like terms to simplify expressions. $4a=a+a+a+a$, and $3a=a+a+a$. So, we have:

$$\begin{aligned} 4a+3a &= (a+a+a+a)+(a+a+a) \\ &= a+a+a+a+a+a+a \\ &= \mathbf{7a}. \end{aligned}$$

— or —

We factor a from each term. This gives

$$\begin{aligned} 4a+3a &= (4+3)a \\ &= \mathbf{7a}. \end{aligned}$$

If we have four a's, and we add three more a's, that makes seven a's all together.

PRACTICE | Simplify each of the following expressions.

53. $5x+4x=$ ______

54. $10y-3y=$ ______

55. $3d+4d+5d=$ ______

56. $s+3s+15s=$ ______

57. $-3w+12w=$ ______

58. $6p+(-p)=$ ______

59. $8c-14c+3c=$ ______

60. $-22g+36g-12g=$ ______

61. $12n-7n-5n=$ ______

62. $93k+47k-92k=$ ______

63. Write a simplified expression for the perimeter of a square with side length s.

63. ______

64. Write a simplified expression for the perimeter of a rectangle with width x and height $3x$.

64. ______

A ***coefficient*** is the number part of a term. For example, the coefficient of $16a$ is 16, and the coefficient of $-3xy$ is -3.

In a **Like Terms Corral**, we draw fences to separate like terms into corrals. The number of square units in each corral is the ***coefficient*** of the sum of the like terms within it. In the example below, the like terms $6x$ and $-2x$ have sum $4x$, so they are fenced into a corral with area 4 square units.

Similarly, the like terms $3y$, $4y$, and $5y$ have sum $12y$, so they are fenced into a corral with area 12 square units.

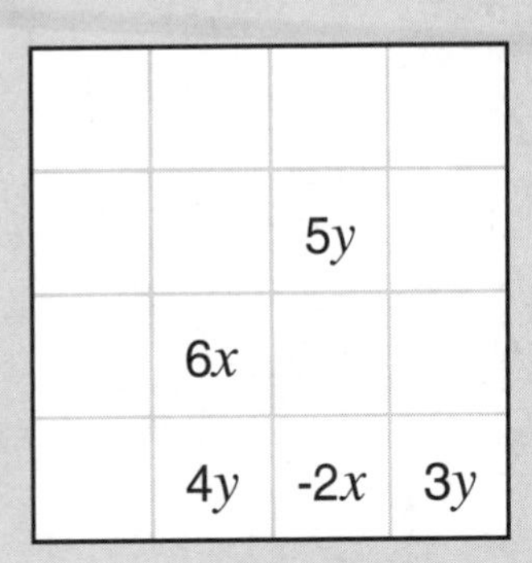

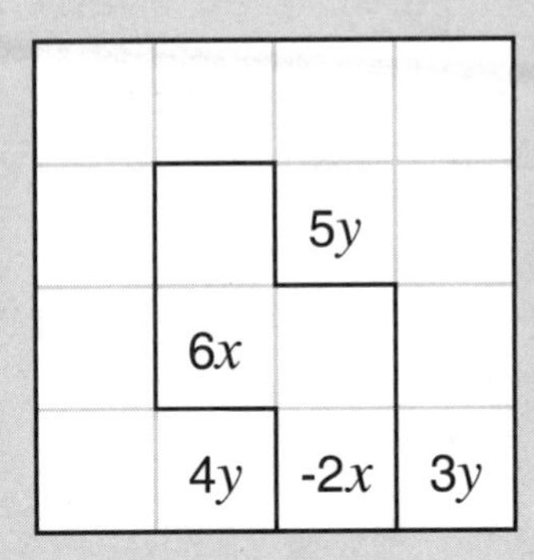

Like terms must all be enclosed within a single corral.

PRACTICE | Solve each Like Terms Corral puzzle below.

65.

a			
$7b$			
			$5a$
			$3b$

66.

$12c$		$2d$	
	$6d$		$-4c$

67.

$6s$				$2r$
	$2r$		$-2s$	
r	$3s$			
				$8r$

68.

	$-3w$			$2v$
		$10w$		
			$3u$	
	u			$7v$

PRACTICE | Solve each Like Terms Corral puzzle below.

69.

	$7m$	$13n$	$5p$	
	$6p$	m	$-7n$	

70.

		$11g$		
$-3f$				$-2g$
		$-3g$		
$-h$				$15f$
		$8h$		

71. ★

				$4x$
		$15y$	$20z$	
	$-11z$			
	$6x$		$-9y$	

72. ★

		$11i$		
	$-7j$		$-2i$	
	$9i$			
			$20j$	
		$-6i$		

73. ★

		$-7b$	$21a$	
	$5c$			
			c	
	$-11a$	$29b$		
				$-8b$

74. ★

	$17r$	$-34s$	$-3t$		
		$-2r$	$49s$	$9t$	

EXAMPLE Simplify the expression $7+6a-2+4b+3a-b$.

We rewrite all subtraction as addition, then rearrange the expression so that like terms are together. Finally, we combine like terms.

$$7+6a-2+4b+3a-b = 7+6a+(-2)+4b+3a+(-b)$$
$$= 6a+3a+4b+(-b)+7+(-2)$$
$$= 9a + 3b + 5$$

The expression simplifies to $\mathbf{9a+3b+5}$. There are no like terms, so the expression cannot be simplified further.

We usually arrange terms that have variables in alphabetical order and put any terms without a variable last.

PRACTICE Simplify the following expressions. If the expression cannot be simplified, circle it.

75. $6a+9a+4+12=$ ________

76. $15+3b+5+18b=$ ________

77. $22-4c-3+6c=$ ________

78. $8d-4+15-12d=$ ________

79. $19e+21f+9e+11f=$ ________

80. $-4g+10+10g-6h+3h=$ ________

81. $12m-5n+11p+7=$ ________

82. $14r+8q-3r-16q+10r=$ ________

83. $3u+7v-2u-v-u=$ ________

84. $7-8s+10t-3s+11t-12=$ ________

EXAMPLE | What is the value of $3x+7x$ when $x=47$?

We replace x with 47, then evaluate the expression:

$$\begin{aligned}3x+7x &= 3(47)+7(47)\\ &= 141+329\\ &= \mathbf{470}.\end{aligned}$$

— or —

We simplify the expression first, then evaluate.
$3x+7x=10x$. When $x=47$, we have $10x=10(47)=\mathbf{470}$.

Simplifying expressions ***before*** evaluating can save time!

PRACTICE | Compute each expression below for $a=7$, $b=$ -17, and $c=29$.

85. $12a+8a=$ ______

86. $5b-4b=$ ______

87. $11c-2c+11c=$ ______

88. $3a-c+2a=$ ______

PRACTICE | Compute each expression below for $r=11$, $s=13$, and $t=$ -5.

89. $r+2t+3r+4t=$ ______

90. $3s-11+t+7s+9=$ ______

91. $2s-6r-5s+19r-8s=$ ______

92. $14(5+s)-13s+17=$ ______

In a **Short Circuit** puzzle, each dot is labeled with an expression. The goal is to draw wires that connect each labeled dot on the left to a dot labeled with an equivalent expression on the right. The wires must not leave the room, cross each other, or pass through walls.

EXAMPLE Complete the Short Circuit puzzle below by connecting the three pairs of equivalent expressions.

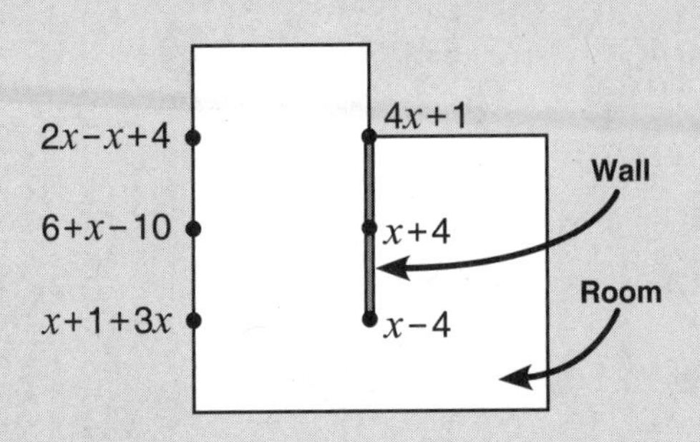

We begin by simplifying each expression on the left. Combining like terms, we have:

$2x-x+4=x+4$
$6+x-10=x+6-10=x+(-4)=x-4$
$x+1+3x=x+3x+1=4x+1$

We connect the dots on the left to the matching dots on the right without crossing wires or passing through walls.

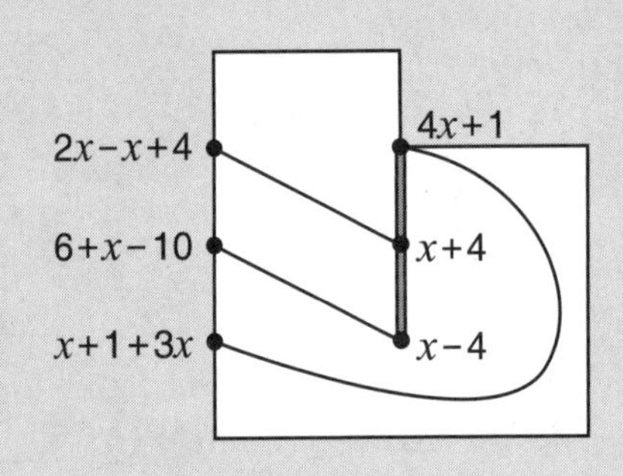

PRACTICE Complete each Short Circuit puzzle below. We recommend using a pencil.

93.

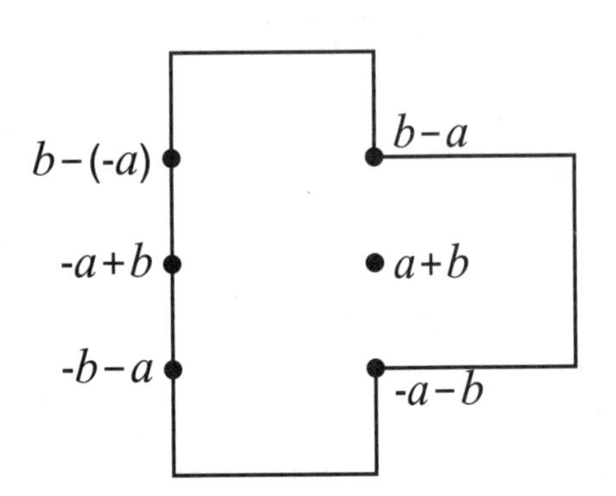

94.

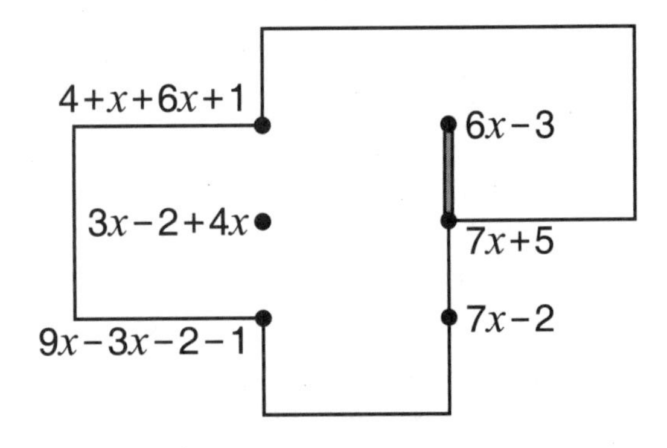

PRACTICE Complete each Short Circuit puzzle below. We recommend using a pencil.

95.

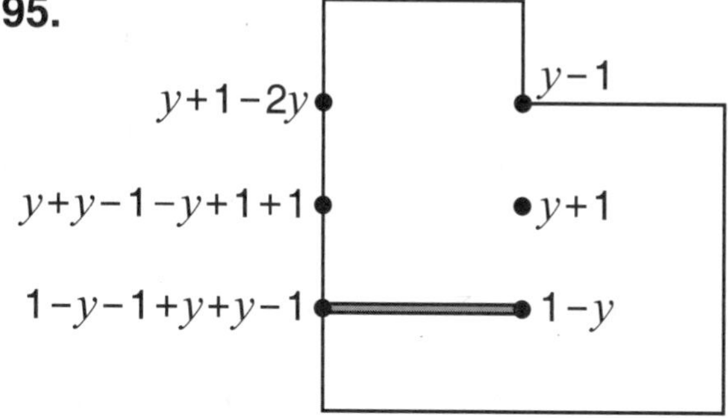

96.

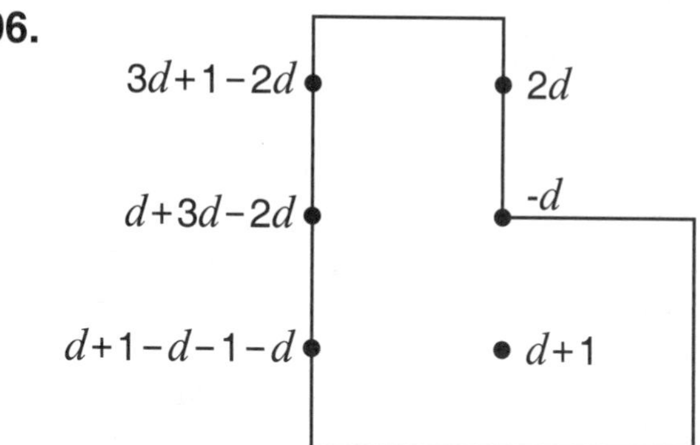

97.

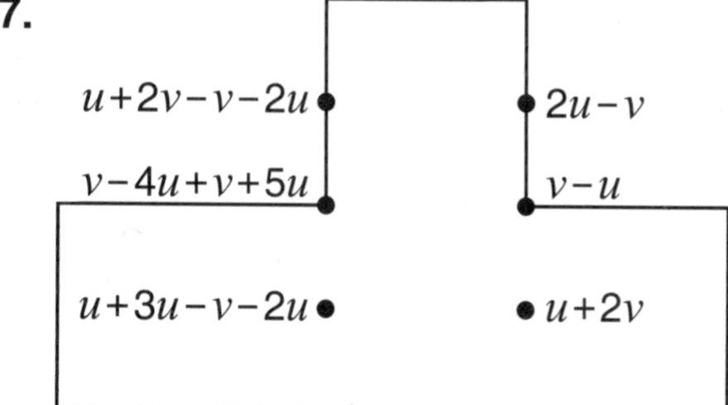

98.

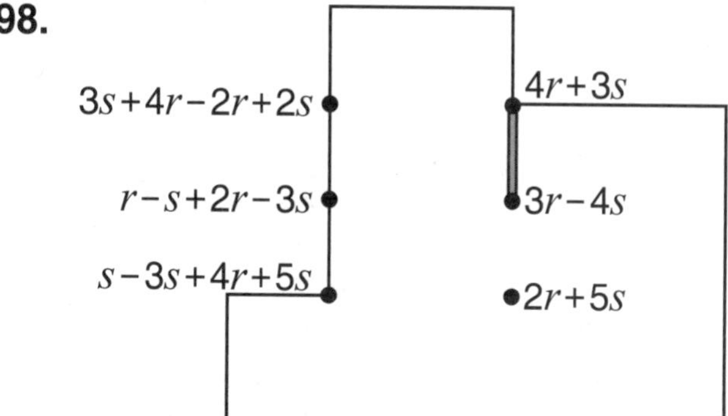

99.

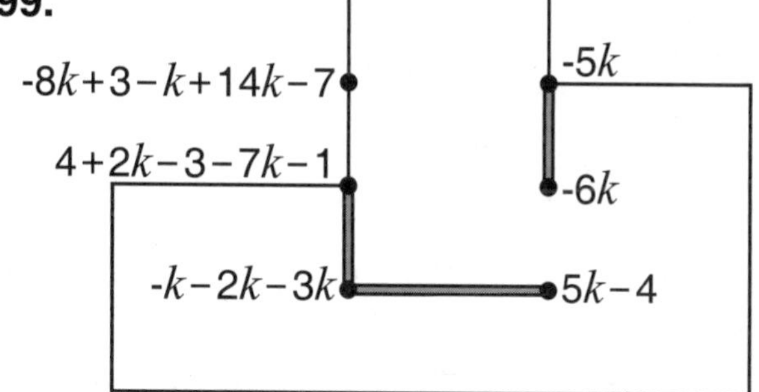

100.

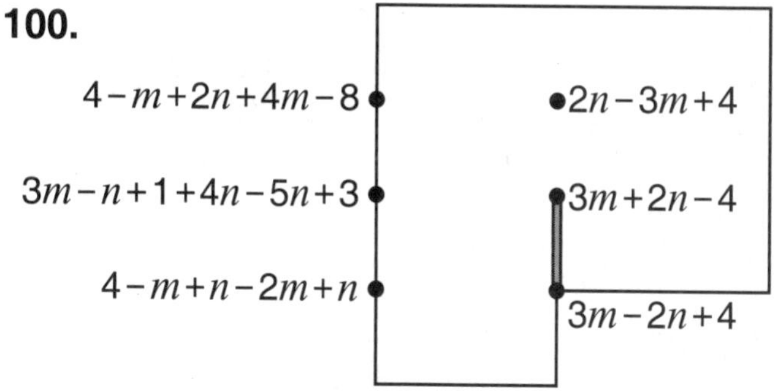

PRACTICE | Answer each question below.

101. What is the value of $6+\frac{3n-12}{4}$ when $n=8$?

101. ________

102. What is the value of $31r-24s+11s+19r+13s$ when $r=13$ and $s=27$?

102. ________

103. Each expression below uses the same ***positive*** value of a. Using the given blanks, write the simplified form of each expression in order from least to greatest.

$4+a-1+a-3$ ________ *least*

$3a+4a-8a$ ________

$5a-2-3a+7$ ________ *greatest*

104. ★ Each expression below uses the same ***negative*** value of b. Using the given blanks, write the simplified form of each expression in order from least to greatest.

$2b-3b-6+8b$ ________ *least*

$-6b+2+3b+4$ ________

$2b+7-5b-7$ ________ *greatest*

105. A rectangle with width w and height h has perimeter $2w+2h$. Two of these rectangles are attached along their widths. Write a simplified expression for the perimeter of the resulting rectangle.

105. ________

106. ★ Simplify the expression $2(x+2(x+2(x+2)))$.

106. ________

Two operations that undo each other are called ***inverse operations***. For example, addition and subtraction are inverse operations.

EXAMPLE | Compute $117+514-514$.

Adding 514 then subtracting 514 is the same as doing nothing.
So, $117+514-514=\mathbf{117}$.

Similarly, multiplication and division are inverse operations.

EXAMPLE | Compute $\frac{647 \cdot 13}{13}$.

Multiplying by 13 then dividing by 13 is the same as doing nothing.

$$\text{So, } \frac{647 \cdot 13}{13}=\mathbf{647}.$$

PRACTICE | Answer each question below.

107. Evaluate each expression below when $a=44$.

a. $a-24+24=$ ______ **b.** $\frac{a}{11} \cdot 11=$ ______ **c.** $\frac{71a}{71}=$ ______

108. Evaluate each expression below when $b=34$.

a. $(b+4)+17-17=$ ______ **b.** $\frac{9(b-7)}{9}=$ ______ **c.** $18-2b-18=$ ______

109. Circle the expressions below that are equal to 87.

$87+46-46$ $\frac{87(13)}{13}$ $34-87+34$ $\frac{-5(87)}{5}$ $9 \cdot \frac{87}{9}$

110. Circle the expressions below that are always equal to x.

$\frac{6x}{x}$ $14+x-14$ $\frac{-11x}{-11}$ $-6+x-6$ $\frac{x}{3} \cdot 3$

EXAMPLE Nyla says, "I'm thinking of a number. If I multiply my number by 7, then add 9, I get 93." What was Nyla's original number?

We work backwards. To find the number Nyla ***added*** 9 to in order to get 93, we ***subtract*** 9 from 93 to get $93-9=84$.

Then, to find the number Nyla ***multiplied*** by 7 to get 84, we ***divide*** 84 by 7 to get $84\div7=12$.

So, Nyla's original number was **12**.

We check our answer:
$12\cdot7=84$, and $84+9=93$. ✓

PRACTICE Solve each mystery number problem below.

111. Worlag chooses a number, multiplies it by 8, then adds 7 to the result. The number he ends up with is 55. What number did Worlag start with?

111. ______

112. Hubert divides his favorite number by 4. He subtracts 10 from this result and gets the number 17. What is Hubert's favorite number?

112. ______

113. Chloe takes her age in years, adds 5 to it, then divides the result by 8 to get 7. How many years old is Chloe?

113. ______

114. On August 1st, Barbara trimmed 11 inches from the length of her ponytail. During the month, her ponytail tripled in length. If Barbara's ponytail was 57 inches long at the end of August, how long was her ponytail before she trimmed it?

114. ______

Many problems can be solved by writing an equation.

EXAMPLE Nyla says, "I'm thinking of a number. If I multiply my number by 7, then add 9, I get 93." What was Nyla's original number?

We use n to represent Nyla's number. Multiplying n by 7, we get $7n$. Adding 9 to the result gives us $7n+9$. We are told that the final result of Nyla's computations is 93. So, we have:

$$7n+9=93.$$

Our goal is to ***isolate the variable n***. We can do this by undoing the steps above in reverse order. To undo adding 9, we subtract 9 from both sides of the equation. $7n+9-9$ equals $7n$, so we have:

$$\begin{array}{r} 7n+9=93 \\ -9\quad -9 \\ \hline 7n=84 \end{array}$$

To undo multiplying by 7, we divide both sides of the equation by 7. Since $\frac{7n}{7}$ equals n, we have:

$$\frac{7n}{7}=\frac{84}{7}$$
$$n=12$$

So, $n=12$, which means Nyla's original number was **12**. We check our answer:

$$7(12)+9=93.\ ✓$$

Review the basics of solving equations in the Variables chapter of Beast Academy 3C!

PRACTICE Solve the following equations by isolating the variable.

Once you have solved each problem below, check your work by replacing the variable in the original equation with your answer.

115. $8w+7=55$

116. $\frac{h}{4}-10=17$

115. $w=$ ________

116. $h=$ ________

117. $\frac{c+5}{8}=7$

118. $3(b-11)=57$

117. $c=$ ________

118. $b=$ ________

PRACTICE | Isolate the variable in each equation below.

119. $5+\frac{x}{2}=-3$

120. $7(b-4)=84$

119. $x=$ ______

120. $b=$ ______

121. $\frac{m+3}{-6}=-11$

122. $-7+13n=58$

121. $m=$ ______

122. $n=$ ______

123. $72=4(r+5)$

124. $19+\frac{v}{15}=49$

123. $r=$ ______

124. $v=$ ______

125. $-3z+8=2$

126. $\frac{k-9}{-3}=13$

125. $z=$ ______

126. $k=$ ______

Find more practice problems at BeastAcademy.com!

EXAMPLE | Solve for x in the equation $9x+6-2x=20$.

First, we simplify the left side of the equation.
Since $9x+6-2x$ equals $7x+6$, the equation simplifies to

$$7x+6=20.$$

To isolate x, we subtract 6 from both sides of the equation to get $7x=14$. Then, we divide both sides by 7 to get $x=\mathbf{2}$.

EXAMPLE | Solve for n in the equation $7n=3n+20$.

To isolate n, we need to get all terms with n on one side of the equation. We can eliminate the $3n$ from the right side of the equation by subtracting $3n$ from both sides:

$$\begin{array}{r} 7n = 3n+20 \\ -3n \quad -3n \\ \hline 4n = 20 \end{array}$$

Then, we divide both sides of $4n=20$ by 4 to get $n=\mathbf{5}$.

We can't always isolate the variable just by undoing operations.
We can use any steps needed to isolate the variable, as long as we do the same thing to both sides.

PRACTICE | Solve each equation below.

127. $16=2x-2+4x$

128. $3-4y+5+11y=85$

127. $x=$ ______

128. $y=$ ______

129. $2n+3(n+6)=43$

130. $9p=6p+21$

129. $n=$ ______

130. $p=$ ______

131. $99-4r=7r$

132. $8t+13=20t-35$

131. $r=$ ______

132. $t=$ ______

133. $24a-31=30a+11$

134. $6(c+5)-2c=56-9c$

133. $a=$ ______

134. $c=$ ______

In a **Circle Sum** puzzle, the number in each circle is the sum of the numbers in the connected circles below it.

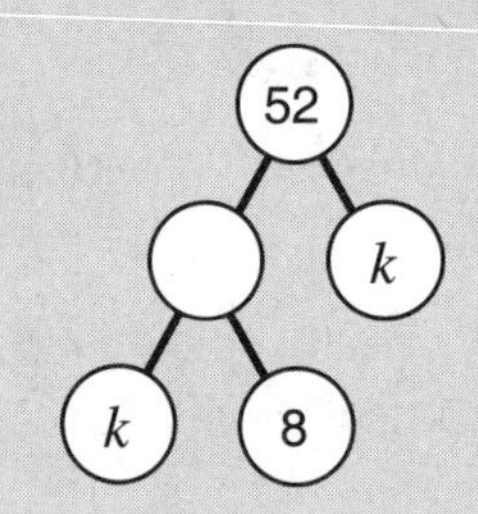

EXAMPLE | Find the value of k in the Circle Sum puzzle to the right.

The blank circle is the sum of the two numbers connected below it, so we can label it $k+8$. Then, we can use the top three circles to write an equation. The sum of $k+8$ and k is 52, so

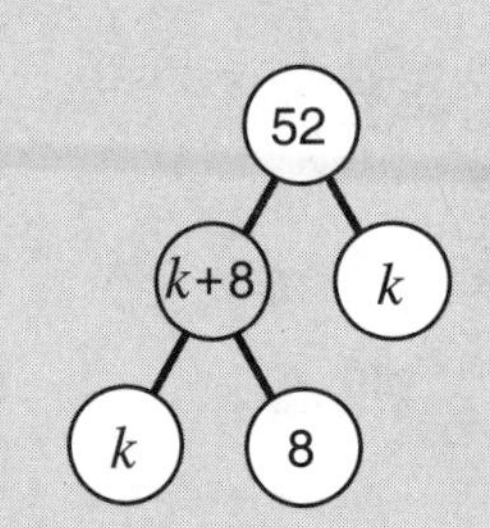

$$(k+8)+k=52.$$

Combining like terms, the equation simplifies to $2k+8=52$.

Subtracting 8 from both sides of the equation gives $2k=44$. Dividing both sides by 2, we have $k=\mathbf{22}$.

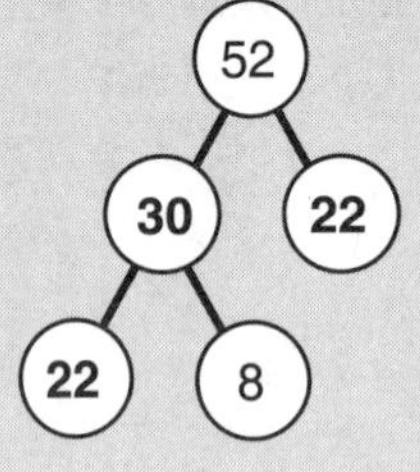

PRACTICE | Find the value of the variable in each diagram below.

135.

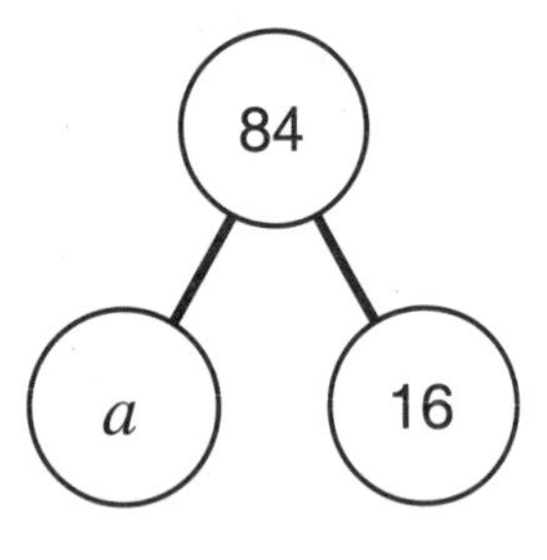

136.

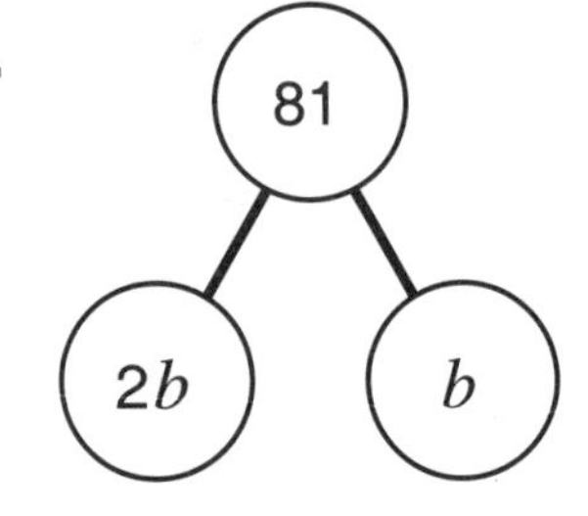

135. $a=$ ________

136. $b=$ ________

137.

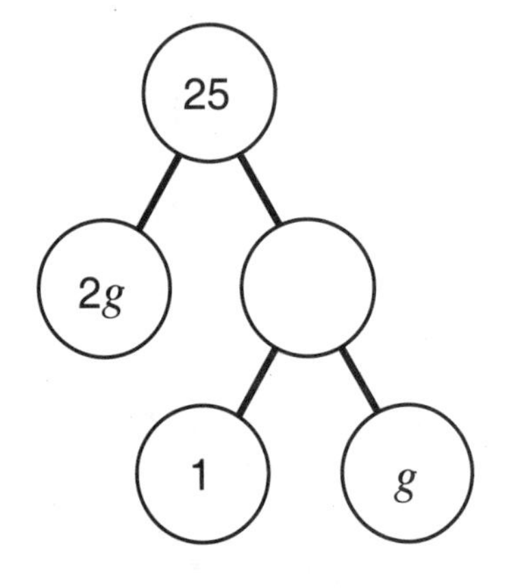

138.

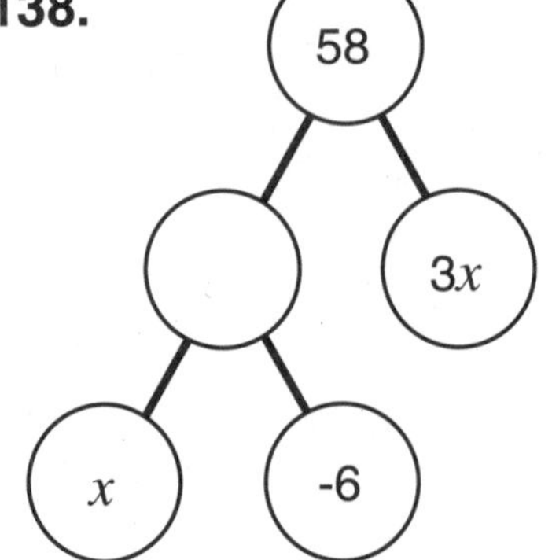

137. $g=$ ________

138. $x=$ ________

139.

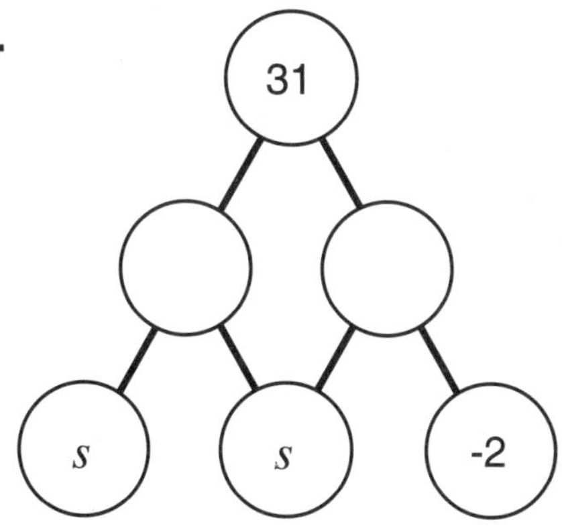

140.

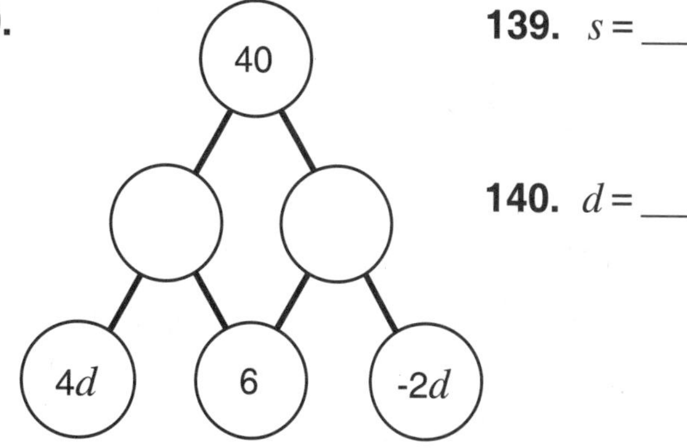

139. $s =$ ________

140. $d =$ ________

141.

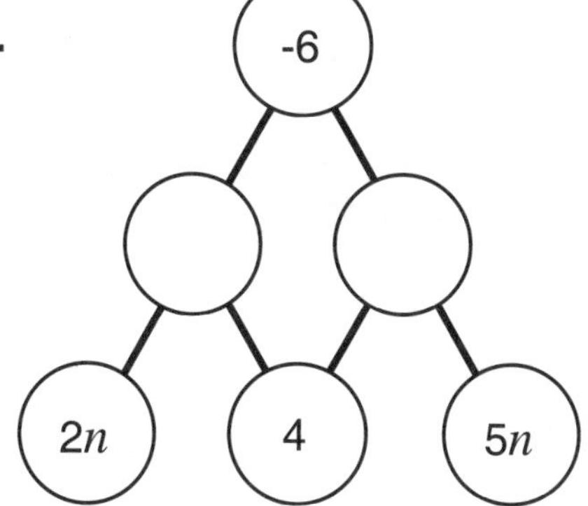

142.

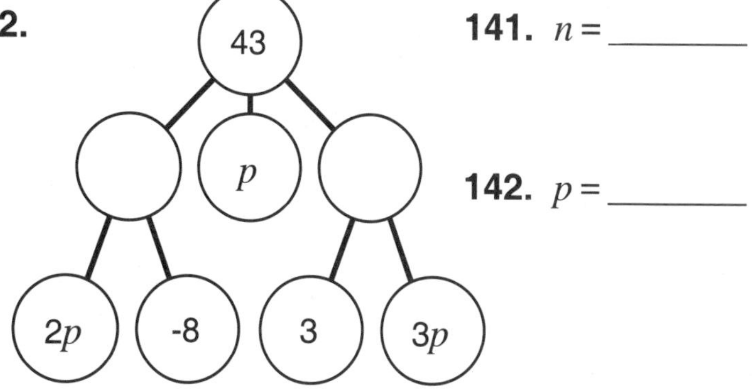

141. $n =$ ________

142. $p =$ ________

143.

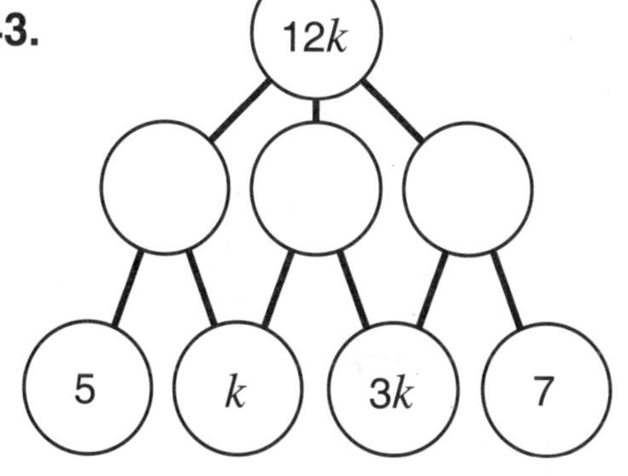

144.

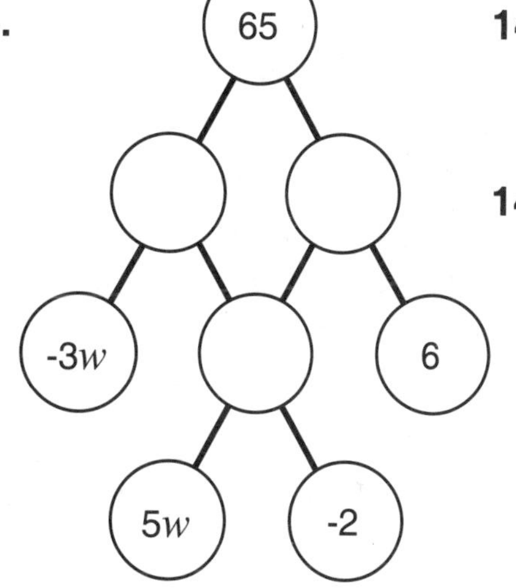

143. $k =$ ________

144. $w =$ ________

PRACTICE | Find the value of the variable in each diagram below.

145.

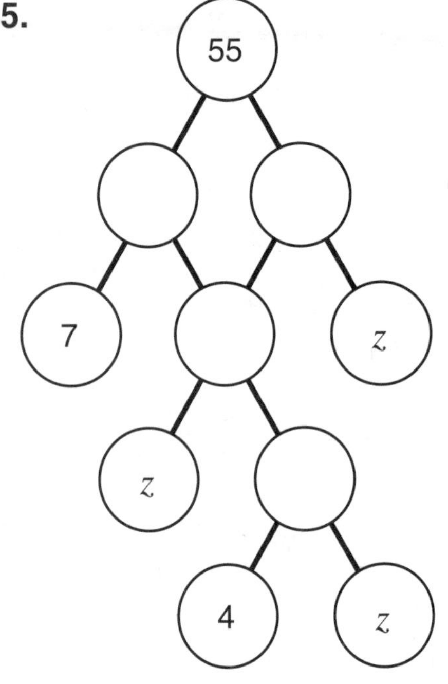

146.

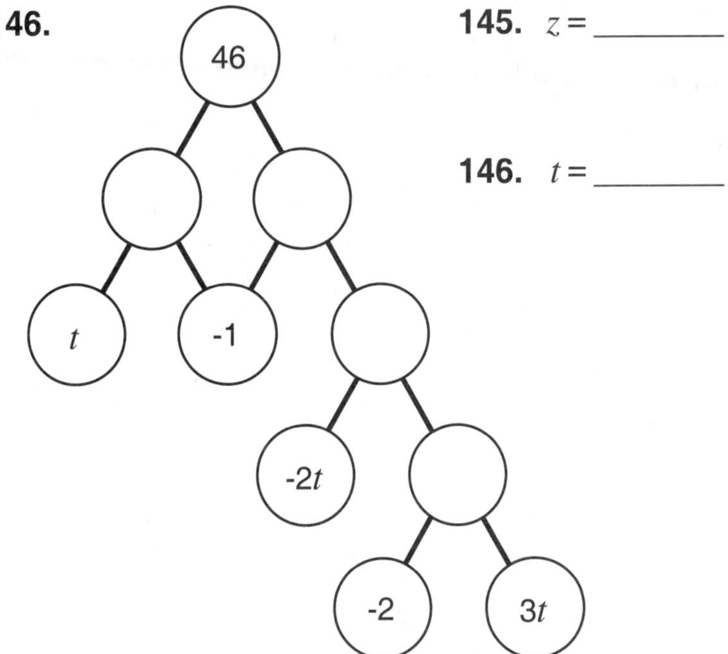

145. $z =$ ________

146. $t =$ ________

147.

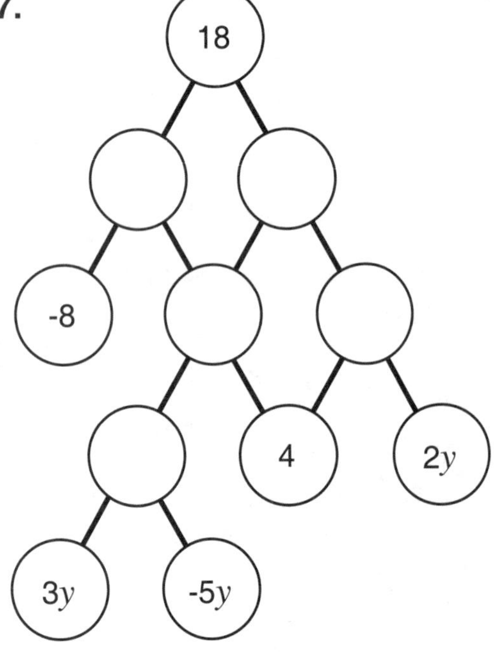

148.

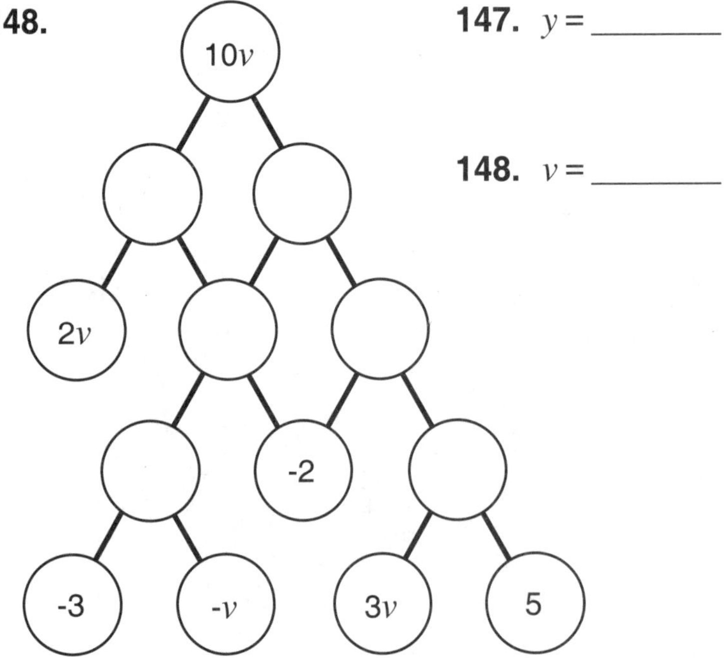

147. $y =$ ________

148. $v =$ ________

EXAMPLE | A rectangle that is 6 cm wider than it is tall has a perimeter of 52 cm. What is the height of the rectangle?

We draw a diagram. If we call the rectangle's height in centimeters h, then its width is $h+6$.

$h+6$

h h

$h+6$

The perimeter is 52 cm, so $h+h+(h+6)+(h+6)=52$.
Combining like terms, the equation simplifies to $4h+12=52$.

Subtracting 12 from both sides of the equation gives $4h=40$.
Dividing both sides by 4, we have $h=10$.

Since h represents the height of the rectangle, the height is **10 cm**.

PRACTICE | For each shape below, use the given perimeter to find the value of the unknown variable. Side lengths are given in inches.

149.

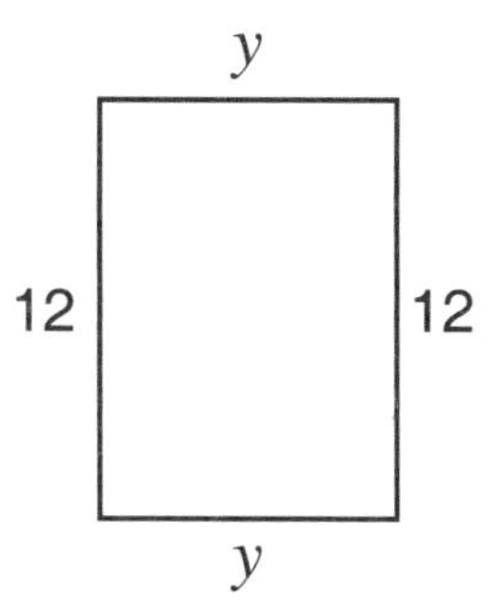

Perimeter: 36 in

$y=$ ________ in

150.

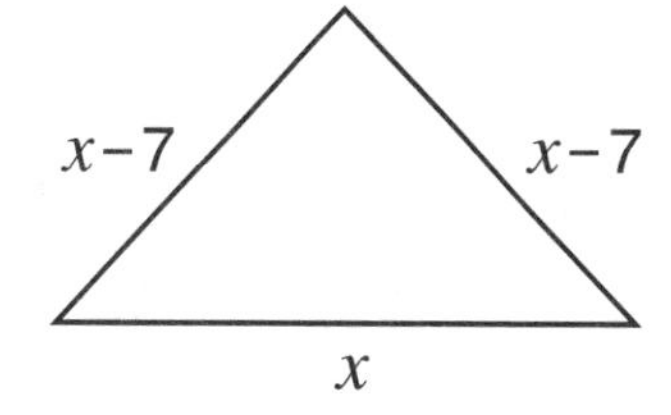

Perimeter: 52 in

$x=$ ________ in

151. A rectangle is 7 inches taller than it is wide. The perimeter of the rectangle is 46 inches. What is its width w?

151. $w=$ ________ in

PRACTICE | Answer each question below. Remember to include units where necessary.

152. The rectangle below has width $k+5$ inches and height $k+2$ inches. The perimeter of the rectangle is $6k$ inches. What is the ***area*** of the rectangle?

152. ________

$k+5$

$k+2$

153. ★ Ralph draws triangle ABC. Side BC is 3 inches longer than side AB, and side AC is 5 inches longer than side BC. The perimeter of the triangle is 35 inches. What is the length of side AB?

153. ________

154. ★ Two congruent octagons each have two opposite sides of length 4 feet and six sides of length s feet. When the octagons are joined along 4-foot sides, the resulting figure has perimeter 80 ft. What is s?

154. $s =$ ________

EXAMPLE

Twelve times the sum of Tim's favorite number and 3 is 84. What is Tim's favorite number?

Writing an equation can help turn a difficult word problem into easier math!

We use t to represent Tim's favorite number.

The sum of Tim's favorite number and 3 is $t+3$.
Twelve times this sum is $12(t+3)$.
This expression is equal to 84, so we have $12(t+3)=84$.

Dividing both sides of $12(t+3)=84$ by 12 gives $t+3=7$.
Subtracting 3 from both sides, we have $t=4$.

Since t represents Tim's favorite number, Tim's favorite number is **4**.

We check our answer:
$12(4+3)=12(7)=84$. ✓

PRACTICE

For each problem below, write and solve an equation to find the unknown value.

155. Three more than the product of a number and seven is 59. What is the number?

155. ________

156. Five less than six times a number is 37. What is the number?

156. ________

157. Borg subtracts 7 from the age of his monkaroo, then multiplies this amount by 5. He gets 20 as a result. How old is Borg's monkaroo?

157. ________

158. Grogg pulls a number of candies out of his pocket. He eats eleven of them, then gives the remaining candies to seven friends. If each friend receives nine pieces of candy, how many candies did Grogg originally pull out of his pocket?

158. ________

Word Problems

EXAMPLE Enrico has six more stamps than his brother Robbie. Together they have 100 stamps. How many stamps does Robbie have?

We let r represent the number of stamps Robbie has.
Enrico has six more stamps than Robbie, so Enrico has $r+6$ stamps.

Together, the brothers have 100 stamps, so $r+(r+6)=100$.
Combining like terms, the equation simplifies to $2r+6=100$.

Subtracting 6 from both sides of $2r+6=100$ gives $2r=94$.
Dividing both sides by 2, we have $r=47$. Since r represents the number of stamps Robbie has, Robbie has **47** stamps.

We check our answer: If Robbie has 47 stamps, Enrico has $47+6=53$ stamps. $47+53=100$. ✓

PRACTICE Write and solve an equation to answer each question below. Remember to include units where necessary.

159. Swillard's age is twice Jorple's age. The sum of their ages is 42 years. How old is Jorple?

159. ________

160. The combined cost of a slingshot and a grapefruit is 23 dollars. If the slingshot is 2 dollars more than six times the cost of the grapefruit, how many dollars does the grapefruit cost?

160. ________

161. Rod weighs seven pounds less than three times his brother Todd's weight. When the two brothers step on a scale together, the scale reads 213 pounds. What is Rod's weight in pounds?

161. ________

162. ★ On Tuesday, Fleet jogged 5 less than 4 times as many miles as his friend Wheezy. If Fleet jogged 22 more miles than Wheezy, how many miles did Fleet jog?

162. ________

EXAMPLE The sum of three consecutive integers is 24. What is the smallest of the three integers?

We let n represent the smallest of the three integers. Since the integers are consecutive, the second integer is one more than n, which is $n+1$.

Similarly, the third integer must be one more than $n+1$, which is $(n+1)+1=n+2$. The sum of all three integers is 24. So, we have:

$$n+(n+1)+(n+2)=24.$$

Combining like terms, this equation simplifies to $3n+3=24$. Subtracting 3 from both sides gives $3n=21$. Dividing both sides by 3, we have $n=7$.

Since n represents the smallest integer, the smallest integer is **7**.

We check our answer: $7+8+9=24$. ✓

PRACTICE Answer each question below.

163. The sum of 5 consecutive integers is 45. What is the smallest of these integers?

163. ________

164. Norbert lists nine consecutive integers ordered from least to greatest. The sum of the numbers in his list is 108. What is the middle number in Norbert's list?

164. ________

165. Grogg skip-counts by 7, starting with x. The sum of the first four numbers he says is 58. What is the value of x?

165. $x=$ ________

166. ★ Alex skip-counts by y, starting with -10. The sum of the first three numbers he says is 3. What is the value of y?

166. $y=$ ________

Think About It

Sometimes, a clever insight can help make a tough problem much easier!

PRACTICE | Answer each question below.

167. If $7x = 3{,}122$, what is $21x$?

167. ________

168. Solve for x in the equation below.

$$-17 = 17x^{101}$$

168. $x =$ ________

169. ★ What is the value of a in the equation below?

$$3a + 4b + 5c + 6 = 6a + 5c + 4b + 3$$

169. $a =$ ________

170. ★ If $x + y = 7$, what is $3x + 3y - 5$?

170. ________

171. ★ If $29x = 6{,}844$ and $17x = 4{,}012$, then what is $12x$?

171. ________

PRACTICE | Answer each question below.

172. What value of n makes the equation below true?

172. $n =$ ________

$$3n+6n+9n+12n+15n+18n+21n+24n+27n=270$$

173. List all values of c that make the equation below true.

173. ____________

$$|6c-30|=12$$

174. ★ Grogg solves the equation $(x-3)^2=49$ and gets $x=10$. Winnie solves the same equation and says, "I found another solution!" What is Winnie's solution?

174. ________

PRACTICE | Answer each question below.

175. ★ In isosceles triangle XYZ, side XY is 6 cm longer than side XZ. If the perimeter of the triangle is 21 cm, what is the length of side YZ?

175. ________

176. ★ The rectangle below has perimeter 54 inches. Grogg cuts the rectangle along the dashed line, and the combined perimeters of the two resulting rectangles is 68 inches. What is the width (w) of the rectangle in inches?

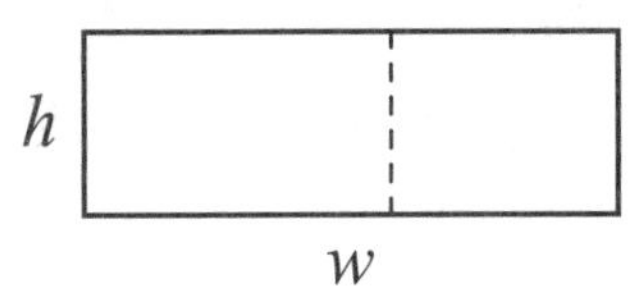

176.  w = ________

177. ★ A rectangular prism with length 4 feet and width 6 feet has a surface area of 138 sq ft. What is the height of the prism in feet?

177. ________ ft

PRACTICE | Answer each question below.

178. ★ Rosie is 6 years older than her sister Suzie, and Suzie is twice as old as her brother Toby. The sum of all three siblings' ages is 31. How old in years is each sibling?

178. Rosie: ________

Suzie: ________

Toby: ________

179. ★ Lizzie skip-counts out loud. The first three numbers she says can be represented by the expressions n, $n+7$, and $2n$, in that order. If Lizzie continues her pattern, what is the next number she will say?

179. ________

180. ★★ Anita and Beth have the same number of trading cards. Anita gives 18 of her cards to Beth. As a result, Beth has twice as many cards as Anita does. How many cards did Anita have before she gave some away?

180. ________

181. ★★ At the Beast Island Zoo, there are two types of creatures in the dobble exhibit: duo-dobbles, which are dobbles with 2 horns, and tri-dobbles, which are dobbles with 3 horns. All together, there are 76 horns in the dobble exhibit. If there are seven more tri-dobbles than duo-dobbles, then how many total dobbles are in the exhibit?

181. ________

HINTS
For Selected Problems

HINTS For Selected Problems

Below are hints to every problem marked with a ★.
Work on the problems for a while before looking at the hints.
The hint numbers match the problem numbers.

11. Draw it!

30. Consider cuts from lots of different angles, not just horizontal cuts and vertical cuts.

31. Consider cuts from lots of different angles, not just horizontal cuts and vertical cuts.

34. What shape is the prism's base?

35. Consider a prism whose bases have 100 sides. How would you compute its numbers of faces, edges, and vertices?

40. What shape is the pyramid's base?

41. Consider a pyramid whose base has 100 sides. How would you compute its numbers of faces, edges, and vertices?

43. Could Phil's polyhedron be a prism? A pyramid?

90. What did you learn from the two previous problems?

91. Is there a move that Player 1 can make that prevents Player 2 from making a cube net with the final square?

93. After the first three squares are drawn, what shapes are possible? Then, consider what you learned from the two previous problems.

96. How can the formulas you found in problems 35 and 41 help you answer this question?

102. What expression represents the area of one *face* of a cube with edge length n cm?

107. What is the area of the shaded region of the prism's net below? How can this help you find the prism's height, h?

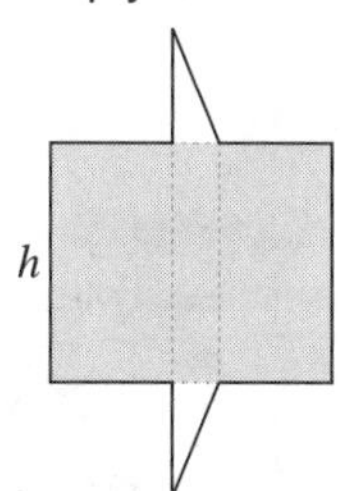

108. How can knowing the perimeter of the prism's bases help you find area of the shaded region of the prism's net below?

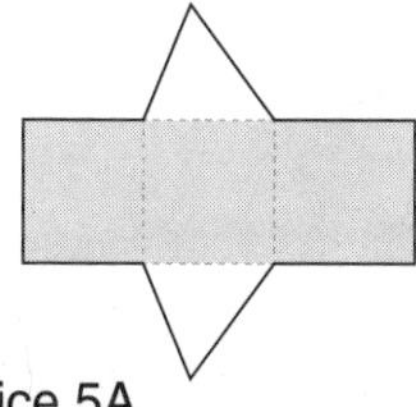

109. b. Find the area of each of the faces you counted in (a). How could you split the pentagonal faces into pieces whose areas you know how to find?

111. The net can be folded into a prism with three different edge lengths. One of them is 5 inches. How could you find the other two?

113. What is the total surface area of the cubes *before* stacking? How much of this surface area is removed as we stack each cube?

— *or* —

Consider the view of the whole sculpture from the bottom, top, front, back, left, and right views. What is the surface area we can see from each view?

114. When we attach six squares to create a cube net, how many sides of these squares make up the perimeter of the net?

115. What is the edge length of the original cube? What is the edge length of the new cube?

— *or* —

How does tripling the edge length of a cube change the area of each of its faces?

116. How does doubling the edge length of a cube change the area of each of its faces?

122. Consider the view of the solid from the bottom, top, front, back, left, and right. What is the surface area we can see from each view?

123. Consider the view of the solid from the bottom, top, front, back, left, and right. What is the surface area we can see from each view?

137. How many cubic feet are in *one* cubic yard?

141. Since Lizzie glues unit cubes together, the edge lengths of her prism must all be whole numbers. What whole number edge lengths give a volume of 165 cubic units?

142. What is the edge length of the original cube? What is the edge length of the new cube?

— *or* —

How does the volume of a cube with edge length n compare to the volume of a cube with edge length $3\times n$?

143. How does the volume of a cube with edge length n compare to the volume of a cube with edge length $2\times n$?

153. Since Jim cuts his block into unit cubes, the edge lengths of his block must all be whole numbers. What whole number edge lengths give a volume of 105 cubic units?

154. What does the fact that 35 cubes have no paint on them tell us about how the blocks are arranged?

155. Test some possibilities for the edge length of the original cube. Can you find a quick way to compute the number of unit cubes with 0 painted faces for any edge length? Can you find a quick way to compute the number of unit cubes with 1 painted face for any edge length?

160. What sizes of rectangles will be attached to create this net? How should these rectangles be attached to create a net with the greatest possible perimeter?

161. Stay organized! There are 11 total faces.

162. How many of the 20 remaining blocks are corner cubes? How many are edge cubes (and not on the corners)? How many faces does each type of cube contribute to the surface area of Cole's figure?

163. Where could we place the 7?

164. Which nets have the same opposite-face pairs as the original net? Which of these nets is a match?

CHAPTER 2

Integers 40

29. How can you organize your work to make sure you find every possible arrangement?

30. What are the least and greatest possible sums? Can we get every integer sum between those?

73. Two blocks below a 1 must have the same sign. What possibilities does that give us for the row with three blocks?

74. Find the missing entry in the row of two blocks. Then, if we assign variables a and b for the two unlabeled blocks in the bottom row, what expressions can we write in terms of a and b for the blocks in the row with three blocks?

81. The first three products are negative.

82. The first three products are negative.

83. The first five products are negative.

84. The first two products are negative.

85. The first four products are negative.

86. The first five products are negative.

102. What is the least possible product for two integers whose absolute value is less than 10? Three integers?

106. Alex has two more -1's to place. Where can Alex place a -1 on this move to guarantee that two rows will have a negative product? Is this a winning move?

107. Alex has two more -1's to place. Where can Alex place a -1 on this move to guarantee that two rows will have a negative product? Is this a winning move?

108. Alex has two more -1's to place. Where can Alex place a -1 on this move to guarantee that two columns will have a negative product? Is this a winning move?

112. How can Grogg play to guarantee that all three columns will have a positive product? Is this strategy enough to win?

113. How can Grogg play to guarantee that all three columns will have a positive product? Is this strategy enough to win?

114. How can Grogg play to guarantee that all three rows will have a positive product? Is this strategy enough to win?

115. Consider a game board with *three* -1's in any row, column, or diagonal. What happens when player 2 adds a fourth -1 on any of these game boards?

116. For the player placing -1's to get 6 points, every product must be negative. Can four -1's be placed on a game board in a way that makes every product negative?

117. In a completed game of Sign Wars, will the combined product of all three rows (row 1)×(row 2)×(row 3) be positive or negative? Will the combined product of all three columns (col 1)×(col 2)×(col 3) be positive or negative?
What does this tell you about the total number of negative products in all of the rows and columns?

126. $660 = 2^2 \times 3 \times 5 \times 11$.

127. $1{,}155 = 3 \times 5 \times 7 \times 11$.

165. Can you work from right to left in a Zig-Zag puzzle? Careful with that leftmost entry!

166. What number must be placed left of the 2 in this part of the puzzle?

2
-12

201. How can we write $(-4)^{30} = 4^{30}$ as a power of 2?

202. Can you evaluate each quotient? This one only *looks* harder than the rest.

205. What do we know about the signs of two integers whose product is positive?

207. **a.** How many of the $2^3 = 8$ possible outcomes of the coin flips give a negative result?

b. How many of the $2^4 = 16$ possible outcomes of the coin flips give a negative result?

c. If the product of the first 999 integers is positive, what is the probability that the product of all 1,000 integers is negative?

If the product of the first 999 integers is negative, what is the probability that the product of all 1,000 integers is negative?

71. Start with the z's.

72. Start with the i's.

73. Start with the a's.

74. Start with the s's.

104. Simplify the expressions first. Which of these expressions will always be positive when b is negative?

106. Stay organized. Start by simplifying $(x+2(x+2))$.

153. Draw triangle ABC. If side AB has length x, how can you label the lengths of the other two sides?

154. Draw the connected octagons. How many sides of length s are part of the perimeter? How many sides of length 4?

162. If we let w represent the number of miles Wheezy jogged, can we write two different expressions for the number of miles that Fleet jogged?

166. Write expressions for Alex's second and third numbers in terms of y.

169. What can you subtract from both sides to simplify this equation?

170. If you know $x+y$, how can you find $3x+3y$?

171. We could solve for x, but we don't need to!
How is $12x$ related to $29x$ and $17x$?

174. Is there a number besides 7 whose square is 49?

175. Since triangle XYZ is isosceles, we know two of its sides have equal length. Which two?

176. What is the length of Grogg's cut?

177. Label the height of the prism h. What expressions can be used to represent the areas of the prism's six faces?

178. Let t represent the age of the youngest sibling, Toby. What is Suzie's age in terms of t? What is Rosie's age in terms of t?

179. What number is Lizzie skip-counting by?

180. If we let c represent the number of cards Anita and Beth each start with, how many cards does Anita have after she gives cards away? How many cards does Beth have after she gets cards from Anita?

181. If we let d represent the number of duo-dobbles in the exhibit, how many tri-dobbles are in the exhibit?
How many horns do the duo-dobbles have all together?
How many horns do the tri-dobbles have all together?

SOLUTIONS

Chapters 1-3

2D Review — page 7

1. The perimeter is $17+13+5+5=$ **40 inches**.
To compute the area of this shape, we split the shape into a right triangle and a square as shown.

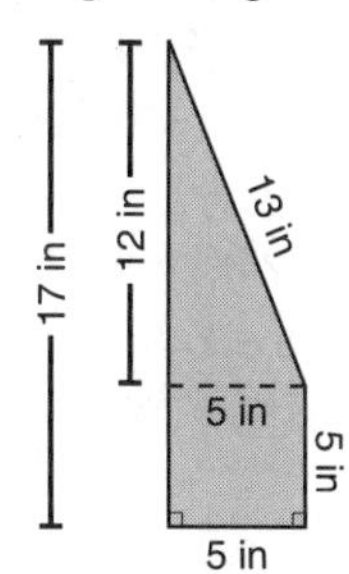

The area of the right triangle is $12\times5\div2=60\div2=30$ sq in.

The area of the square is $5\times5=25$ sq in.

So, the area of the larger shape is $30+25=$ **55 sq in**.

2. The perimeter is $13+4+15=$ **32 in**. The area of any triangle is equal to half the product of its base and height, so the area is $(4\times12)\div2=48\div2=$ **24 sq in**.

3. A regular pentagon has 5 sides of equal length. So, the perimeter of the pentagon is $5\times35=$ **175 cm**.

4. The area of a right triangle is half the product of its leg lengths: $7\times13\div2=91\div2=$ **$45\frac{1}{2}$ sq mm**.

5. The area of the large rectangle is $5\times6=30$ sq m. The area of the square is $4\times4=16$ sq m. No matter where the square is cut out, the area of the remaining shape is $30-16=$ **14 sq m**.

6. Since the perimeter of each triangle is 16 ft, the side length of each triangle is $16\div3=5\frac{1}{3}$ ft. The six sides of the hexagon are triangle sides. So, the perimeter of the hexagon is $6\times5\frac{1}{3}=$ **32 ft**.

— or —

The sides of the hexagon are six triangle sides. Each triangle has three sides. So, the perimeter of the hexagon is twice the perimeter of one triangle: $16\times2=$ **32 ft**.

Three Dimensions — 8

7. There are **6 faces**,

12 edges,

and **8 vertices**.

8. There are **5 faces**,

9 edges,

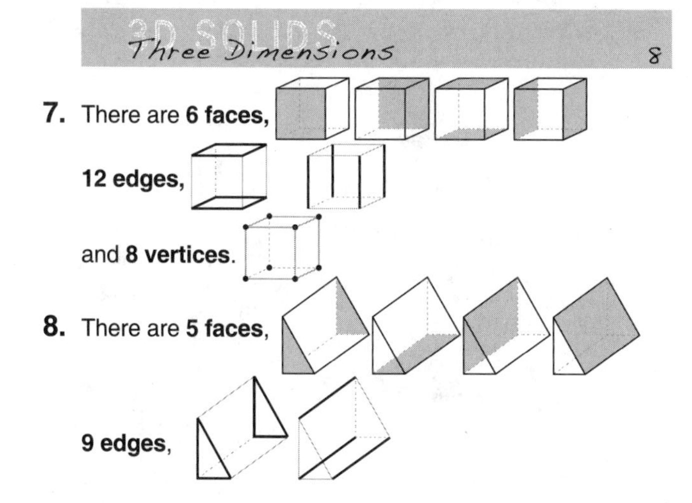

and **6 vertices**.

9. There are **7 faces**,

12 edges,

and **7 vertices**.

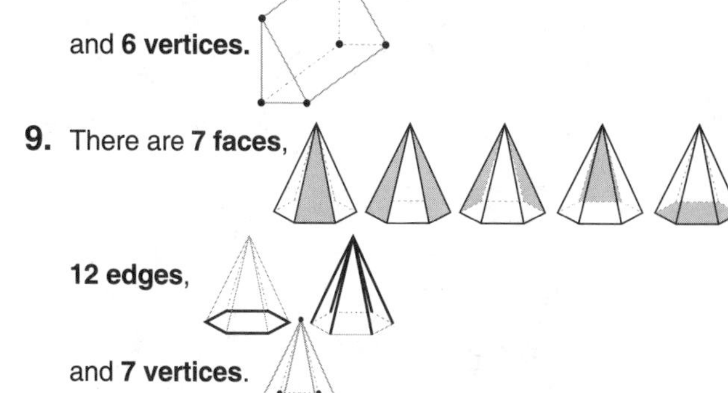

10. There are **8 faces**,

12 edges,

and **6 vertices**.

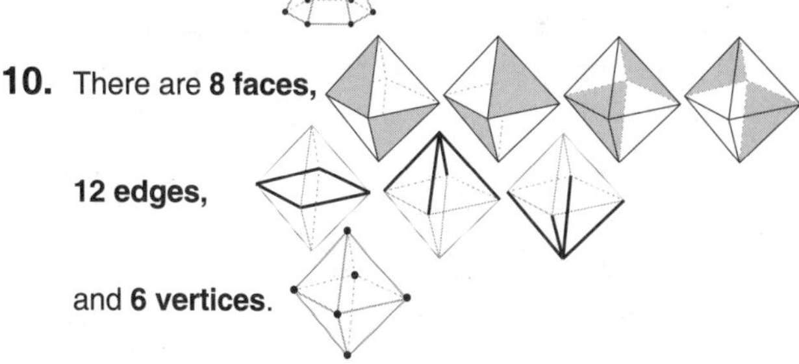

11. To create this polyhedron, each triangular face must share an edge with each of the three other triangles.

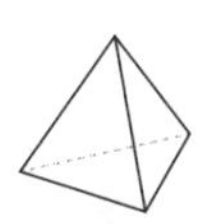

This polyhedron has **6 edges** and **4 vertices**.

Prisms — 9

12. The bases of this prism are octagons, so this is an **octagonal prism**.

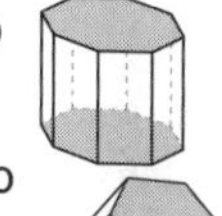

13. The bases of this prism are hexagons, so this is a **hexagonal prism**.

14. The bases of this prism are decagons, so this is a **decagonal prism**.

15. The bases of this prism are triangles, so this is a **triangular prism**.

16. The bases of this prism are pentagons, so this is a **pentagonal prism**.

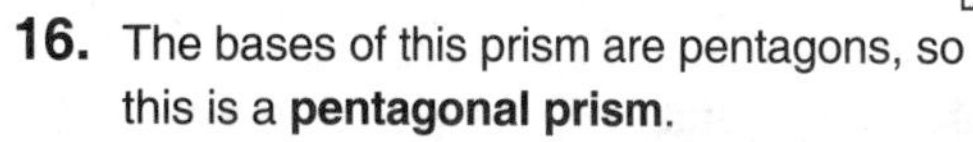

17. In this prism, each pair of opposite faces is a pair of congruent rectangles. So, this is a **rectangular prism**. Any pair of opposite faces can be used as the bases of a rectangular prism, so you may have shaded any one of the three pairs of opposite faces:

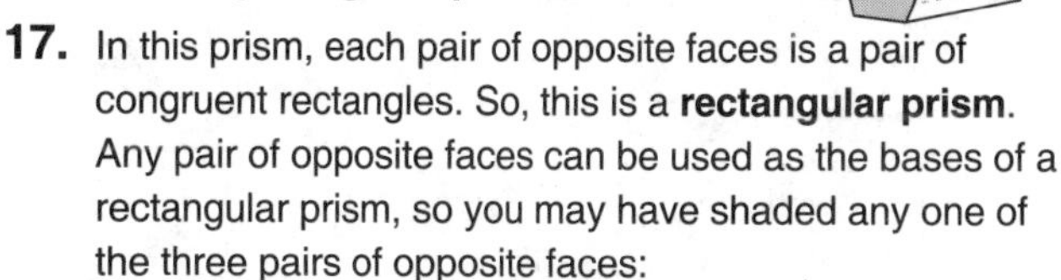

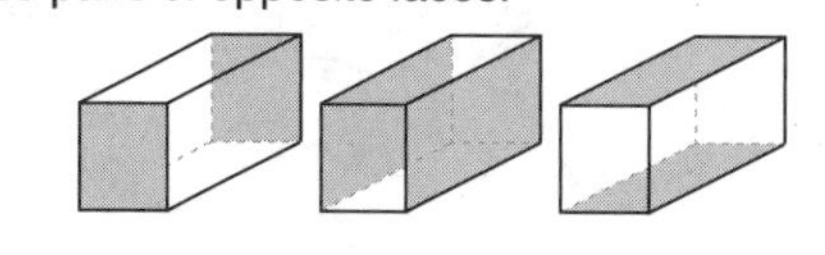

Below is the correct matching for #12-17:

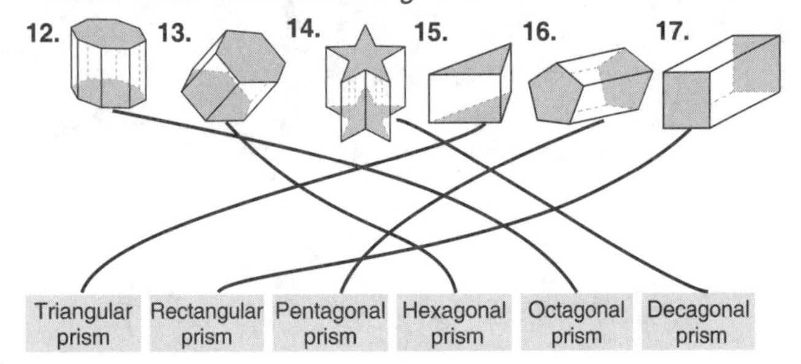

18. We count faces, edges, and vertices, then fill in the table as shown.

Prism	Faces	Edges	Vertices
Triangular prism	5	9	6
Rectangular prism	6	12	8
Pentagonal prism	7	15	10
Hexagonal prism	8	18	12
Octagonal prism	10	24	16
Decagonal prism	12	30	20

19. A nonagonal prism has two congruent nonagon bases and nine lateral faces that connect each side of one base to a side of the other base.

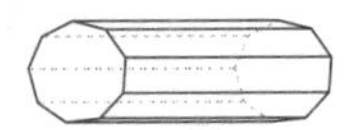

So, there are $2+9=$ **11 faces**.

There are 9 edges on each base and 9 edges that connect the bases. So, there are $9+9+9=$ **27 edges**.

There are 9 vertices on each base of the prism. There are no other points where edges meet. So, there are $9+9=$ **18 vertices**.

20. The base of this pyramid is a pentagon, so this is a **pentagonal pyramid**.

21. Every face of this pyramid is a triangle, so this is a **triangular pyramid**. Any face of a triangular pyramid can be used as the base, so you may have shaded any one of the four faces.

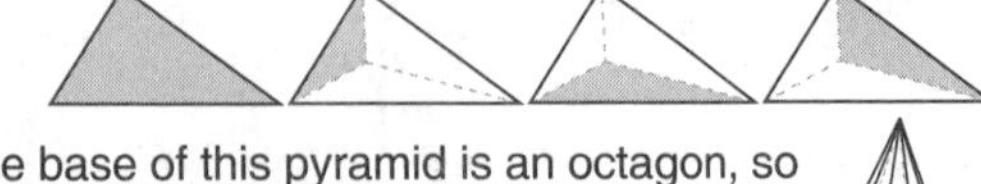

22. The base of this pyramid is an octagon, so this is an **octagonal pyramid**.

23. The base of this pyramid is a square, so this is a **square pyramid**.

24. The base of this pyramid is a hexagon, so this is a **hexagonal prism**.

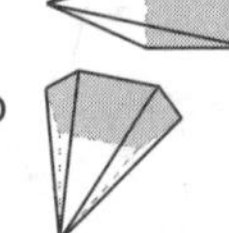

Below is the correct matching for #20-24:

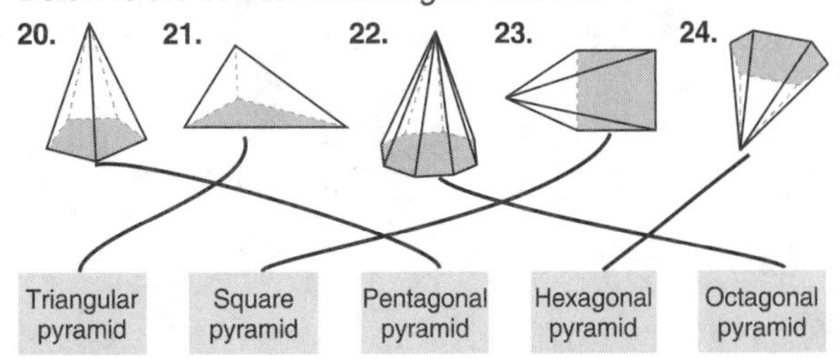

25. We count faces, edges, and vertices, then fill in the table as shown.

Pyramid	Faces	Edges	Vertices
Triangular pyramid	4	6	4
Square pyramid	5	8	5
Pentagonal pyramid	6	10	6
Hexagonal pyramid	7	12	7
Octagonal pyramid	9	16	9

26. A heptagonal pyramid has one heptagonal base and seven triangular faces. So, there are $1+7=$ **8 faces**.

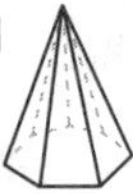

There are 7 edges on the base and 7 edges that connect the base to the apex. So, there are $7+7=$ **14 edges**.

There are 7 vertices on the base of the pyramid and there is 1 apex, so there are $7+1=$ **8 vertices**.

27. An icosagonal prism has 20 edges on each base and 20 more edges that connect the bases. So, an icosagonal prism has $20+20+20=60$ edges.

An icosagonal pyramid has 20 edges on the base and 20 edges connecting the base to the apex. So, an icosagonal pyramid has $20+20=40$ edges.

So, an icosagonal prism has $60-40=$ **20** more edges than an icosagonal pyramid.

28. A regular tetrahedron is a special pyramid, and a cube is a special rectangular prism. **Since all pyramids and prisms are polyhedra, the regular tetrahedron and cube are both polyhedra.**

Spheres, cylinders, and cones all have curved surfaces. **Since polyhedra have no curved surfaces, the sphere, cylinder, and cone are not polyhedra.**

29. We consider a few cuts we could make:

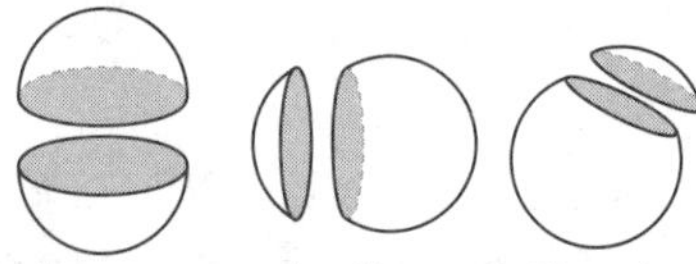

No matter where we make the cut, the shape of the new flat sides of both pieces is a circle.

30. With one horizontal cut, the new flat faces are circles. With one vertical cut, the new flat faces are rectangles.

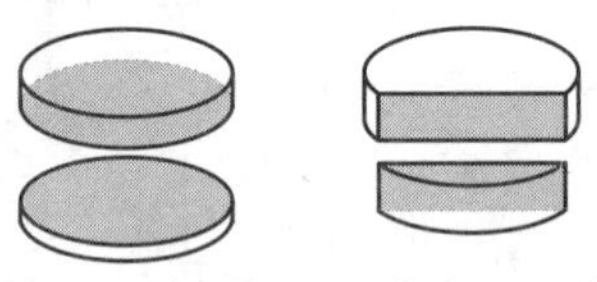

There are just three other types of shapes that could be the new flat faces: an oval (ellipse), a shape with two straight and two curved sides, and a shape with one curved side and one straight side.

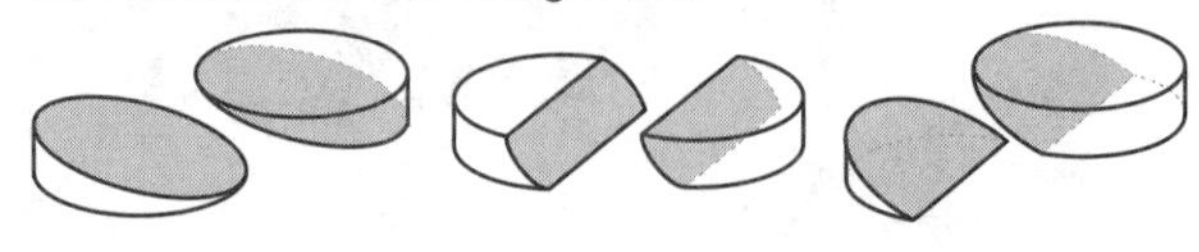

The flat faces cannot be parallelograms, octagons, or triangles. We circle the shapes that the new flat faces could be:

31. With one horizontal cut, the new flat faces are circles. With one vertical cut from the apex of the cone, the new flat faces are triangles.

With a different vertical or angled cut, the new flat faces are shapes with one curved and one flat side.

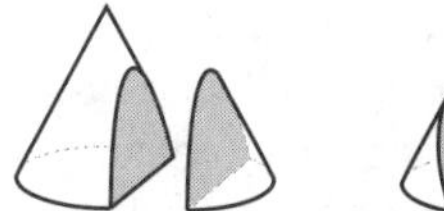
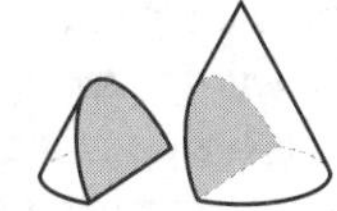

There is just one other type of shape that could be the new flat faces: an oval (ellipse).

The flat faces cannot be rectangles or parallelograms. We circle the shapes that the new flat faces could be:

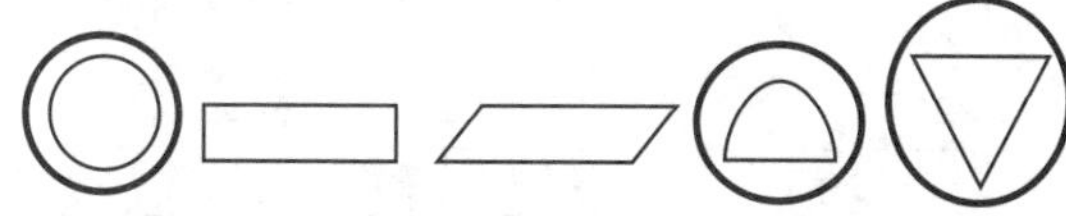

32. Two faces of the prism are bases, which leaves $10-2=8$ lateral faces. Each lateral face connects a side of one base to a side of the other base. Since there are 8 lateral faces, the bases must be polygons with 8 sides each. An 8-sided polygon is an **octagon**.

33. There is one edge along each side of each base. So, if the base has n sides, there are $n+n$ edges on the two bases. There are also n edges that connect the bases. So, there are $n+n+n=3\times n$ edges.

So, the number of edges on a prism is three times the number of sides on its bases. Therefore, a prism with 18 edges has a base with $18\div3=6$ sides.

A 6-sided polygon is a **hexagon**.

34. First, we identify the prism's base shape. All of the vertices of a prism are on its bases. The prism has two congruent bases, so a prism with 22 vertices has a base with $22\div2=11$ vertices and 11 sides.

A prism with an 11-sided base has 2 bases and 11 lateral faces for a total of $2+11=$ **13 faces**.

35. Faces: A prism has 2 bases. There is a lateral face connecting each side of one base to a side of the other base. So, if the base has n sides, the prism has n lateral faces. Therefore, the prism has $\mathbf{n+2}$ **faces**.

Edges: There is one edge along each side of each base. The base has n sides, so there are $n+n$ edges on the two bases. There are also n edges that connect the bases. So, there are $n+n+n=\mathbf{3\times n}$ **edges**.

Vertices: There are n vertices on each base of the prism, and every vertex is on exactly one base. So, there are $n+n=\mathbf{2\times n}$ **vertices.**

36. The height of each lateral face is equal to the height of the prism: 24 inches. The width of each lateral face is equal to the side length of the regular heptagonal bases. Since the perimeter of each base is 28 inches, the side length of each base is $28\div7=4$ inches.

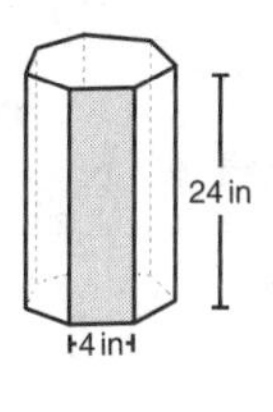

So, the area of one lateral face is $24\times4=$ **96 sq in**.

37. The original cube is a rectangular prism, so it had 6 faces, 12 edges, and 8 vertices. Cutting out the hole did not remove any faces, edges or vertices, but it created 4 new faces, 12 new edges, and 8 new vertices.

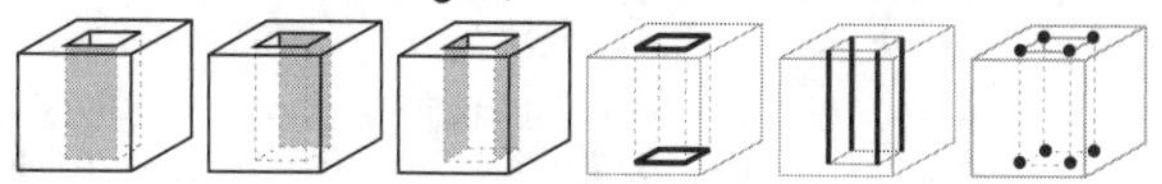

So, the new solid has $6+4=$ **10 faces**, $12+12=$ **24 edges**, and $8+8=$ **16 vertices**.

38. One face of the pyramid is its base, which leaves $10-1=9$ triangular faces. Each triangular face is attached to one side of the base, so the base must have 9 sides. A 9-sided polygon is a **nonagon**.

39. If there are n edges on the base of the pyramid, then there are n edges that connect the base to the apex, which give us a total of $n+n$ edges.

So, the number of edges on a pyramid is twice the number of sides on its base. Therefore, a pyramid with exactly 20 edges has a base with $20\div2=10$ sides.

A 10-sided polygon is a **decagon**.

40. First, we identify the pyramid's base shape. The pyramid's apex is one vertex. The remaining vertices are on its base. Therefore, a pyramid with 22 vertices has a base with $22-1=21$ vertices and 21 sides. A pyramid with a 21-sided base has 1 base and 21 triangular faces for a total of $1+21=$ **22 faces**.

41. Faces: A pyramid has 1 base. There is a triangular face attached to each side of the base. So, if the base has n sides, the pyramid has n triangular faces that meet at the apex. Therefore, the pyramid has $\mathbf{n+1}$ **faces**.

Edges: There are n edges on the base of the pyramid and n edges that connect the base to the apex. So, the pyramid has $n+n=\mathbf{2\times n}$ **edges**.

Vertices: A pyramid with an n-sided base has n vertices on its base. The only other vertex is the apex, so the pyramid has $\mathbf{n+1}$ **vertices**.

42. Each edge of the pyramid is a side of at least one of the triangular faces. Since each triangular face of the pyramid is an equilateral triangle with perimeter 15 cm, the length of each edge of the pyramid is $15\div3=5$ cm.

A pentagonal pyramid has $2\times5=10$ edges. So, the sum of the lengths of all the edges is $5\times10=$ **50 cm**.

43. A prism has three times as many edges as the number of sides of its bases. So, a prism with 12 edges has a base with $12 \div 3 = 4$ sides (rectangular prism). A rectangular prism has 6 faces, 12 edges, and 8 vertices.

A pyramid has two times as many edges as the number of sides of its bases. A pyramid with 12 edges has a base with $12 \div 2 = 6$ sides (hexagonal pyramid). A hexagonal pyramid has 7 faces, 12 edges, and 7 vertices.

We found two different polyhedra that each have 12 edges but different numbers of faces and vertices. So, we cannot tell how many faces or vertices Phil's polyhedron has without more information.

In fact, there are many different polyhedra with 12 edges but different numbers of faces and vertices, such as the solid shown in problem 10 (a regular octahedron).

3D SOLIDS — Nets 14–16

44. The 12 pentagonal faces of this net can be folded into a **dodecahedron**.

45. The 6 square faces of this net can be folded into a **cube**.

46. The 4 equilateral triangle faces of this net can be folded into a **regular tetrahedron**.

47. The 5 faces of this net (1 square and 4 triangles) can be folded into a **square pyramid**.

48. The 5 faces of this net (2 congruent triangles and 3 rectangles) can be folded into a **triangular prism**.

49. The 8 equilateral triangle faces of this net can be folded into a **regular octahedron**.

50. The rectangle of this net can be wrapped around both circles to create a **cylinder**.

The correct matching for #44-50 is shown below.

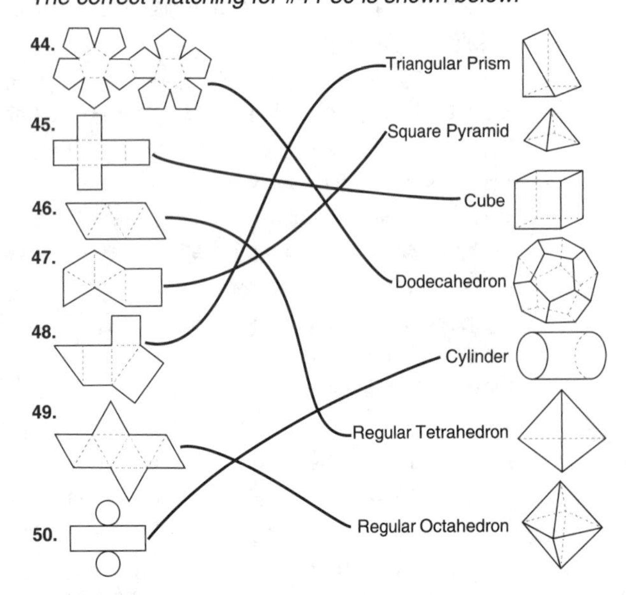

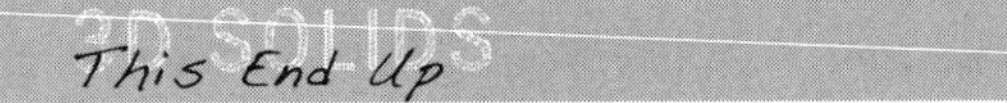

3D SOLIDS — This End Up 17

51. We number the squares of the net as shown.

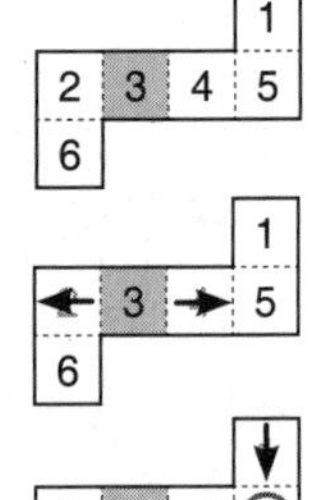

Since face 3 is the bottom of the cube, faces 2 and 4 must have arrows that point away from face 3.

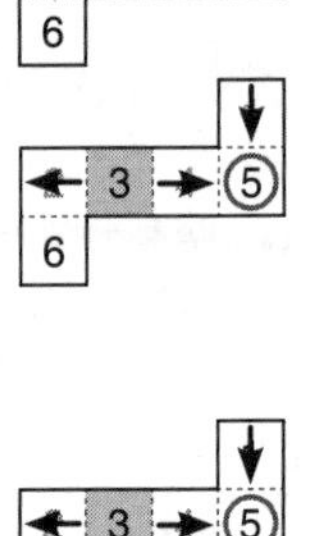

The arrow on face 4 points to face 5. So, face 5 is the top of the cube. We place a circle on face 5. Then, we place an arrow on face 1 that points to face 5.

Finally, the arrow on face 2 does not point toward or away from face 6. So, face 6 is a side of the cube (not top or bottom). It must therefore contain an arrow. This arrow must point in the same direction as the arrow on face 2.

52. Step 1: Step 2: Final:

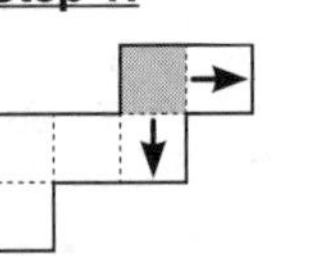

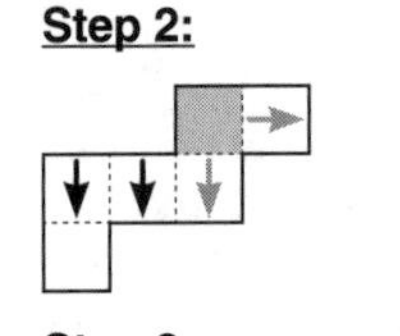

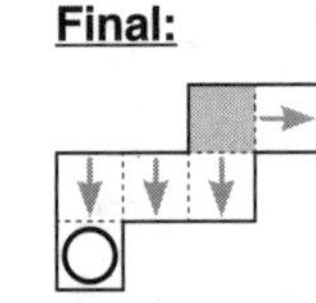

53. Step 1: Step 2: Final:

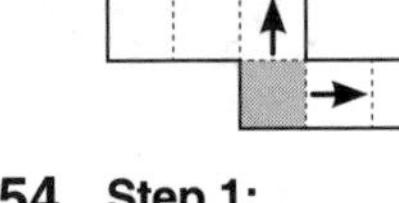

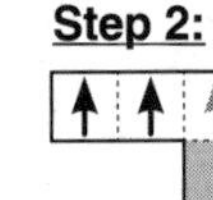

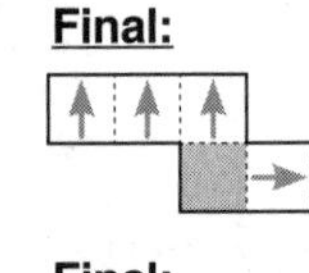

54. Step 1: Step 2: Final:

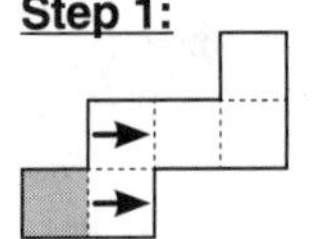

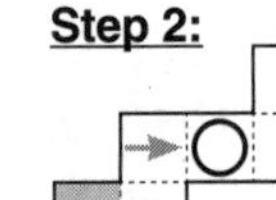

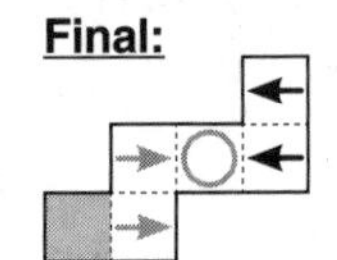

55. Step 1: Step 2: Final:

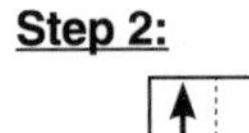

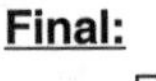

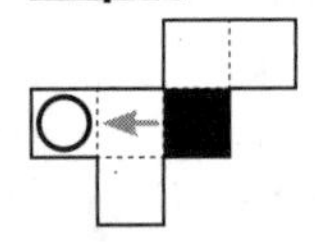

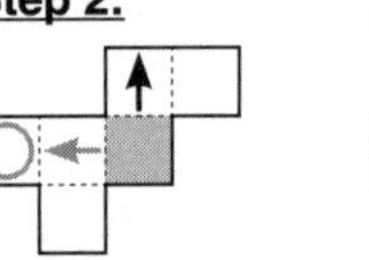

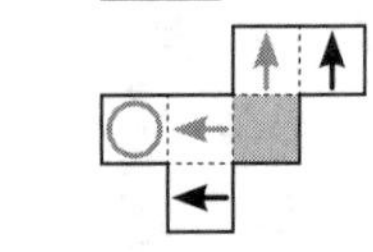

56. Step 1: Step 2: Final:

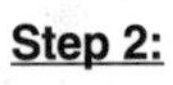

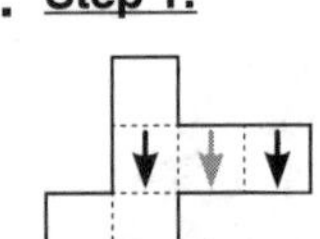

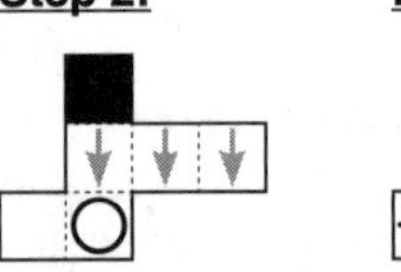

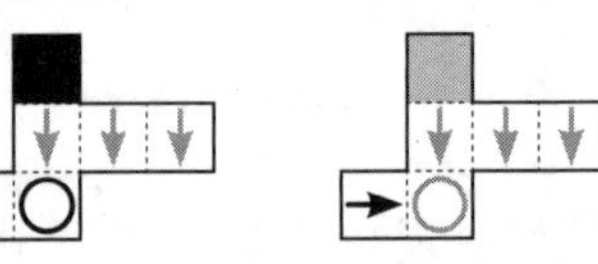

3D SOLIDS — Die Net Puzzles 18

57. We label the blank faces A, B, and C. Then, we visualize folding the net. Below is one way to visualize the net being folded into a cube.

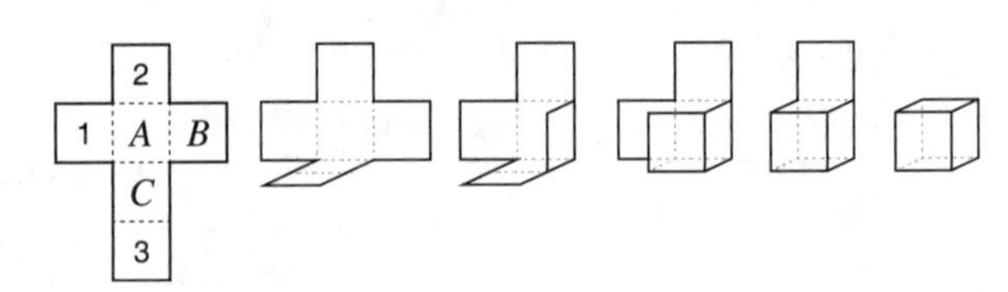

When the net is folded as shown, face 2 is the top, face 1 is the left side, face A is the back, face B is the right side, face C is the bottom, and face 3 is the front. You may have folded differently, but the pairs of opposite faces will be the same.

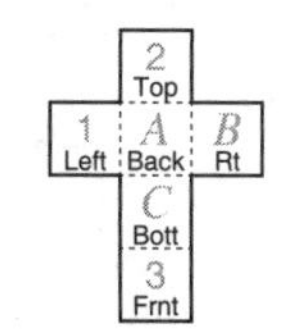

So, face A is opposite face 3, face B is opposite face 1, and face C is opposite face 2. We label the faces as shown:
$A=7-3=\mathbf{4}$, $B=7-1=\mathbf{6}$, and $C=7-2=\mathbf{5}$.

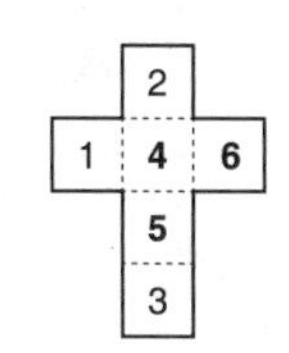

For each problem below, we use the strategy discussed above to label the blank faces in each die net.

58.

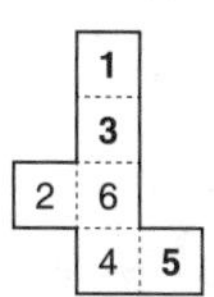

59.

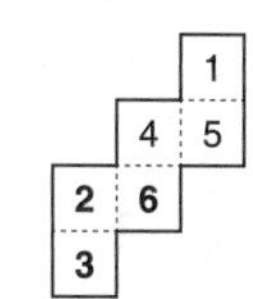

60.

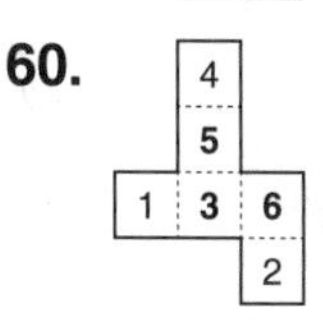

61.

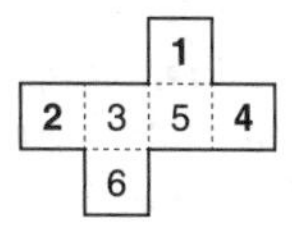

62.

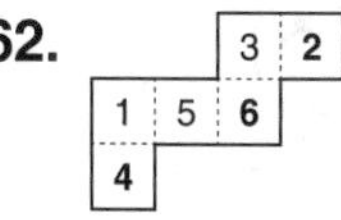

63. When the net is folded, the blank square shares an edge with each of face 4 and face 1. So, the blank square cannot be opposite either face 4 or face 1. In a die net, the opposite of face 4 is face 3, and the opposite of face 1 is face 6. Therefore, the blank square must contain either a 2 or a 5, which is another opposite pair! Their sum is $2+5=\mathbf{7}$.

In a cube net, any three squares that are arranged in an L-shape must be part of three different opposite pairs.

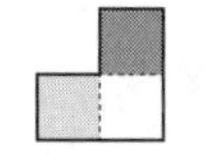
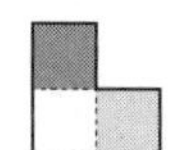
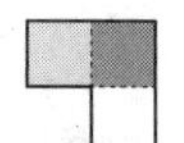
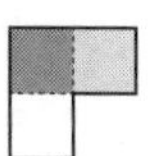

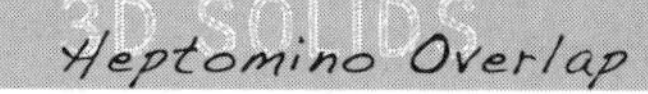

3D SOLIDS — Heptomino Overlap 19

64. Below is one way to visualize folding this heptomino into a cube.

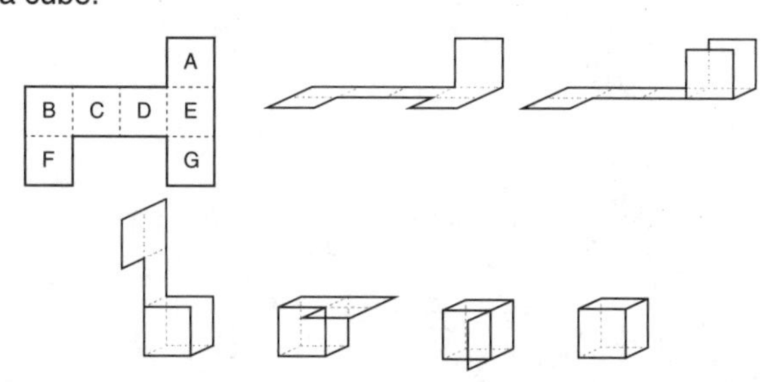

When we fold the cube as shown, face E is the bottom, face A is the back, face G is the front, face D is the left side, face C is the top, face B is the right side, and face F is the front.

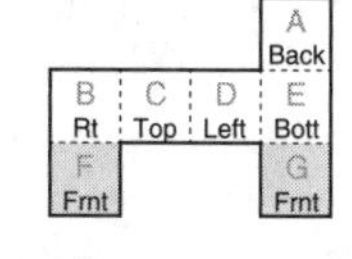

Faces F and G are both the front of the cube when face E is the bottom. So, when this heptomino is folded into a cube, **faces F and G** will overlap.

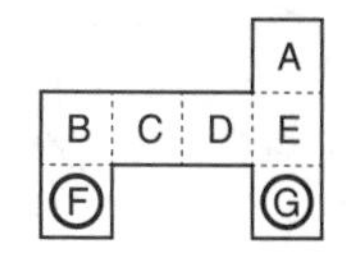

We visualize folding each heptomino into a cube to find the two overlapping squares.

65. (D and E)

66.

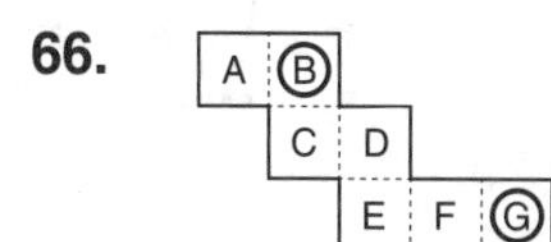

67.

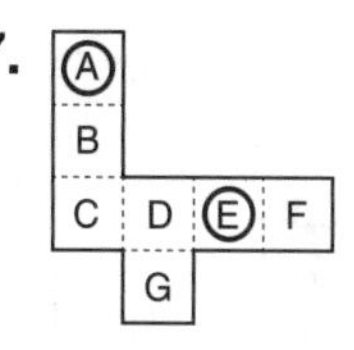

68.

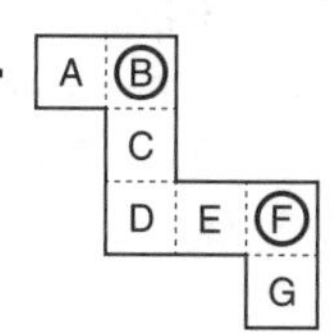

69.

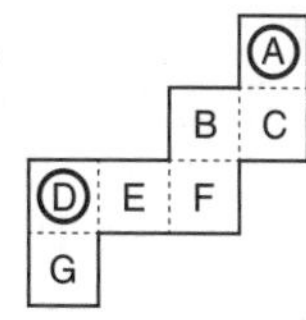

3D SOLIDS — Dot Cube Puzzles 20–21

70. We label each square with a number as shown.

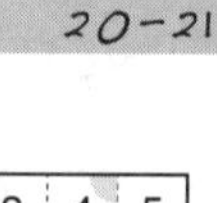
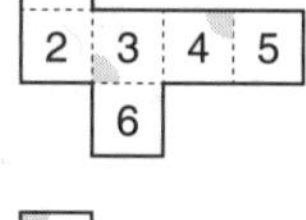

One of the painted vertices is the bottom-left corner of face 3. Faces 2, 3, and 6 meet at this vertex. So, the bottom-right corner of face 2 and the top-left corner of face 6 must also be painted to complete the spot on this vertex.

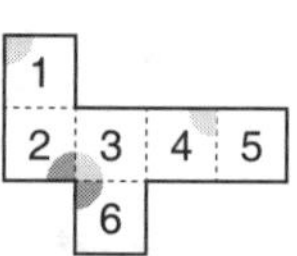

The other spot is on the vertex where faces 1, 4, and 5 meet. The top-left corner of face 5 must also be painted to complete the spot on this vertex.

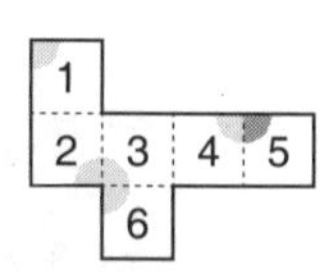

71. The bottom-right corner of face 3 and the top-right corner of face 5 must be painted to complete the spot on the vertex where faces 3, 4, and 5 meet.

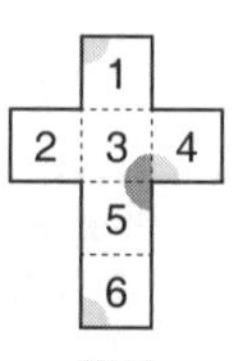

There is already some paint on the top-left corner of face 1. When we fold this net into a cube, the top-left corner of face 2 will be attached to the top-left corner of face 1 (and the bottom-left corner of face 6). We paint this corner to complete the dot on this vertex.

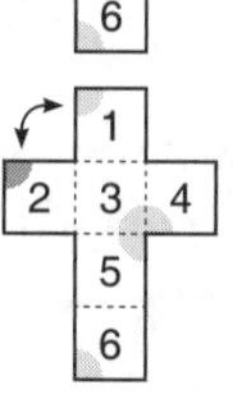

72. **Step 1:**

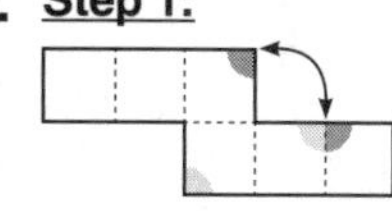

Final:

73. Step 1: Final:

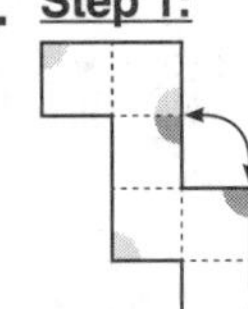

74. We label each square with a number and use the strategies described earlier to place paint as shown. So, one painted vertex is the point where faces 2, 3, and 4 meet. The other painted vertex is the point where faces 1, 5, and 6 meet.

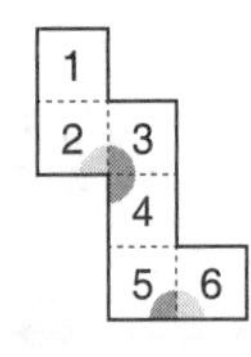

Every square has some paint on one corner, so we look for the painted corner of face 1. Since the bottom-left and bottom-right corners of face 1 touch face 2, those cannot be painted. Similarly, the top-right corner of face 1 will be attached to the top-right corner of face 3 when the net is folded into a cube. So, the top-right corner of face 1 cannot be painted either.

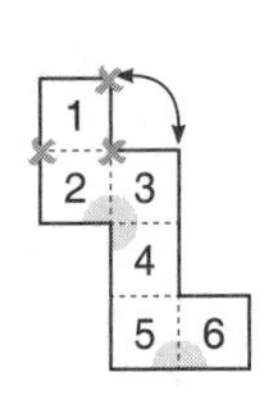

This leaves the top-left corner of face 1 to be the vertex that is attached to faces 5 and 6. We paint the final corner as shown.

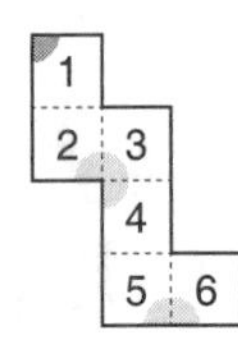

75. We label each square with a number. Then, we complete the spot on the vertex where faces 4, 5, and 6 meet.

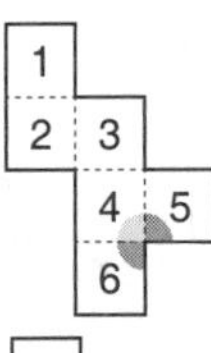

Since one painted vertex is the point where faces 4, 5, and 6 meet, the other painted vertex must be the point where faces 1, 2, and 3 meet.

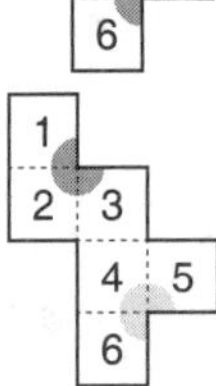

76. We label each square with a number and use the strategies described earlier to paint face 2 as shown.

Since faces 1 and 2 are part of one painted vertex, their opposite faces (5 and 4) are part of the *other* painted vertex.

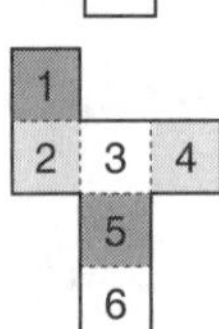

Also, from the location of the paint on faces 1 and 2, we see that the vertex shared by faces 1, 2, and 3 is not painted.

So, one dot is on the vertex where faces 3, 4, and 5 meet. The other is on the vertex where faces 1, 2, and 6 meet.

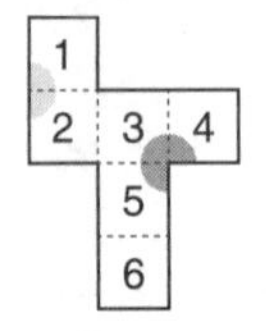

The two corners of face 6 that touch face 5 cannot be painted.

Also, the bottom-right corner of face 6 will be attached to face 4 when we fold this net into a cube. So, this corner cannot be painted either.

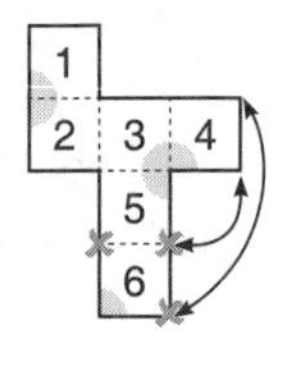

This leaves the bottom-left corner of face 6 to be attached to faces 1 and 2. We paint this corner as shown.

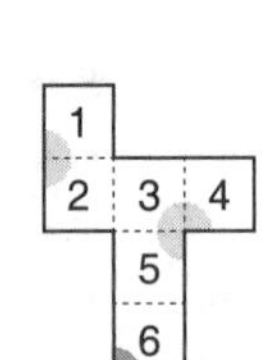

We visualize and use the strategies discussed above to complete each puzzle as shown.

77.

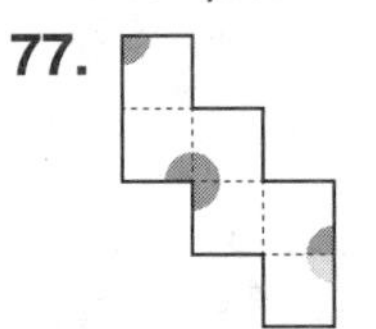

78.

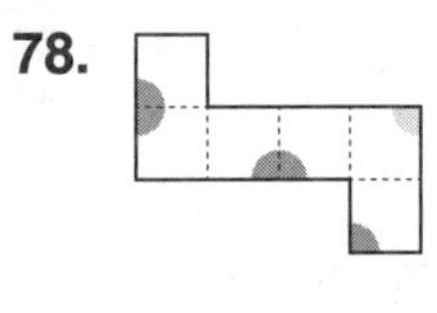

79. When folded as shown, we have the back face and the left face of a cube. The previously-painted corner is part of the *back-top-right* painted vertex. The other painted vertex is the opposite vertex: the *front-bottom-left* vertex. The **bottom-left corner of the blank square** is part of this front-bottom-left vertex, so it also is painted.

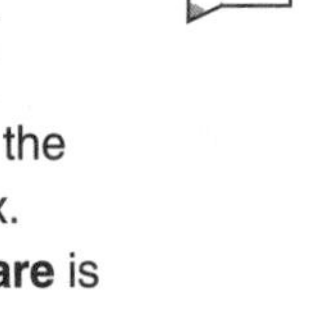

Consider how this observation for any pair of squares would allow you to quickly solve the previous puzzles!

3D SOLIDS Grid Net 22–24

80. Player 2 has four possible moves.

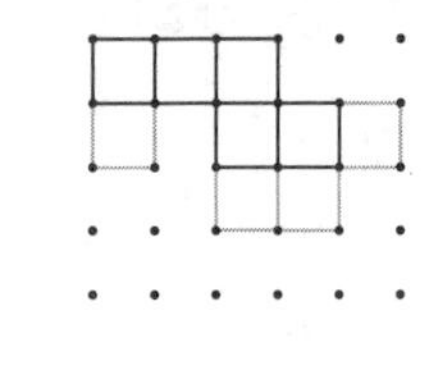

To help us complete the cube net, we consider which pairs of these squares will be opposites when the net is folded (white, gray, and dark gray).

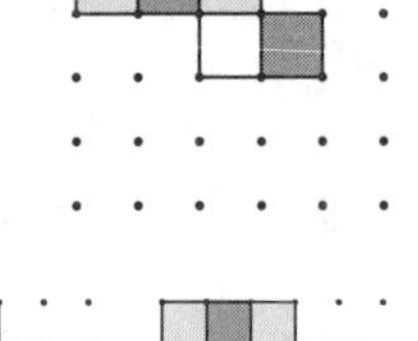

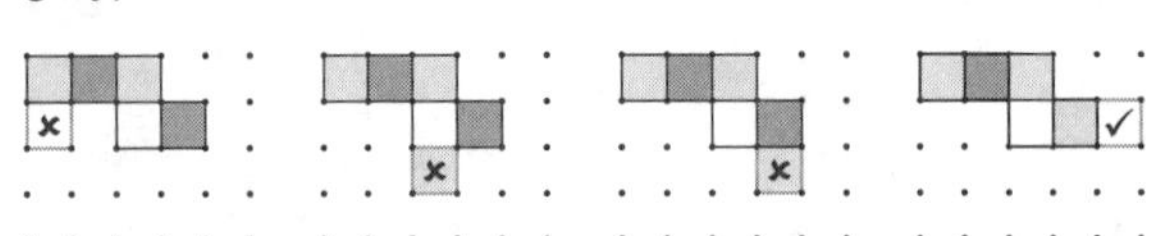

In the first arrangement, the two white squares will overlap when we fold it, so this cannot be a cube net. In the two middle arrangements, we do not get three pairs of opposite faces, so they cannot be cube nets.

Only one move gives us a cube net. This is Player 2's winning move.

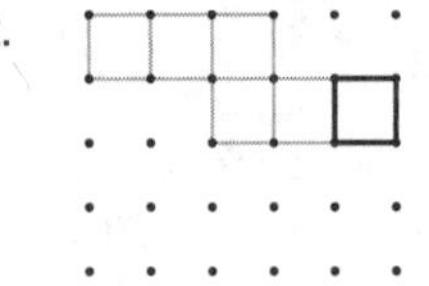

81. Below are all of Player 2's possible moves, with opposite faces marked.

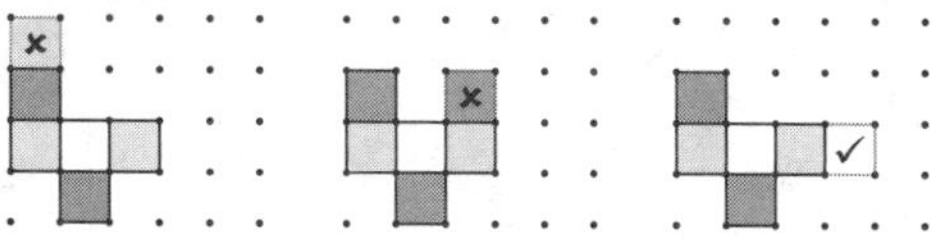

Only one move gives us a cube net. This is Player 2's winning move.

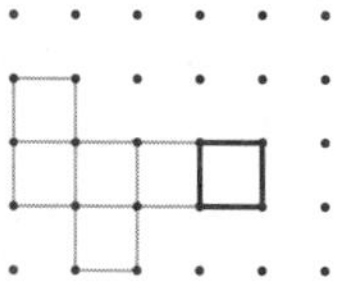

82. Below are all of Player 2's possible moves, with opposite faces marked.

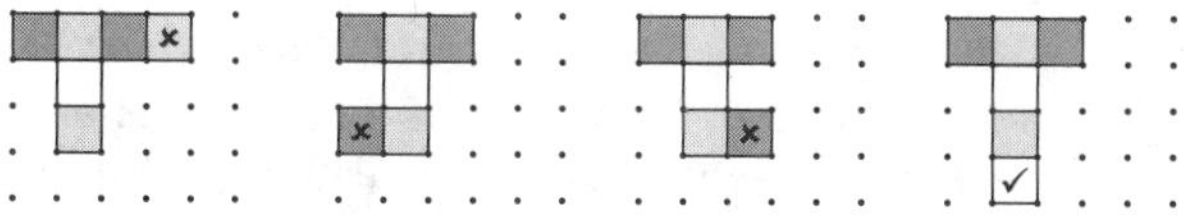

Only one move gives us a cube net. This is Player 2's winning move.

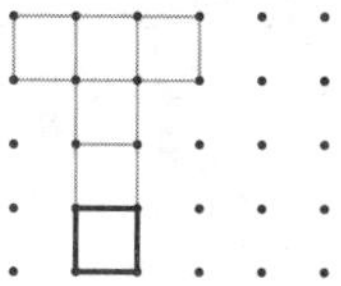

83. Below are all of Player 2's possible moves, with opposite faces marked.

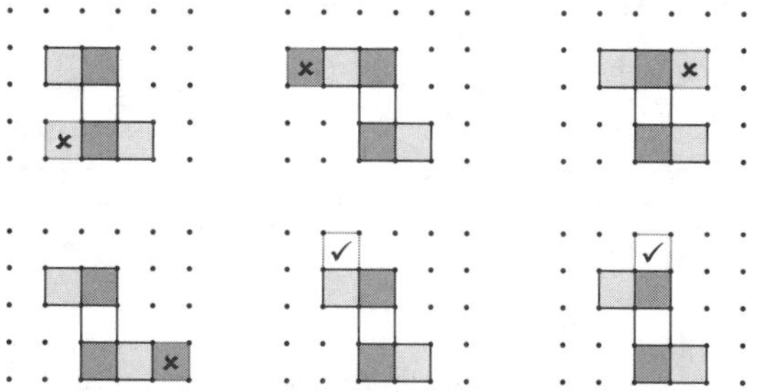

Only two moves give us a cube net. These are Player 2's two winning moves.

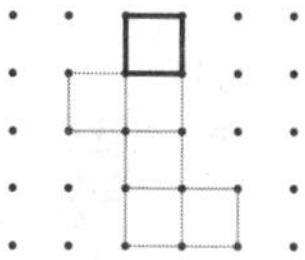

and

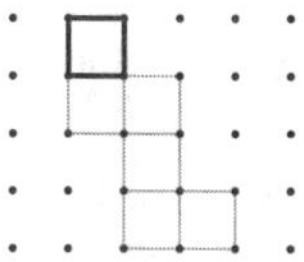

84. We number the squares and consider pairs of opposite faces, as shown.

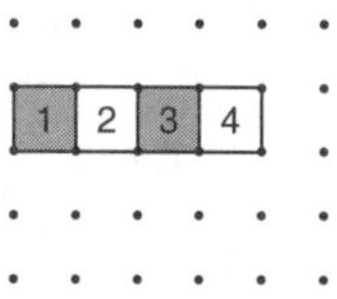

1 2 3 4

If Player 1 plays square 5 as shown, then faces 1 and 5 will both be opposite face 3. So, faces 1 and 5 will always overlap.

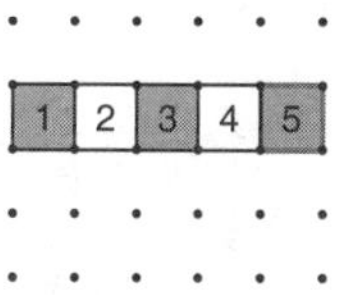

1 2 3 4 5

Therefore, if Player 1 makes this move, Player 2 cannot make a cube net.

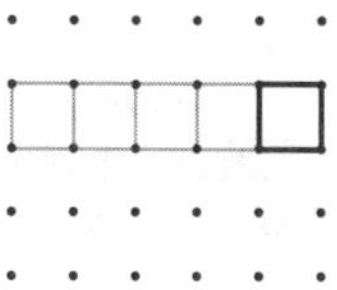

For any other placement of the fifth square, Player 2 can create a cube net with the sixth square.

85. Player 1 has two possible moves.

If Player 1 plays square 5 as shown, then Player 2 will have a move that creates a cube net (square 6).

So this move does not guarantee that Player 2 cannot make a cube net with the final move.

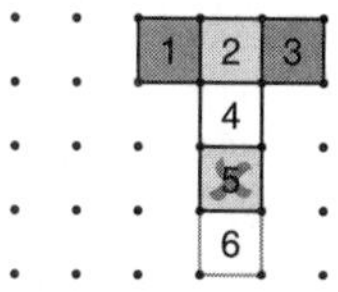

If Player 1 plays square 5 as shown, then Player 2 cannot attach a sixth square that will be opposite face 4.

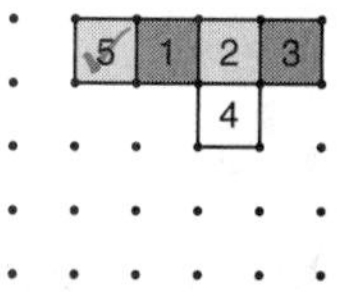

So, this move guarantees that Player 2 will not be able to create a cube with the final move.

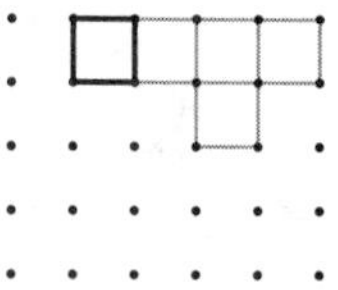

86. Player 1 has 7 possible moves.

If Player 1 plays square 5 in any one of the six ways shown below, then Player 2 will have a move that creates a cube net (square 6).

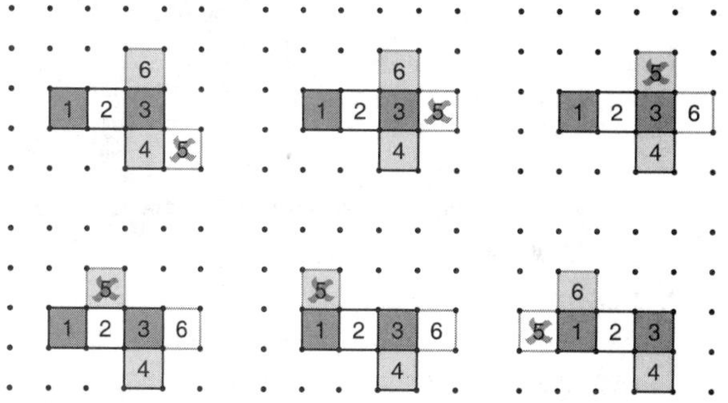

So, none of the moves above guarantee that Player 2 cannot make a cube net with the final move.

If Player 1 plays square 5 as shown, then faces 4 and 5 will overlap when the arrangement is folded.

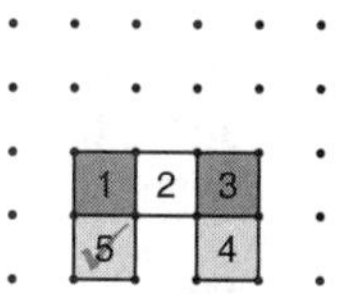

So, this move guarantees that Player 2 will not be able to create a cube net with the final move.

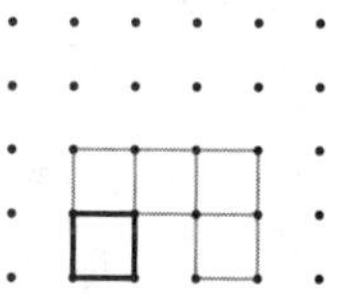

87. Player 1 has three possible moves.

If Player 1 plays square 5 as shown, then Player 2 will have a move that creates a cube net (square 6).

So, this move does not guarantee that Player 2 cannot make a cube net with the final move.

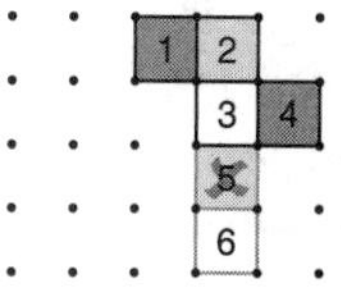

If Player 1 plays square 5 as shown in either diagram below, then Player 2 cannot attach a sixth square that will be opposite face 3.

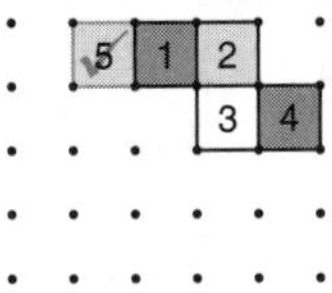

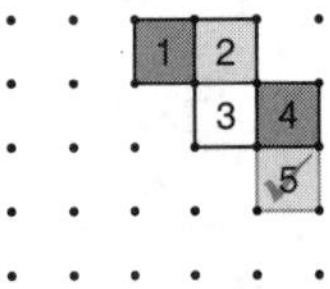

Below are the two winning moves for Player 1.

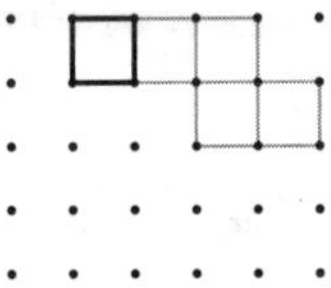

and

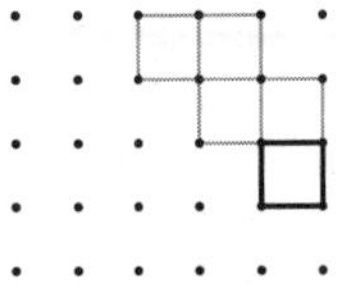

88. Player 1 has three possible moves.

If Player 1 places square 5 as shown, then Player 2 cannot attach a sixth square that will be opposite face 4.

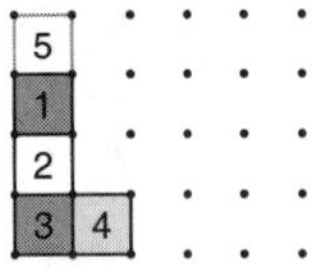

If Player 1 places square 5 as shown, then faces 4 and 5 will overlap when the arrangement is folded.

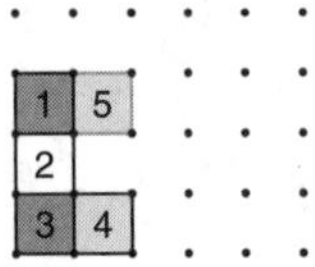

If Player 1 places square 5 as shown, then faces 1 and 5 will overlap when the arrangement is folded.

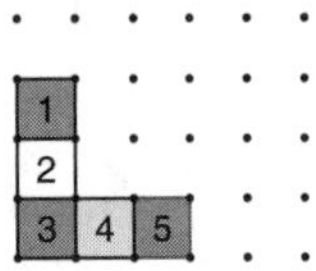

So, Player 1 will win this game no matter what his or her move is.

89. Player 2 must add one of the two squares below, creating an L-shaped arrangement of 4 squares (tetromino) in the corner of the game board.

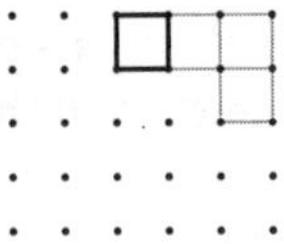

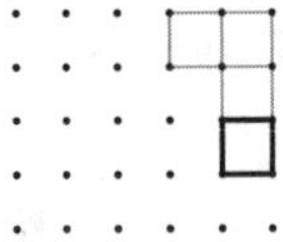

In the previous problem, we determined that if there is an L-shaped tetromino in the corner of the game board, then Player 1 will win.

So, it does not matter where Player 2 draws the fourth square. Player 1 will always win if there is an L-shaped arrangement of 3 squares (triomino) in the corner of the game board.

90. From the previous problems, we see that Player 1 wins whenever the game board has an L-shaped arrangement of 3 squares (triomino) in the corner of the board.

Player 1 can guarantee an L-triomino in a corner by placing the very first square in a corner.

After Player 2's move, the board will look like one of the arrangements below (possibly in a different corner):

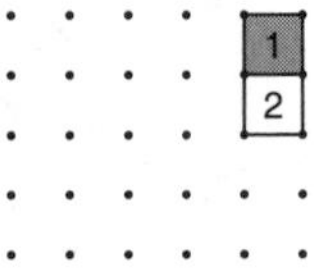

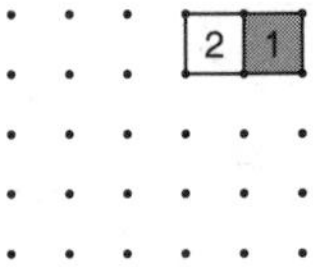

No matter which corner these arrangements are in, Player 1 can always make an L-triomino.

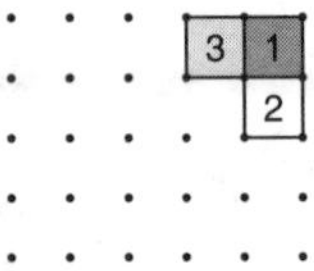

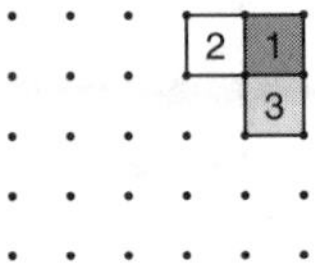

Therefore, on a 5-by-6 dot grid, Player 1 can always win by drawing the first square in a corner and then making an L-shaped triomino with the third square.

91. Player 1 has six possible moves on this board. For all six of these moves, Player 2 can place a sixth square and make a cube net. An example of Player 2's winning move is given on each board below.

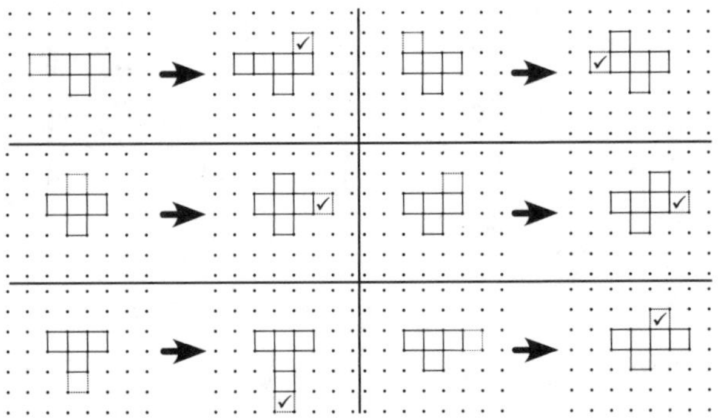

So, Player 2 will win this game, no matter what Player 1's move is.

92. We look back at our work in the previous problems, where possible moves were not restricted by the edges of the grid. We saw that if there is a T-shaped arrangement of 4 squares (tetromino) on such a grid, then no matter where Player 1 plays, Player 2 can make a net and win.

On this game board, Player 2 can add one of the two squares below to create a T-shaped tetromino. Either of these moves guarantees that Player 2 will have a winning move on his or her next turn, regardless of Player 1's next move.

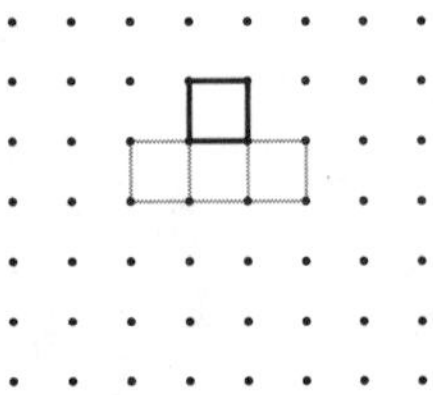

or

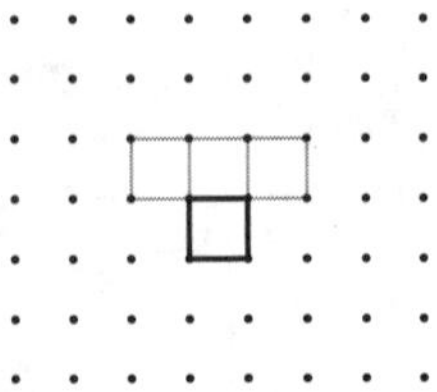

93. After the first three moves, only two shapes are possible. We can rotate the first three squares played on any unlimited dot grid to look like one of the following shapes:

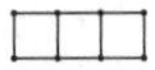

or 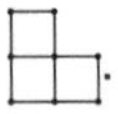 .

Player 2 will then add a fourth square to one of these arrangements and create a tetromino. As we saw in previous problems, Player 2 always has a winning move if he or she creates a T-tetromino on an unlimited grid.

Player 2 can always add a square to a triomino to create a T-tetromino, as shown below.

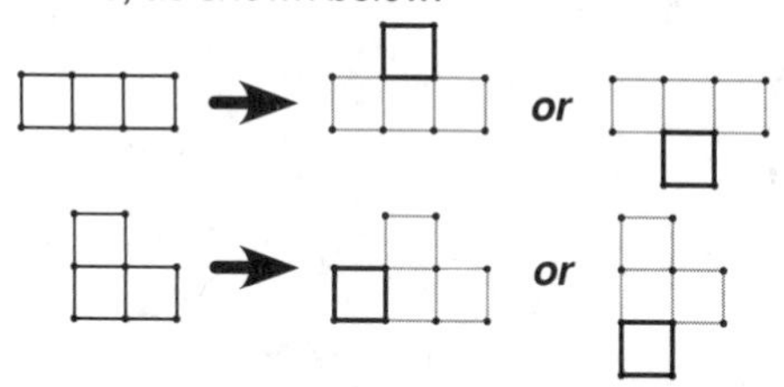

So, Player 2's winning strategy is to create a T-tetromino when placing the fourth square. Then, no matter what Player 1's next move is, Player 2 will always be able to create a cube net.

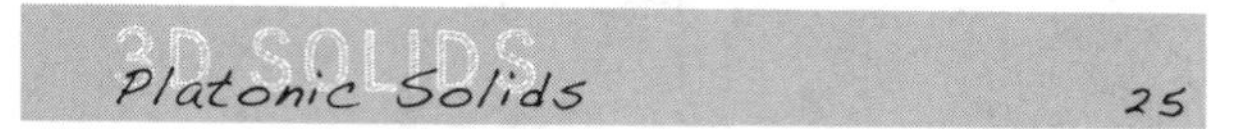

94. We look at the face shapes and count faces, edges, and vertices to fill out the table as shown below.

Platonic Solid	Face Shape	# of Faces	# of Edges	# of Vertices
Regular Tetrahedron	Equilateral triangle	4	6	4
Cube	**Square**	6	12	8
Regular Octahedron	**Equilateral triangle**	8	12	6
Regular Dodecahedron	**Regular pentagon**	12	30	20
Regular Icosahedron	**Equilateral triangle**	20	30	12

If you have trouble visualizing the last two platonic solids, you can use their nets and some reasoning to count the faces, vertices, and edges.

Regular dodecahedron: From the net, we see it is made of **12 regular pentagons**. Each pentagon has 5 sides for a total of $12\times5=60$ sides. Two pentagon sides meet at each edge of the dodecahedron, so this polyhedron has $60\div2=$ **30 edges**. Similarly, each pentagon has 5 corners for a total of $12\times5=60$ corners. Three pentagon corners meet at each vertex of the dodecahedron, so this polyhedron has $60\div3=$ **20 vertices**.

Regular icosahedron: From the net, we see it is made of **20 equilateral triangles**. Each triangle has 3 sides for a total of $20\times3=60$ sides. Two triangle sides meet at each edge of the icosahedron, so this polyhedron has $60\div2=$ **30 edges**. Similarly, each triangle has 3 corners for a total of $20\times3=60$ corners. Five triangle corners meet at each vertex of the icosahedron, so this polyhedron has $60\div5=$ **12 vertices.**

95. We compute the sum of the numbers of faces and vertices and organize our work in a table:

Platonic Solid	# of Faces + # of Vertices	# of Edges
Regular Tetrahedron	$4+4=$ **8**	**6**
Cube	$6+8=$ **14**	**12**
Regular Octahedron	$8+6=$ **14**	**12**
Regular Dodecahedron	$12+20=$ **32**	**30**
Regular Icosahedron	$20+12=$ **32**	**30**

We see that **the sum of the numbers of faces and vertices of a platonic solid is always two more than the number of its edges**.

We could also write an equation for this relationship: $\boldsymbol{f+v=e+2}$, or equivalently, $\boldsymbol{f+v-2=e}$.

96. The tables we completed in problems 18 and 25 show that this relationship is true for those pyramids and prisms. For example, a hexagonal prism has 8 faces, 12 vertices, and 18 edges, and $8+12=18+2$. Similarly, a square pyramid has 5 faces, 5 vertices and 8 edges, and $5+5=8+2$.

We use the formulas we found in problems 35 and 41 to show that it is also true for all prisms and pyramids.

A prism with an n-sided polygon base has $2+n$ faces, $n+n$ vertices, and $n+n+n$ edges. The sum of the number of faces and vertices is

$$(2+n)+(n+n)=2+n+n+n,$$

which is two more than the number of its edges $(n+n+n)$.

A pyramid with an n-sided polygon base has $n+1$ faces, $n+1$ vertices, and $n+n$ edges. The sum of the number of faces and vertices is

$$(n+1)+(n+1)=n+n+1+1=n+n+2,$$

which is two more than the number of its edges $(n+n)$.

Yes, the relationship described in problem 95 is also true for all pyramids and prisms.

This relationship is called Euler's formula, and it is true for all polyhedra!

97. The prism has six rectangular faces. Two faces are 6-by-11, two faces are 6-by-15, and two faces are 11-by-15.

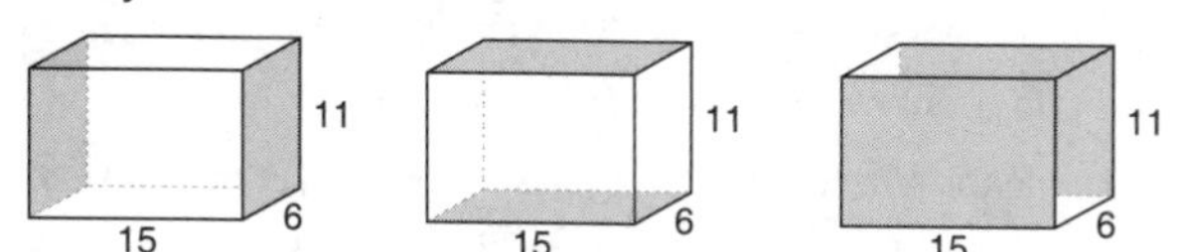

So, the surface area of the prism is

$$\begin{aligned}&2\times(6\times11)+2\times(6\times15)+2\times(11\times15)\\&=2\times66+2\times90+2\times165\\&=132+180+330\\&=\textbf{642 sq cm}.\end{aligned}$$

98. The prism has two 3-by-4 faces, two 3-by-12 faces, and two 4-by-12 faces. So, the surface area is

$$2\times(3\times4)+2\times(3\times12)+2\times(4\times12)=2\times12+2\times36+2\times48$$
$$=24+72+96$$
$$=\mathbf{192\text{ sq cm}}.$$

99. The prism has two 2-by-7 faces, two 2-by-16 faces, and two 7-by-16 faces. So, the surface area is

$$2\times(2\times7)+2\times(2\times16)+2\times(7\times16)=2\times14+2\times32+2\times112$$
$$=28+64+224$$
$$=\mathbf{316\text{ sq cm}}.$$

100. The prism has two 8-by-10 faces, two 8-by-20 faces, and two 10-by-20 faces. So, the surface area is

$$2\times(8\times10)+2\times(8\times20)+2\times(10\times20)$$
$$=2\times80+2\times160+2\times200$$
$$=160+320+400$$
$$=\mathbf{880\text{ sq cm}}.$$

101. The six faces of a cube are congruent squares, so each face of this cube is a 9-by-9-foot square. The surface area is $6\times(9\times9)=6\times81=$ **486 sq ft**.

102. Each of the six faces of the cube is a square with an area of $n\times n=n^2$ sq cm. So, the surface area of a cube with side length n is $\mathbf{6\times n^2}$ **sq cm**.

103. This triangular prism has five faces: two right triangles and three rectangles. We compute the area of each.

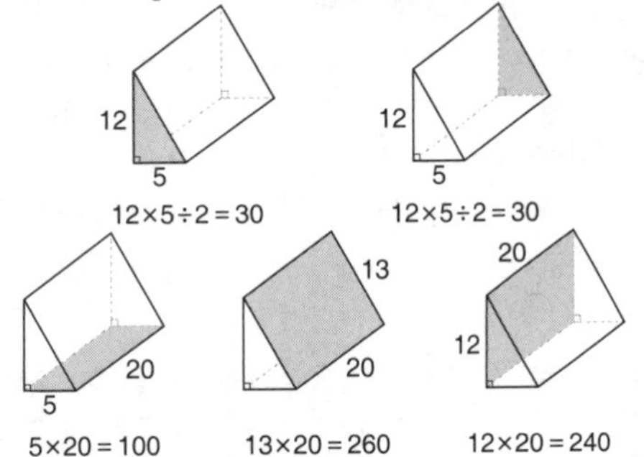

The surface area of the prism is
$30+30+100+260+240=$ **660 sq in**.

104. The leg lengths of each right triangle are 20 and 15, so the area of each right triangle face is $15\times20\div2=150$ sq in. The areas of the rectangular faces are $25\times30=750$ sq in, $15\times30=450$ sq in, and $20\times30=600$ sq in.

So, the surface area of the prism is
$150+150+750+450+600=$ **2,100 sq in**.

105. The area of each right triangle face is $9\times12\div2=54$ sq in. The areas of the rectangular faces are $9\times10=90$ sq in, $12\times10=120$ sq in, and $15\times10=150$ sq in.

The surface area of the prism is
$54+54+90+120+150=$ **468 sq in**.

106. Each triangular face has a height of 12 inches when we consider the 21-inch side as the base. So, the area of each triangle face is $21\times12\div2=126$ sq in.
The rectangular faces are $13\times10=130$ sq in, $20\times10=200$ sq in, and $21\times10=210$ sq in.

The surface area of the prism is
$126+126+130+200+210=$ **792 sq in**.

107. The surface area of the prism is made up of two triangular faces and three rectangular faces. We consider a net of the prism and let h represent the height of the prism.

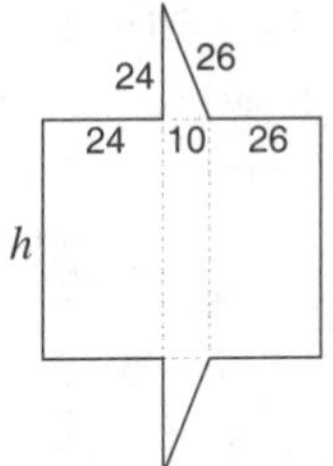

The area of each right-triangle face is $10\times24\div2=120$ sq m. The remaining surface area is $3{,}240-120-120=3{,}000$ sq m.
The remaining surface area of the prism is made up of the three rectangular (lateral) faces with height h and a combined width of $24+10+26=60$ meters.

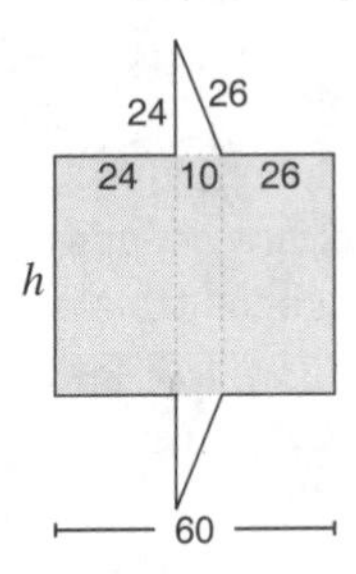

So, the combined area of these three rectangles is $60\times h=3{,}000$ sq m. Since $60\times\boxed{50}=3{,}000$, we have $h=50$. Therefore, the height of the prism is **50 meters**.

108. To find the combined area of the lateral faces, we consider the net of this triangular prism. We use a, b, and c to represent the three side lengths of the triangle.

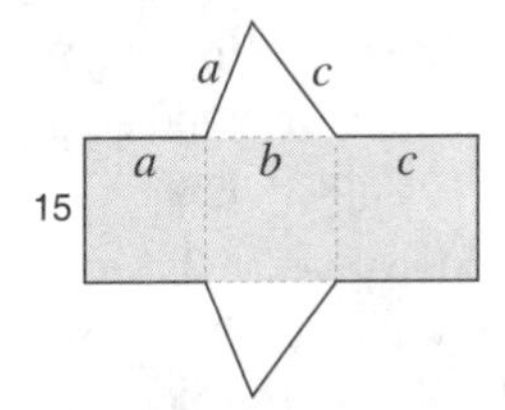

The combined area of the lateral faces is equal to the area of the shaded rectangle in the middle of this net, which has height 15 and width $a+b+c$.

Since a, b, and c are the sides of the triangle base, $a+b+c$ is equal to the perimeter of the triangle: 42 cm. So, the shaded rectangle's area is $15\times42=630$ sq cm.

We are told the area of each triangular face: 168 sq cm.

Therefore, the surface area of the solid is
$168+168+630=$ **966 sq cm**.

We can compute the total area of the lateral faces of ***any*** *prism in the way we discussed above. The total area of the lateral faces of a prism is equal to the product of the base's perimeter and prism's height!*

109. a. We count two pentagonal faces, three square faces, and two rectangular faces. So, the solid has $2+3+2=$ **7** faces.

b. Each pentagonal face is made by attaching a right triangle to a square. So, we can find the area of each pentagon by splitting it into two pieces.

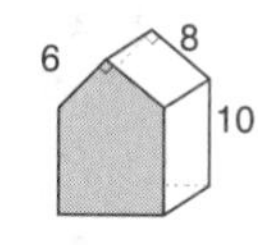

The area of the right triangle is $6\times8\div2=24$ sq cm. The area of the square is $10\times10=100$ sq cm. So, the area of each pentagonal face is $24+100=124$ sq cm.

The area of each of the three square faces is $10\times10=100$ sq cm.

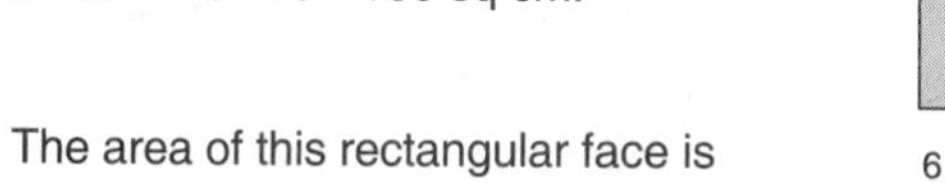

The area of this rectangular face is $6\times10=60$ sq cm.

The area of this rectangular face is $8\times10=80$ sq cm.

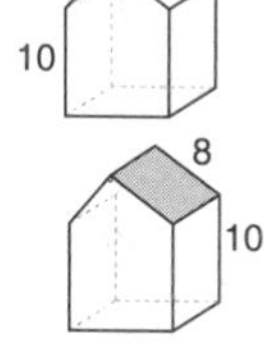

So, the surface area of the solid is

$$124\times2+100\times3+60+80=248+300+60+80$$
$$=\mathbf{688\ sq\ cm}.$$

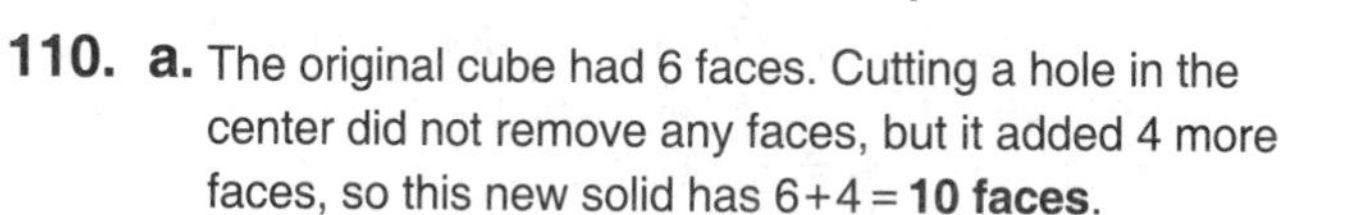

110. a. The original cube had 6 faces. Cutting a hole in the center did not remove any faces, but it added 4 more faces, so this new solid has $6+4=$ **10 faces**.

b. The entire surface of the original cube is red. The surface area of a cube with side length 8 is $6\times(8\times8)=6\times64=$ **384 sq ft**.

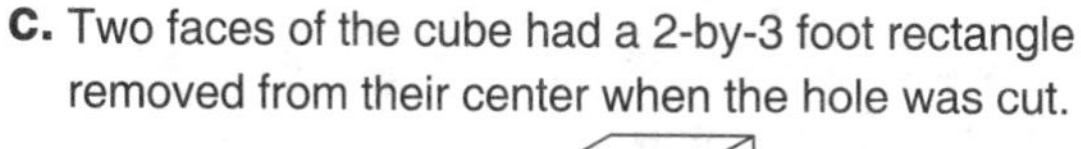

c. Two faces of the cube had a 2-by-3 foot rectangle removed from their center when the hole was cut.

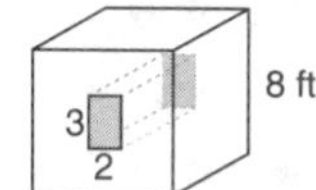

So, cutting the hole removed $2\times(2\times3)=12$ sq ft of red surface. This leaves $384-12=$ **372 sq ft** of red surface on the new solid.

d. The hole created four new unpainted faces. Two of these faces are 2-by-8 rectangles. The other two faces are 3-by-8 rectangles.

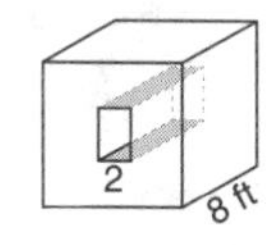
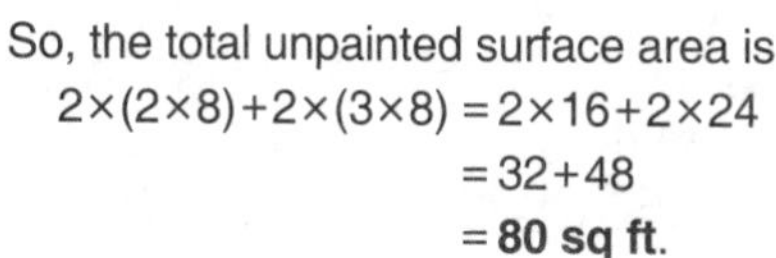

So, the total unpainted surface area is

$$2\times(2\times8)+2\times(3\times8)=2\times16+2\times24$$
$$=32+48$$
$$=\mathbf{80\ sq\ ft}.$$

e. The surface area consists of 372 sq ft of red surface and 80 sq ft of unpainted surface. So, the surface area of the new solid is $372+80=$ **452 sq ft**.

111. The total area of the four shaded rectangles on the right is $22\times5=110$ sq in.

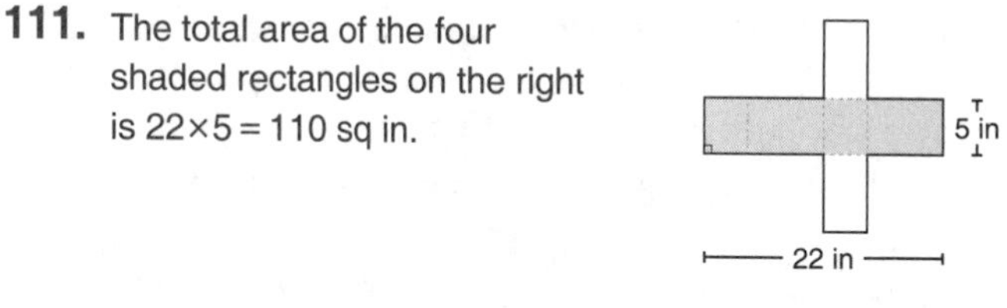

The remaining (white) rectangles are congruent. The height of the entire figure is 19 inches, and the rectangle in the middle is 5 inches tall.

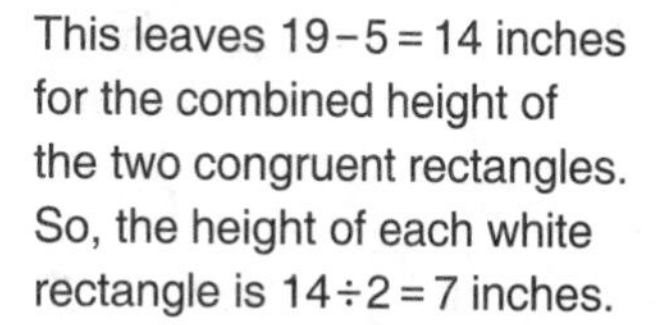

This leaves $19-5=14$ inches for the combined height of the two congruent rectangles. So, the height of each white rectangle is $14\div2=7$ inches.

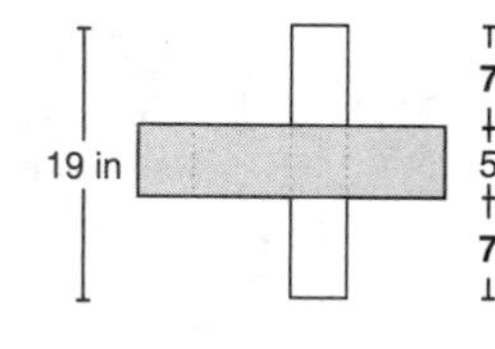

Opposite sides of a rectangle are equal. Also, sides that will be attached during folding are equal. So, the 8 labeled segments below have the same length (7 in):

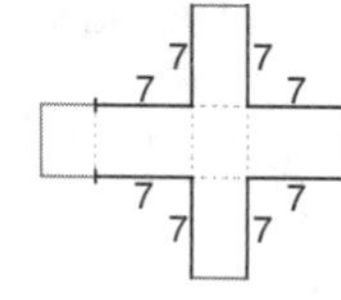

The width of the figure is 22 inches, and two of the rectangles are each 7 inches wide. This leaves $22-7-7=8$ inches for the total width of the two other congruent rectangles. So, the width of each rectangle is $8\div2=4$ inches.

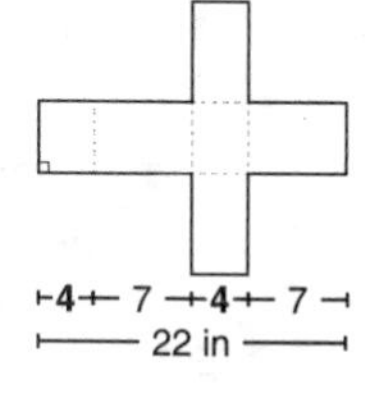

So, the shaded rectangles are 4 inches by 7 inches and each have an area of $4\times7=28$ sq in.

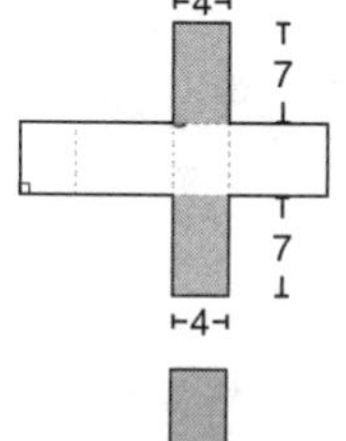

Therefore, the total surface area of the prism is $110+28+28=$ **166 sq in**.

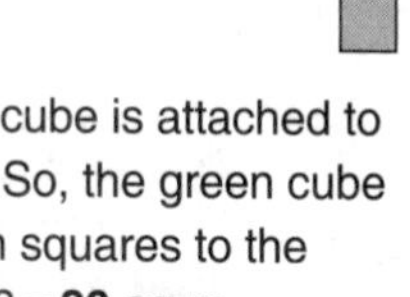

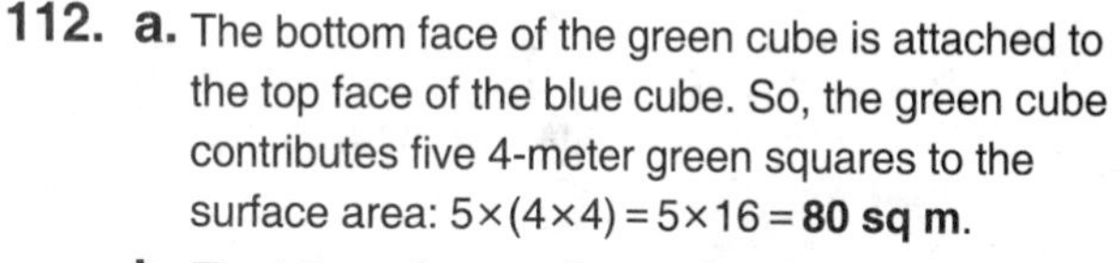

112. a. The bottom face of the green cube is attached to the top face of the blue cube. So, the green cube contributes five 4-meter green squares to the surface area: $5\times(4\times4)=5\times16=$ **80 sq m**.

b. The blue cube contributes five 8-meter blue squares plus one 8-meter blue square missing a 4-meter square (shown below).

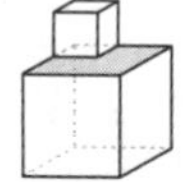

So, the total blue surface area is:

$$5\times(8\times8)+(8\times8-4\times4)=5\times64+(64-16)$$
$$=320+48$$
$$=\mathbf{368\ sq\ m}.$$

c. Since the green and blue cubes are the only two figures that contribute to the surface area, we can add our answers to the previous two questions: $80+368=$ **448 sq m**.

113. The edge lengths of the cubes from bottom to top are 9, 8, 7, 6, 5, 4, and 3 inches. Before stacking, the seven cubes have a total surface area of

$$\begin{aligned}&6\times(3\times3)+6\times(4\times4)+6\times(5\times5)+6\times(6\times6)+6\times(7\times7)\\&\quad+6\times(8\times8)+6\times(9\times9)\\&=6\times(3\times3+4\times4+5\times5+6\times6+7\times7+8\times8+9\times9)\\&=6\times(9+16+25+36+49+64+81)\\&=6\times280\\&=1{,}680\text{ sq in.}\end{aligned}$$

Then, the surfaces where two cubes touch are not part of the final solid's surface area. For example, when the 8-inch cube is stacked on the 9-inch cube, two 8-inch squares are covered: one 8-inch square on the top face of the 9-inch cube and the entire bottom face of the 8-inch cube.

So, we subtract two 8-inch squares, two 7-inch squares, two 6-inch squares, two 5-inch squares, two 4-inch squares, and two 3-inch squares from the total surface area of the cubes.

$1{,}680-64-64=1{,}552$.
$1{,}552-49-49=1{,}454$.
$1{,}454-36-36=1{,}382$.
$1{,}382-25-25=1{,}332$.
$1{,}332-16-16=1{,}300$.
$1{,}300-9-9=1{,}282$.

The surface area of the sculpture is **1,282 sq in**.

— or —

Consider the view of the whole sculpture from all six sides (bottom, top, front, back, left, and right).

The bottom of the sculpture is simply the bottom of the 9-inch cube: a 9-inch square with area $9\times9=81$ sq in.

Next, we consider the top view of the sculpture. Our view of the top of sculpture looks something like this:

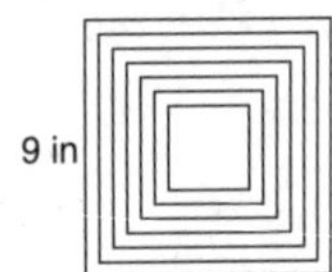

So, the combined tops of all the sculpture layers make up one complete 9-inch square for a total of $9\times9=81$ sq in.

Then, on each of the four remaining sides of the sculpture (front, back, left, and right), we have one complete face of each cube:
$3\times3+4\times4+5\times5+6\times6+7\times7+8\times8+9\times9$
$=9+16+25+36+49+64+81$
$=280$ sq in.

3 in

9 in

So, the total surface area of the sculpture is $81+81+4\times280=162+1{,}120=$ **1,282 sq in**.

114. Six squares have a total of $6\times4=24$ sides.

No cube net has a 2-by-2 square of faces (⊞) because this arrangement prevents us from creating a cube, so all attachments to create a cube net require exactly one side of two different squares. So, to connect six squares into a cube net, we will make a total of 5 attachments. Those 5 attachments exclude $5\times2=10$ square sides from the perimeter of the net. Therefore, the perimeter of any cube net includes $24-10=14$ sides of congruent squares.

Examples:

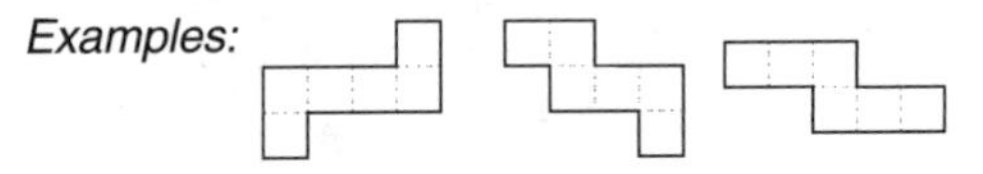

So, a cube net with a perimeter of 56 inches is made up of squares with side length $56\div14=4$ inches. So, the edge length of this cube is 4 inches and its surface area is $6\times(4\times4)=6\times16=$ **96 sq in**.

115. A cube has six identical square faces. Since this cube has a surface area of 96 sq in, the area of each face is $96\div6=16$ sq in. Since $4\times4=16$, the edge length of the original cube is 4 inches.

A cube with edges 3 times as long has $4\times3=12$-inch edges and surface area $6\times(12\times12)=6\times144=$ **864 sq in**.

— or —

When we triple the edge length of a cube, we triple the side length of its square faces. Tripling the side length of a square multiplies its area by $3\times3=9$. Multiplying the area of each face of a cube by 9 makes the surface area of the cube 9 times greater.

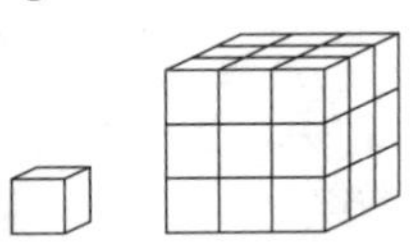

Since the surface area of the original cube is 96 sq in, the surface area of a cube with edges three times as long is $96\times9=$ **864 sq in**.

116. When we double the edge length of a cube, we double the side length of all the faces. Doubling the side length of a square makes its area $2\times2=4$ times as large.

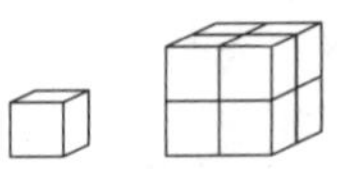

The area of each new face is 4 times the area of a face of the original cube. So, if the edge lengths of a cube are doubled, the surface area of the larger cube is **4** times the surface area of the original cube.

117. You may have used any methods shown in the examples to find that the surface area of this figure is **22 square units**.

118. We count the visible faces from each of the six views.

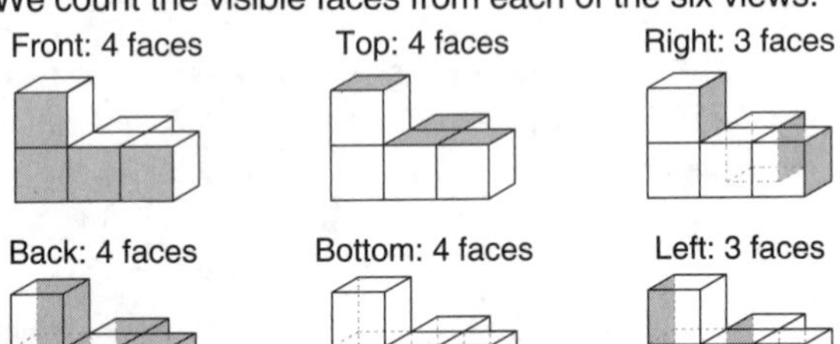

So, the total surface area of the stack is $4+4+3+4+4+3=$ **22 square units**.

— or —

Before the cubes were attached, each cube had 6 visible faces for a total of $5\times6=30$ faces. Attaching the cubes hid some of these faces.

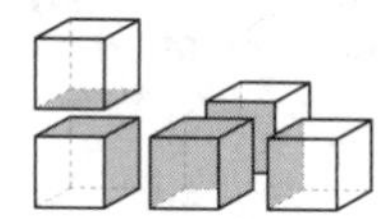

Since 8 of the faces are hidden, $30-8=22$ faces are still visible. Therefore, the surface area of the stack is **22 square units**.

— or —

Three cubes each have five faces showing. One cube has four faces showing. One cube has three faces showing.

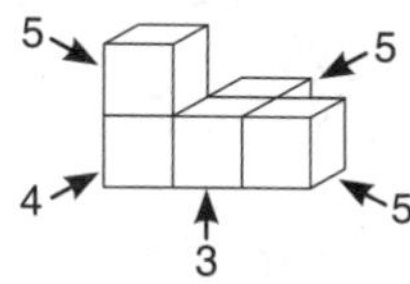

All together, there are $(5\times3)+(4\times1)+(3\times1)=22$ cube faces showing, so the surface area of the stack is **22 square units**.

We can use any of the three approaches discussed above to find the surface area of each stack.
For Method 1, we will always show the sum as front+back+top+bottom+right+left.
For Method 2, we will always show the difference as (total # of faces)−(# of faces hid by attachments).
For Method 3, we will always show each product as (# faces)×(# cubes with that many faces showing).

119. The surface area of the stack is

Method 1	Method 2	Method 3
$4+4+6+6+3+3$	$36-10$	$(5\times2)+(4\times4)$

$=$ **26 square units**.

Be careful if you use the "views" approach on this problem. When counting the number of faces from the left or right "views", don't forget about the faces that are inside the "C"-shape of the stack!

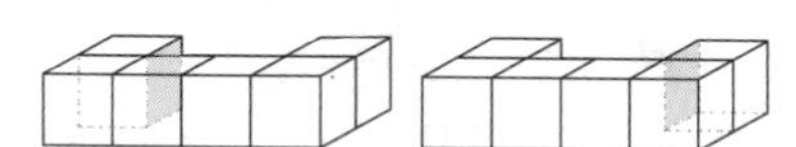

120. The surface area of the stack is

Method 1	Method 2	Method 3
$6+6+5+5+4+4$	$48-18$	$(5\times2)+(4\times3)$ $+(3\times2)+(2\times1)$

$=$ **30 square units**.

121. The surface area of the stack is

Method 1	Method 2	Method 3
$6+6+5+5+4+4$	$42-12$	$(5\times4)+(4\times2)+(2\times1)$

$=$ **30 square units**.

122. From each of the left, right, front, and back, the solid looks like this, with four visible faces:

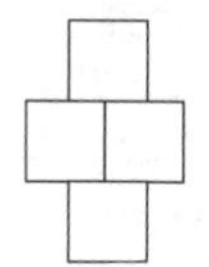

From the top or the bottom, the solid looks like this:

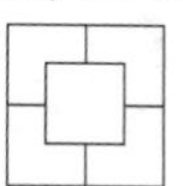

The surface area visible from the top and from the bottom is the same as the surface area of a 2-by-2 square: $2\times2=4$ square units.

From each of the 6 views, we see 4 square units of the solid's surface area. So, the total surface area is $6\times4=$ **24 square units**.

123. Four faces are visible from each of the bottom, left, and right sides. Ten faces are visible from each of the front and back sides. The view from the top of the stack looks like this:

The combined tops of all the cubes make one complete 4-by-1-unit rectangle for a total of $4\times1=4$ square units.

So, the surface area of the stack is $4+4+4+10+10+4=$ **36 square units**.

3D SOLIDS

Volume 32-35

124. The base of the prism is a 15-by-6 rectangle. We can cover the base using $15\times6=90$ cubes. The prism is 11 inches tall, so we can stack 11 layers inside the prism.

Eleven layers of 90 cubes is $11\times90=990$ cubes. So, the volume of the prism is $15\times6\times11=$ **990 cubic inches**.

125. The volume of the prism is $4\times3\times12=$ **144 cubic inches**.

126. The volume of the prism is $16\times7\times3=$ **336 cubic inches**.

127. The volume of the prism is $40\times16\times20=$ **12,800 cubic inches.**

128. The cube is a rectangular prism with a 5-mm square base and height 5 mm. So, the volume of the cube is $5\times5\times5=$ **125 cubic mm**.

129. The cube is a rectangular prism with a n-ft square base and height n ft. So, the volume of the cube is $\mathbf{n\times n\times n}$ **cubic ft** $=\mathbf{n^3}$ **cubic ft**.

130. The area of the base is $(9\times12)\div2=108\div2=54$ sq in. The height of the prism is 10 inches, so the volume of the prism is $54\times10=$ **540 cubic inches**.

— or —

Cutting a 9-by-12-by-10-inch rectangular prism in half diagonally gives us two copies of this triangular prism. So, the volume of the triangular prism is half the volume of a 9-by-12-by-10-inch rectangular prism:

$(9\times12\times10)\div2=1{,}080\div2=$ **540 cubic inches**.

131. $(7\times24\div2)\times10=84\times10=$ **840 cubic inches**.

— or —

$(7\times24\times10)\div2=1{,}680\div2=$ **840 cubic inches**.

132. $(15\times8\div2)\times20=60\times20=$ **1,200 cubic inches**.

— or —

$(15\times8\times20)\div2=2{,}400\div2=$ **1,200 cubic inches**.

133. Each triangular face has a height of 12 inches when we consider the 14-inch side as the base. So, the area of each triangle is $12\times14\div2=84$ sq in. The height of the prism is 25 inches, so the volume of the prism is $84\times25=(21\times4)\times25=21\times(4\times25)=$ **2,100 cubic inches**.

134. We let h represent the height of this prism, and we write an equation for the volume:

$$50\times h=600 \text{ cubic ft.}$$

Since $50\times\boxed{12}=600$, we have $h=12$. So, the height of the prism is **12 feet**.

135. A cylinder is like a prism, but with a circular base. The area of each base of the cylinder is 3 square meters, and the diagram shows that the cylinder's height is 12 meters. So, the volume of the cylinder is (*area of base*)$\times$*height* $=3\times12=$ **36 cubic meters**.

136. The area of the base of the prism is $3\times w$ sq cm, and its height is 6 cm. We write an equation for the volume:

$$(3\times w)\times6=126 \text{ cubic cm.}$$

Since $(3\times w)\times6=3\times6\times w=18\times w$, we have $18\times w=126$ cubic cm. Since, $18\times\boxed{7}=126$, we have $w=$ **7**.

— or —

With a rectangular prism, we can consider any parallel pair of faces as the bases. So, we let the 6-by-3 faces be the bases, and we write an equation for the volume: $(6\times3)\times w=126$ cubic cm. Since $6\times3=18$, we have $18\times w=126$ cubic cm. Then, $18\times\boxed{7}=126$, so $w=$ **7**.

137. One yard is equal to three feet. A cube with edge length 1 yard has a volume of $1\times1\times1=1$ cubic yard. That same cube has an edge length of 3 feet so its volume is $3\times3\times3=27$ cubic feet. So, 1 cubic yard is equal to 27 cubic feet, and 7 cubic yards is equal to $7\times27=$ **189 cubic feet**.

138. The volume of the original cube is $10\times10\times10=1{,}000$ cubic cm. The volume of the removed triangular prism is $(6\times8\div2)\times10=240$ cubic cm. So, the volume of the remaining solid is $1{,}000-240=$ **760 cubic cm**.

139. The volume of a cube with 20-cm edges is $20\times20\times20=8{,}000$ cubic cm. The volume of a cube with 40-cm edges is $40\times40\times40=64{,}000$ cubic cm. Together, the volume of these two cubes is $8{,}000+64{,}000=$ **72,000 cubic cm**.

No matter where the smaller cube is placed on top of the larger cube, their combined volume is always the same.

140. A cube has six identical square faces. Since this cube has a surface area of 150 square meters, the area of each face is $150\div6=25$ square meters.

Since $5\times5=25$, the edge length of the original cube is 5 meters. The volume of a cube with edge length 5 meters is $5\times5\times5=$ **125 cubic meters**.

141. If we use l and w to represent the length and width of the base rectangle and h to represent the height of Lizzie's prism, then we can write an equation for the volume: $(l\times w)\times h=165$ cubic units.

Since Lizzie glues together unit cubes, we know that l, w, and h are all whole numbers. There are five ways to write 165 as the product of three whole numbers:

$165=1\times1\times165$, $165=1\times3\times55$, $165=1\times5\times33$, $165=1\times11\times15$, and $165=3\times5\times11$.

For each of the first four factorizations, we cannot use two of those lengths to make a rectangular base with a perimeter of 28 units.

For the fifth factorization ($3\times5\times11$), we could only have a base with a perimeter of 28 units if the base is a 3-by-11 rectangle. So, the height of the prism is **5 units**.

142. The volume of a cube with side length n inches is $n\times n\times n$ cubic inches, and the volume of this cube is 1,000 cubic inches. Since $10\times10\times10=1{,}000$, the edge length of this cube is 10 inches.

A cube with edges three times as long has edge length $10\times3=30$ inches and volume $30\times30\times30=$ **27,000 cubic inches**.

— or —

The volume of a cube with side length n inches is $n\times n\times n$ cubic inches. A cube with edges three times as long has edge length $3\times n$ inches and volume $(3\times n)\times(3\times n)\times(3\times n)$ cubic inches.

We rearrange and simplify this volume expression: $(3\times n)\times(3\times n)\times(3\times n)=3\times3\times3\times n\times n\times n=27\times(n\times n\times n)$.

So, the volume of the larger cube is *twenty-seven times* the volume of the original cube.

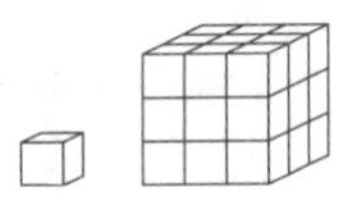

The volume of the original cube in this problem is 1,000 cubic inches, so the volume of a cube with edges three times as long is $27\times1{,}000=$ **27,000 cubic inches**.

143. The volume of a cube with side length n inches is $n\times n\times n$ cubic inches. A cube with edges two times as long has edge length $2\times n$ inches and volume $(2\times n)\times(2\times n)\times(2\times n)$ cubic inches.

We rearrange and simplify this volume expression: $(2\times n)\times(2\times n)\times(2\times n)=2\times2\times2\times n\times n\times n=8\times(n\times n\times n)$.

So, the volume of the larger cube is *eight times* the volume of the original cube.

Therefore, if the edge lengths of a cube are doubled, the volume of the larger cube is **8** times the volume of the original cube.

3D SOLIDS Painted Cubes 36–37

144. The block is divided into $8\times4\times5=$ **160** unit cubes.

145. The **8** corner cubes are the only cubes that have exactly three faces painted.

146. The unit cubes with exactly two faces painted are on the original block's edges, but are not corner cubes.

On an edge with length 8 units, 6 cubes have exactly two faces painted. On an edge with length 5 units, 3 cubes have exactly two faces painted. On an edge with length 4 units, 2 cubes have exactly two faces painted.

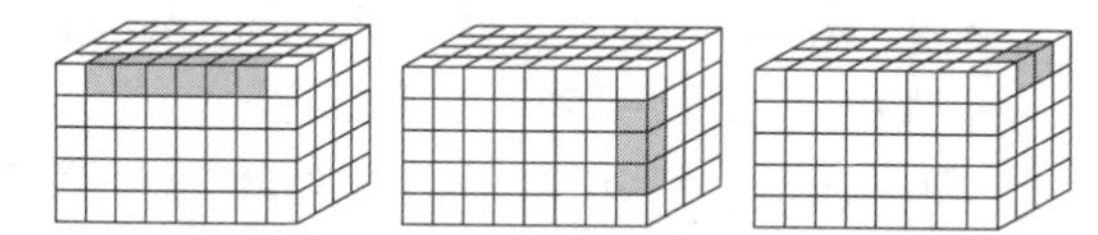

There are 12 edges on the original block: four of each length. So, there are $4\times6+4\times3+4\times2=24+12+8=$ **44** unit cubes with exactly two faces painted.

147. The unit cubes with exactly one face painted are on the original block's faces, but are not on edges or corners. On a 5-by-8 face, $3\times6=18$ cubes have exactly one face painted. On a 4-by-5 face, $2\times3=6$ unit cubes have exactly one face painted. On a 4-by-8 face, $2\times6=12$ cubes have exactly one face painted.

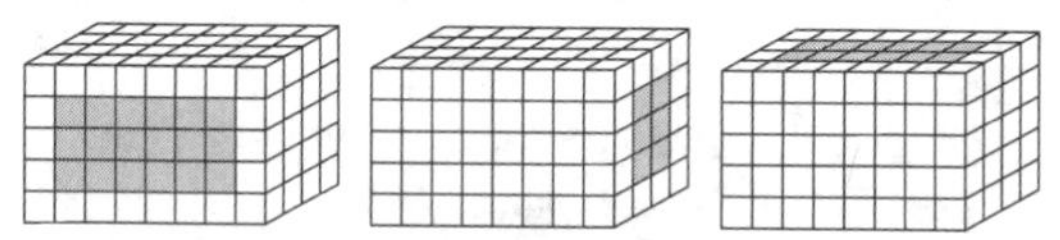

There are 6 faces on the original block: two of each size. So, there are $2\times18+2\times6+2\times12=36+12+24=$ **72** unit cubes with exactly one face painted.

148. No cube is on more than three faces of the original block, so **0** unit cubes have more than three faces painted.

149. In previous problems, we counted the number of cubes with 1, 2, 3, and more than three faces painted. The remaining unit cubes must be those without any paint on them: $160-8-44-72=$ **36** unit cubes.

— or —

There is one layer of cubes on each of the top, bottom, left, right, front, and back sides of the block. If we remove these layers, then all the remaining cubes have no paint on them.

After removing the outer layer of painted cubes, we are left with a block of unpainted unit cubes that has width $8-2=6$, length $4-2=2$, and height $5-2=3$.

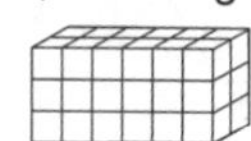

So, $6\times2\times3=$ **36** unit cubes do not have any paint on them.

150. The 5-by-8 and 4-by-8 faces are painted green, so every unit cube on these faces has at least one green face. Each 5-by-8 face has $5\times8=40$ cubes. Then, there are $2\times8=16$ other cubes on each 4-by-8 face that are *not* on a 5-by-8 face.

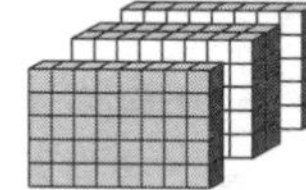

All together, $40+40+16+16=$ **112** cubes have at least one face painted green.

— or —

In problems 145 to 148, we determined that a total of $8+44+72=124$ unit cubes have paint on them.

Each 4-by-5 face has $2\times3=6$ cubes with *only* red paint.

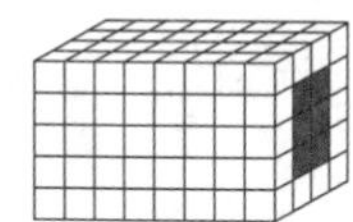

All other painted cubes have at least one green face. So, there are $124-6-6=$ **112** cubes with at least one face painted green.

151. The volume of a cube with side length n units is $n\times n\times n$ cubic units, and the volume of this cube is 125 cubic units. Since $5\times5\times5=125$, the edge length of Lizzie's large cube is 5 units.

There is one painted layer of cubes on each of the top, bottom, left, right, front, and back sides of the block. If we remove these layers, then all the remaining cubes have no paint on them.

After removing the outer layer of painted cubes, we are left with a block of unpainted unit cubes that has width $5-2=3$, length $5-2=3$, and height $5-2=3$.

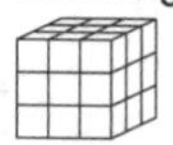

So, $3\times3\times3=$ **27** unit cubes do not have any paint on them.

152. Grogg cuts the block into $5\times6\times7=210$ unit cubes. Only the 8 corner cubes have three painted faces. So, the probability that Alex's cube has exactly three faces painted is

$$\frac{\text{\# of corner cubes}}{\text{total \# of cubes}}=\frac{8}{210}=\mathbf{\frac{4}{105}}.$$

153. If we use l, w, and h to represent the dimensions of Jim's original block, then we can write an equation for the volume: $(l\times w)\times h=105$ cubic units.

Jim cuts the block into a whole number of unit cubes, so the dimensions of the original block must be whole units. Therefore, l, w, and h are all whole numbers. The prime factorization of 105 is $3\times5\times7$.

There are two ways that a unit cube could have exactly three faces painted. It could be on the edge of a block with one dimension equal to 1 and the others larger, or it could be the corner cube of a block whose dimensions are all greater than 1.

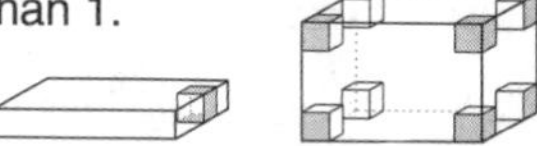

We first consider blocks with one dimension equal to 1. We can use 105's prime factorization to find three ways to write 105 as the product of 1 and two larger numbers: $1\times3\times35$, $1\times5\times21$, and $1\times7\times15$.

However, just one 35-unit, 21-unit, or 15-unit edge has many more than 8 edge cubes:

So, all the dimensions of the original block are greater than 1. Since 105 has only 3 prime factors, those three factors are the dimensions of the block: 3 by 5 by 7 units.

Next, we count the number of cubes with exactly two faces painted. Only the edge cubes that are not corner cubes have exactly two faces painted. On an edge with length 3 units, 1 cube has exactly two faces painted. On an edge with length 5 units, 3 cubes have exactly two faces painted. On an edge with length 7 units, 5 cubes have exactly two faces painted.

A 3-by-5-by-7-unit block has four 3-unit edges, four 5-unit edges, and four 7-unit edges. So, there are $(4\times1)+(4\times3)+(4\times5)=4+12+20=$ **36** unit cubes with exactly two faces painted.

154. As in the previous problem, the dimensions of the original block are all whole numbers and are therefore factors of 189. The prime factorization of 189 is $189=3\times3\times3\times7$.

The fact that some cubes have no paint on them tells us that all of the dimensions of the original block are greater than 2. There are two ways to write 189 as a product of three factors greater than 2: $3\times3\times21$ and $3\times7\times9$.

We then count the cubes with no paint in each size block. After removing the outer layer of painted cubes, we are left with a block that has width, length, and height that are each 2 less than the original block.

So, a 3-by-3-by-21 block has $(3-2)\times(3-2)\times(21-2)$ $=1\times1\times19=19$ cubes with no paint.

Similarly, a 3-by-7-by-9 block has $(3-2)\times(7-2)\times(9-2)=1\times5\times7=35$ cubes with no paint.

So, the original block was 3 by 7 by 9 units. The longest dimension of this original block is **9 units**.

— or —

After removing the outer layer of painted cubes, we are left with a block that has width, length, and height that are each 2 less than the original block.

This inner block must also have whole-number dimensions. The prime factorization of 35 is $35=5\times7$. So the block of unpainted cubes is either 1-by-1-by-35 or 1-by-5-by-7. Therefore, the original block is either 3-by-3-by-37 or 3-by-7-by-9.

A 3-by-3-by-37 block would be cut into $3\times3\times37=333$ unit cubes, while a 3-by-7-by-9 block would be cut into $3\times7\times9=189$ unit cubes.

So, the original block was 3 by 7 by 9 units. The longest dimension of this original block is **9 units**.

155. A cube with edge length 3 units is the smallest cube with at least 1 unpainted unit cube. We count the number of unit cubes that have exactly 0 painted faces and those with 1 painted face, starting with edge length 3.

Edge length 3: One unit cube on each face of the original cube has exactly 1 face painted, for a total of $1\times6=6$ cubes.

However, after removing the outer layer of painted unit cubes, we are left with just 1 unpainted cube. ✗

Edge length 4: $2\times2=4$ unit cubes on each face of the original cube have exactly 1 face painted, for a total of $6\times(2\times2)=6\times4=24$ cubes.

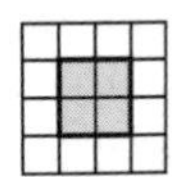

Then, after removing the outer layer of painted cubes, we are left with an unpainted block whose edges are all $4-2=2$ units, giving us $2\times2\times2=8$ unpainted unit cubes.

So, there are 24 unit cubes with exactly one face painted and 8 unpainted unit cubes. ✗

Edge length 5: $6\times(3\times3)=54$ unit cubes with exactly 1 face painted and $3\times3\times3=27$ unpainted unit cubes. ✗

Edge length 6: $6\times(4\times4)=96$ cubes with exactly 1 face painted and $4\times4\times4=64$ unpainted unit cubes. ✗

Edge length 7: $6\times(5\times5)=150$ unit cubes with exactly 1 face painted and $5\times5\times5=125$ unpainted unit cubes. ✗

Edge length 8: $6\times(6\times6)=216$ unit cubes with exactly 1 face painted and $6\times6\times6=216$ unpainted unit cubes. ✓

So, the edge length of the original cube is 8 units, and it was cut into $8\times8\times8=$ **512** unit cubes.

— or —

We find a formula for the number of unit cubes with exactly 1 or 0 faces painted.

In a cube with edge length n units, every face has $(n-2)\times(n-2)$ cubes with exactly one face painted, for a total of $6\times(n-2)\times(n-2)$ cubes. Also, after removing the outer layer of painted cubes, we are left with an unpainted block of $(n-2)\times(n-2)\times(n-2)$ cubes.

The problem tells us that these two numbers are equal, so we write an equation:

$$\underline{6}\times(n-2)\times(n-2)=\underline{(n-2)}\times(n-2)\times(n-2).$$

The only difference between the expressions on the left and right is underlined. If $n-2=6$, both products are $6\times6\times6=216$. Since $\boxed{8}-2=6$, we have $n=8$. So, the side length of the original cube is 8 units, and it was cut into $8\times8\times8=$ **512** unit cubes.

Notice that $6\times(n-2)\times(n-2)=(n-2)\times(n-2)\times(n-2)$ is also true when both sides equal 0. However, this would mean that $n-2=0$ and $n=2$. A cube with edge length 2 has no unpainted unit cubes, but we are told that Cammie has *some* unpainted cubes.

3D SOLIDS Challenge Problems 38–39

156. Four congruent triangle bases together are 8 inches, so each base is $8\div4=2$ inches. The height of each triangle is 3 inches. So, the area of each triangle is $2\times3\div2=3$ sq in.

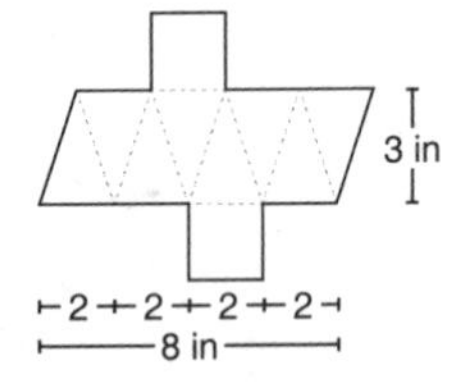

The side length of each square is equal to the base of each triangle: 2 inches. So, the area of each square is $2\times2=4$ sq in.

There are 8 triangles and 2 squares, so the surface area of this solid is $(8\times3)+(2\times4)=24+8=$ **32 sq in**.

157. We can find the number of four-inch cubes it takes to fill the length, width, and height of the prism. One foot is equal to 12 inches, so we can place $12\div4=3$ four-inch cubes along a 1-foot edge.

So, we can place $3\times3=9$ four-inch cubes along each 3-foot edge, $5\times3=15$ four-inch cubes along each 5-foot edge, and $12\times3=36$ four-inch cubes along each 12-foot edge.

Therefore, we can fill the prism with $9\times15\times36=$ **4,860** four-inch cubes.

— *or* —

One foot is equal to 12 inches, so we can place $12\div4=3$ four-inch cubes along a 1-foot edge. So, we can fill 1 cubic foot with $3\times3\times3=27$ four-inch cubes.

The volume of the prism is $3\times5\times12=180$ cubic feet, and we fill each cubic foot with 27 four-inch cubes. So, we can fill the prism with $180\times27=$ **4,860** four-inch cubes.

158. The volume of the water by itself is $30\times30\times20=18{,}000$ cubic cm. The volume of the water and the rocks is $30\times30\times22=19{,}800$ cubic cm. So, the total volume of the rocks is $19{,}800-18{,}000=$ **1,800 cubic cm**.

— *or* —

Adding the rocks increased the water height by 2 cm. So, the total volume of the rocks is $2\times30\times30=$ **1,800 cubic cm**.

Recall the project from the Beast Academy 3C Practice book, where we explored sinking unusually-shaped objects into water to determine their volume! This is related to the Archimedes Principle.

159. When John removes each corner cube, he removes 1 square unit from 3 faces of the prism. He also adds 3 new 1-unit square faces to the new solid.

So, when John removes all eight corner cubes, he removes a total of $3\times8=24$ square units from faces of the prism, but he also creates $3\times8=24$ new square faces, which adds 24 square units to the surface area. Therefore, the surface area of the new solid is the same as the surface area of the original solid.

$$2\times(5\times6)+2\times(5\times7)+2\times(6\times7)=2\times30+2\times35+2\times42$$
$$=60+70+84$$
$$=\mathbf{214\ square\ units}.$$

160. The net of a 2-by-5-by-10-meter rectangular prism consists of two 2-by-5 rectangles, two 2-by-10 rectangles, and two 5-by-10 rectangles. To join all six rectangles into a net, we must make 5 attachments.

To make the greatest possible perimeter, we want to attach the rectangles along their shortest sides, leaving the longest sides to add to the perimeter of the net.

We first attach the 2-by-5 and 2-by-10 meter rectangles along their 2-meter sides as shown, so that congruent rectangles will be opposite faces.

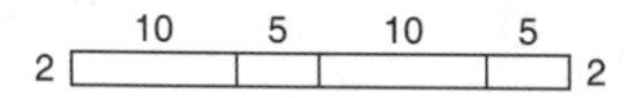

Then, we must attach the 5-by-10 rectangles to complete the net. Since the 5-meter sides are the shortest sides of these rectangles, we attach the 5-by-10 rectangles to 2-by-5 rectangles along their 5-meter sides.

Below are the four ways we could do this to create a prism net:

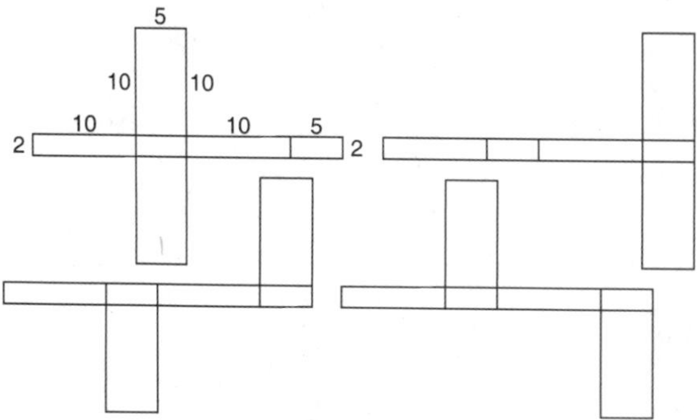

Each of these nets has the same perimeter.

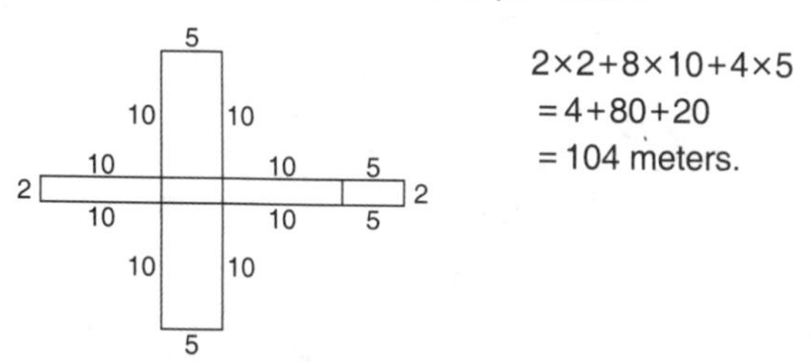

$2\times2+8\times10+4\times5$
$=4+80+20$
$=104$ meters.

So, **104 meters** is the greatest possible perimeter of a net for this prism.

161. The surface area of the original cube is $6\times10\times10=600$ sq ft. When the prism is cut out, a 3-by-4 rectangle is removed from the top face of the cube.

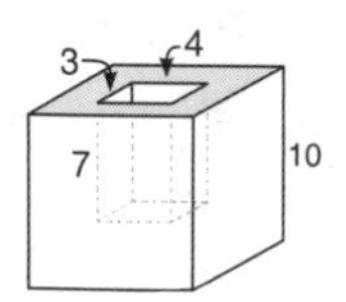

Also, because the cut-out prism did not make a hole through the entire cube, we added *five* new faces: two 3-by-7 faces, two 4-by-7 faces, and one 3-by-4 face.

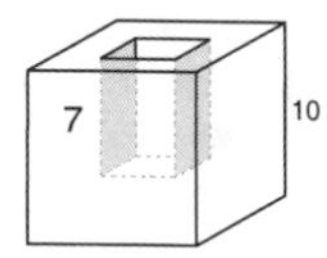

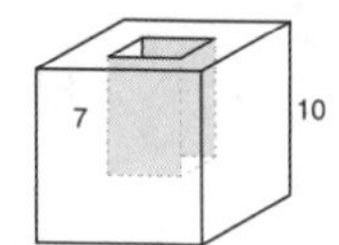

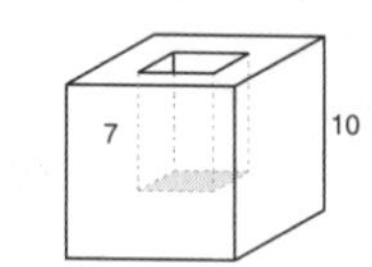

So, the surface area of the remaining solid is
$600-(3\times4)+2\times(3\times7)+2\times(4\times7)+(3\times4)$
$=600-12+2\times21+2\times28+12$
$=600-12+42+56+12$
$=600-12+12+42+56$
$=600+42+56$
$=$ **698 sq ft**.

The volume of the original cube is $10\times10\times10=1{,}000$ cubic feet. The volume of the removed prism is $3\times4\times7=84$ cubic feet. So, the volume of the remaining solid is $1{,}000-84=$ **916 cubic feet**.

162. After Cole removes seven blocks, 20 blocks remain: 8 corner cubes and 12 edge cubes that are not corner cubes.

Each corner cube contributes three 1-unit square faces to the surface area of the new figure. Each edge cube contributes four 1-unit square faces to the surface area of the new figure.

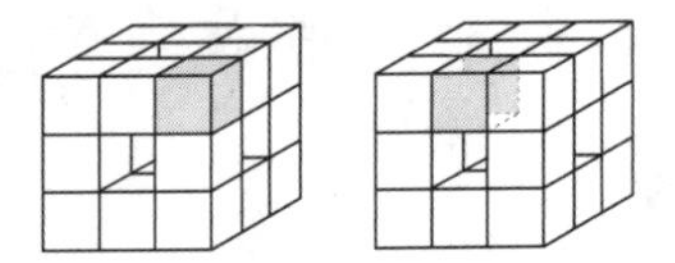

Therefore, the surface area of the remaining solid is $8\times3+12\times4=24+48=$ **72 square units**.

The volume of the original cube is 27 cubic units. Cole removed a total of 7 cubes. So, the volume of the remaining solid is $27-7=$ **20 cubic units**.

163. We use arrows to indicate the sides of the net that will be attached when the net is folded into an octahedron:

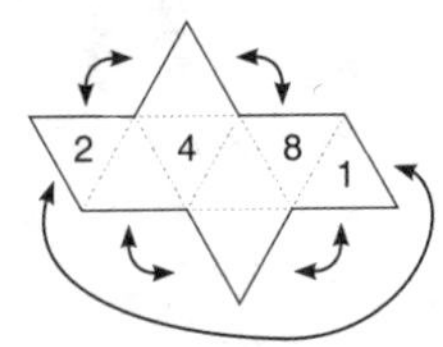

Face 1 is already next to face 2. There are two faces that we could label "7" so that face 7 touches face 8:

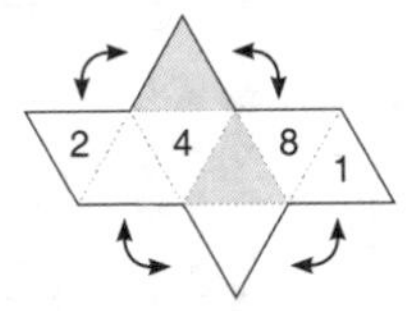

However, if we place the 7 in the upper triangle, then there are no empty faces that share an edge with 7. So, no face could be labeled "6."

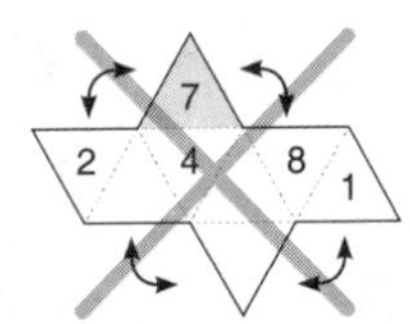

Therefore, face 7 must be the lower triangle.

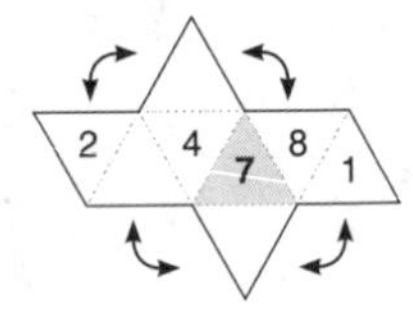

Then, there is only one unlabeled triangle that touches face 7. So, that must be face 6.

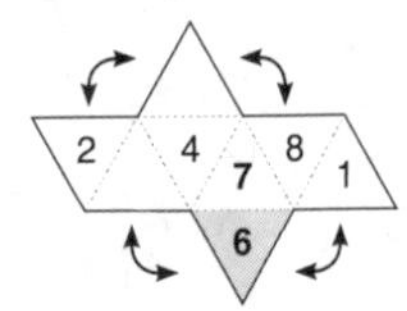

The remaining triangles must be labeled "3" and "5", with face 3 touching faces 2 and 4 and with face 5 touching faces 4 and 6.

164. We first identify opposite face pairs on the original cube:

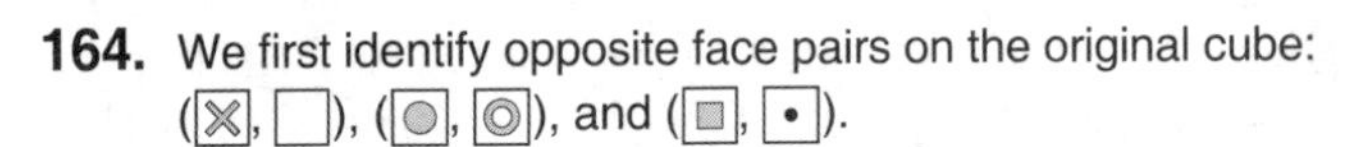

We can eliminate the four nets below because they do not have the same three opposite-face pairs:

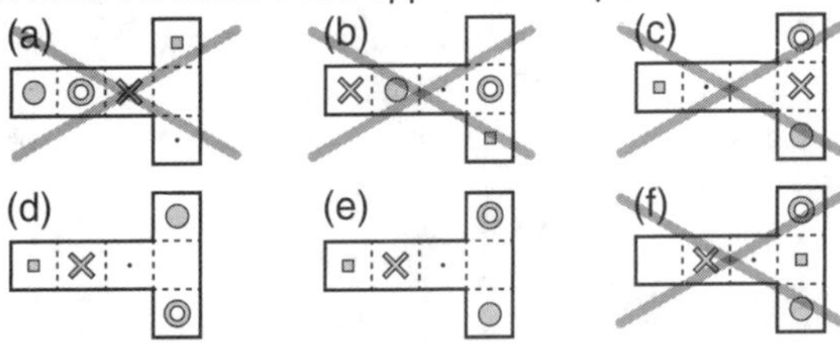

Then, to help us compare the two remaining nets (d and e) to the original, we look at the vertex where the ring, dot, and blank faces meet.

When we fold the nets below so that the shapes are on the outside of the cube, these faces appear as shown:

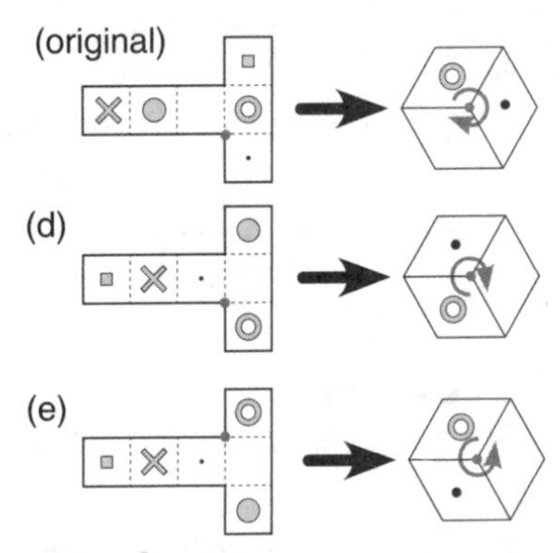

Notice that the original net and net (d) have the faces ◎ → • → □ clockwise around their shared vertex.

Net (e) has the faces ◎ → • → □ counter-clockwise around their shared vertex. So, net (d) is the only choice that makes a cube identical to the first:

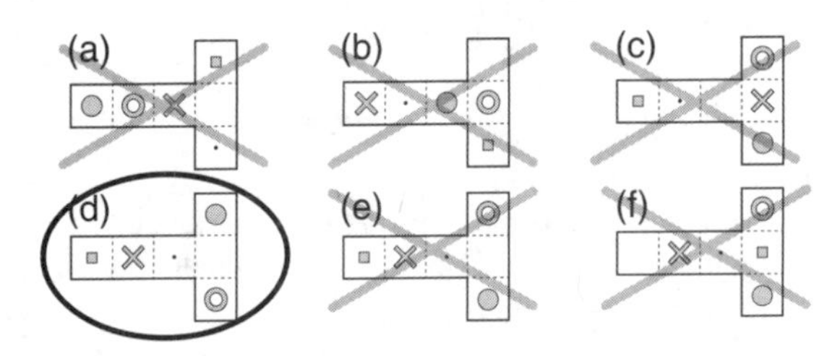

You can also print these nets out from BeastAcademy.com and fold them up into cubes to see that the circled net is the only possibility.

We say that two cubes which are different, but have the same opposite-face pairs, have different ***chiralities****.*

INTEGERS Review pages 41–43

1. To add 7+(-2), we start at 7 and move 2 units to the left.

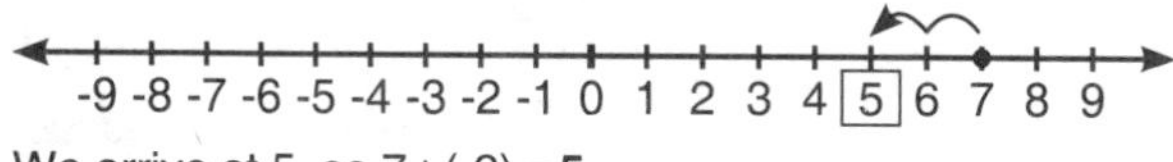

We arrive at 5, so 7+(-2) = **5**.

2. 3+(-5) = **-2**.
3. -6+(-8) = **-14**.
4. -9+6 = **-3**.
5. To add 5+(-6)+(-7), we start at 5, move 6 units left to -1, then 7 more units left to -8. So, 5+(-6)+(-7) = **-8**.
6. -4+9+(-5) = **0**.
7. To add -36+(-42), we start at -36 and move 42 units left to **-78**.

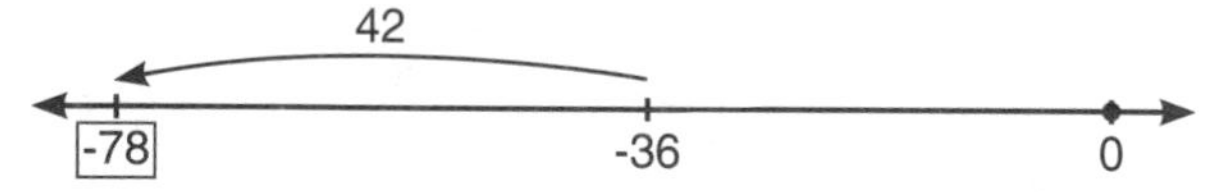

8. -94+41 = **-53**.
9. 67+(-28) = **39**.
10. 120+(-234) = **-114**.
11. -654+789 = **135**.
12. -75+94+(-26) = **-7**.
13. To subtract -7−9, we start at -7 and move 9 units left to **-16**.

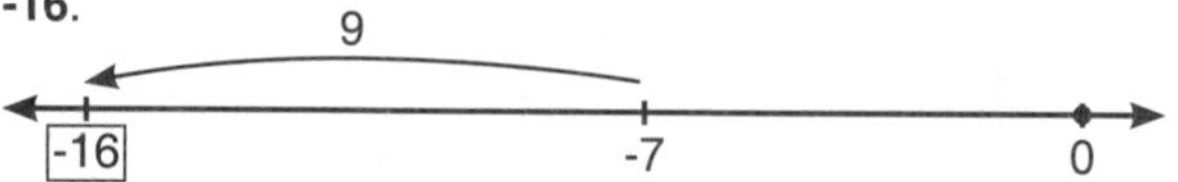

14. 6−10 = **-4**.
15. 3−(-12) = 3+12 = **15**.
16. Subtracting a number from itself always gives us zero. -8−(-8) = -8+8 = **0**.
17. 8−(-2)−12 = 8+2−12 = 10−12 = **-2**.
18. -9−4−(-5) = -9−4+5 = -13+5 = **-8**.
19. 34−(-57) = 34+57 = **91**.
20. 101−117 = **-16**.
21. -58−24 = **-82**.
22. 314−256 = **58**.
23. -470−(-840) = -470+840 = **370**.
24. -67−(-24)−46 = -67+24−46 = -43−46 = **-89**.
25. We first write all subtraction as addition.

-198+299 600+(-803) -943+(-500) -345+435 -639+738+(-1,000)

Then, we consider the sums one at a time.

198−(-299) = -198+299: If we start 198 units to the left of zero and move 299 units to the right, then we cross zero and continue moving right. So, 198−(-299) is positive.

600+(-803): If we start 600 units to the right of zero and move 803 units to the left, then we cross zero and continue moving left. So, 600+(-803) is negative.

-943−500 = -943+(-500): We start at a negative number and move left. So, -943−500 is negative.

-345+435: If we start 345 units to the left of zero and move 435 units to the right, then we cross zero and continue moving right. So, -345+435 is positive.

-639+738−1,000 = -639+738+(-1,000): Because addition is associative, we first consider 738+(-1,000). If we start 738 units to the right of zero and move 1,000 units left, we will cross zero and continue moving left. So, 738+(-1,000) is negative. Adding any negative number to this sum will give a negative result.
So, -639+(738+(-1,000)) = -639+738−1,000 is negative.

We circle the three negative results:

-198−(-299) (600+(-803)) (-943−500) -345+435 (-639+738−1,000)

26. The difference between Grogg's and Alex's integers will be greatest when they are farthest apart.

The greatest integer Grogg could pick is 999.
The least integer Alex could pick is -999.

So, the greatest possible difference between Grogg's and Alex's integers is 999−(-999) = 999+999 = **1,998**.

27. We notice that (-7)+(-7)+(-7) = -21. By changing one -7 to -8 and another -7 to -6, we get three consecutive integers whose sum is -21:

(-8)+(-7)+(-6) = -21.

So, the integers Winnie circles are -6, -7, and -8. The number farthest to the left on the number line is **-8**.

28. The Polar Yeti lives where the temperature is within 25 degrees of -13°F:

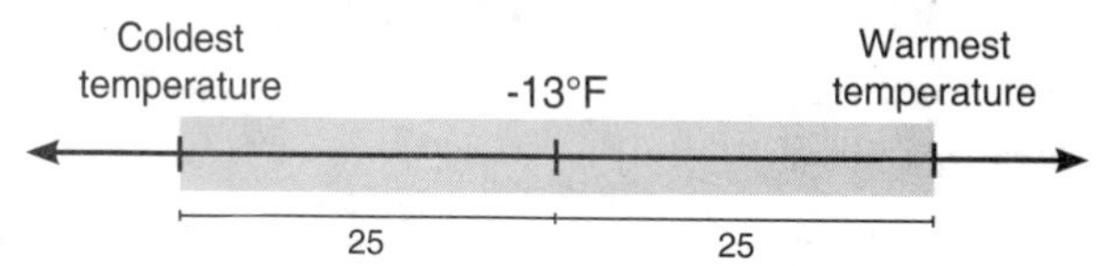

So, the coldest temperature where the Polar Yeti lives is -13−25 = **-38**°F. The warmest temperature where the Polar Yeti lives is -13+25 = **12**°F.

29. We can fill three blanks with three different numbers in 3! = 3×2×1 = 6 ways. We list each of the 6 possible ways and compute each result.

-2−(3−5) = -2−(-2) = -2+2 = 0. -2−(5−3) = -2−2 = -4.
3−(-2−5) = 3−(-7) = 3+7 = 10. 3−(5−(-2)) = 3−7 = -4.
5−(-2−3) = 5−(-5) = 5+5 = 10. 5−(3−(-2)) = 5−5 = 0.

So, **3** different values (0, -4, and 10) can be made.
Remember that the order in which we subtract numbers matters. Subtraction is neither commutative nor associative!

30. The least possible sum of two different numbers in the list is $-20+(-19)=-39$. Similarly, the greatest possible sum is $19+20=39$.

We check that it is possible to create a sum equal to *every* integer from -39 to 39.

To get every integer from -39 to 0, we can add -20 to each other number on this list. For example, $-20+(-19)=-39$, $-20+(-18)=-38$, and $-20+20=0$.

To get every integer from 1 to 39, we can add 20 to each other number on this list, except -20. For example, $20+(-19)=1$, $20+(-18)=2$, and $20+19=39$.

So, we can make 39 positive sums, 39 negative sums, and zero. This gives us $39+39+1=$ **79** different sums.

INTEGERS Multiplication, Part 1 — 44–45

31. To multiply $3\times(-2)$, we can add 3 copies of -2:

$$3\times(-2)=(-2)+(-2)+(-2) = -6.$$

So, $3\times(-2)=$ **-6**.

32. $5\times(-3)=(-3)+(-3)+(-3)+(-3)+(-3)=$ **-15**.

33. $2\times(-9)=(-9)+(-9)=$ **-18**.

34. $6\times(-11)=(-11)+(-11)+(-11)+(-11)+(-11)+(-11)=$ **-66**.

35. To multiply $13\times(-4)$, we add 13 copies of -4. Every copy of -4 takes us 4 units further to the left of zero. So, adding 13 copies of -4 takes us $13\times4=52$ units to the left of zero, to -52. So, $13\times(-4)=$ **-52**.

36. $7\times(-1)=$ **-7**.

37. $150\times(-6)=$ **-900**.

38. $140\times(-20)=$ **-2,800**.

39. $3\times4\times(-5)=12\times(-5)=$ **-60**.

40. $7\times8\times(-2)=56\times(-2)=$ **-112**.

41. $-3\times2=2\times(-3)=(-3)+(-3)=$ **-6**.

— or —

Two numbers with opposite signs always have a negative product. So, $-3\times2=$ **-6**.

42. $-5\times4=$ **-20**.

43. $-9\times6=$ **-54**.

44. $-8\times12=$ **-96**.

45. $-15\times20=$ **-300**.

46. $-30\times22=$ **-660**.

47. $-25\times24=$ **-600**.

48. $-4\times444=$ **-1,776**.

49. $6\times(-5)\times8=-30\times8=$ **-240**.

50. $-4\times9\times2=-36\times2=$ **-72**.

INTEGERS Multiplication, Part 2 — 46–47

51. As we move down the list, the products increase by 6.

$-6\times2=-12$ (+6)
$-6\times1=-6$ (+6)
$-6\times0=0$
$-6\times(-1)=$ ___
$-6\times(-2)=$ ___
$-6\times(-3)=$ ___

We continue the pattern to find the missing products, as shown below.

$-6\times2=-12$ (+6)
$-6\times1=-6$ (+6)
$-6\times0=0$ (+6)
$-6\times(-1)=$ **6** (+6)
$-6\times(-2)=$ **12** (+6)
$-6\times(-3)=$ **18**

52. We already know how to compute the first three products:

$-4\times2=$ **-8**
$-4\times1=$ **-4**
$-4\times0=$ **0**
$-4\times(-1)=$ ___
$-4\times(-2)=$ ___
$-4\times(-3)=$ ___

As we move down the list, the products increase by 4. We continue the pattern to find the missing products, as shown below.

$-4\times2=$ **-8** (+4)
$-4\times1=$ **-4** (+4)
$-4\times0=$ **0** (+4)
$-4\times(-1)=$ **4** (+4)
$-4\times(-2)=$ **8** (+4)
$-4\times(-3)=$ **12**

In each problem below, we find and continue the pattern to complete each list of products as shown.

53.
$-7\times2=$ **-14** (+7)
$-7\times1=$ **-7** (+7)
$-7\times0=$ **0** (+7)
$-7\times(-1)=$ **7** (+7)
$-7\times(-2)=$ **14** (+7)
$-7\times(-3)=$ **21** (+7)
$-7\times(-4)=$ **28**

54.
$-9\times2=$ **-18** (+9)
$-9\times1=$ **-9** (+9)
$-9\times0=$ **0** (+9)
$-9\times(-1)=$ **9** (+9)
$-9\times(-2)=$ **18** (+9)
$-9\times(-3)=$ **27** (+9)
$-9\times(-4)=$ **36**

55. The product of two negatives is positive, so $-2\times(-6)=$ **12**.

56. The product of two negatives is positive, so $-13\times(-1)=$ **13**.

57. The product of a negative and a positive is negative, so $-4\times7=$ **-28**.

58. The product of two negatives is positive, so $-6\times(-3)=$ **18**.

59. The product of a positive and a negative is negative, so $7\times(-5)=$ **-35**.

60. The product of two negatives is positive, so $-9\times(-9)=$ **81**.

61. The product of two negatives is positive, so $-60\times(-8)=$ **480**.

62. The product of two positives is positive, so $4\times19=4\times(20-1)=80-4=\mathbf{76}$.

63. The product of two negatives is positive, so $-18\times(-5)=\mathbf{90}$.

64. The product of two negatives is positive, so $-130\times(-30)=\mathbf{3{,}900}$.

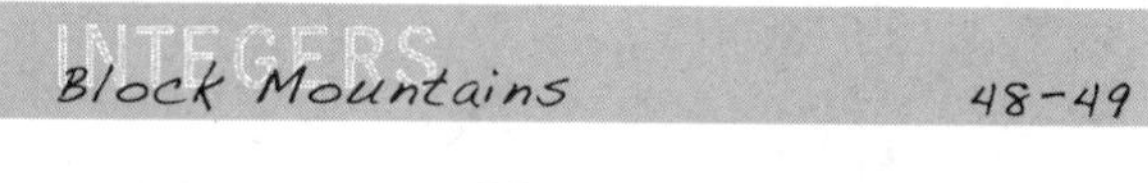

65. **66.**

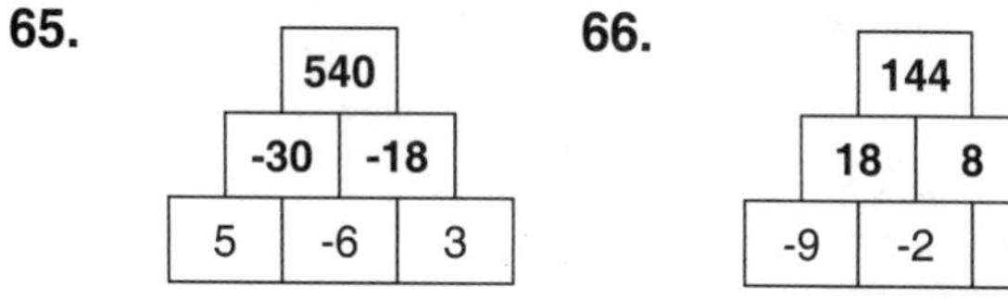

67. **68.**

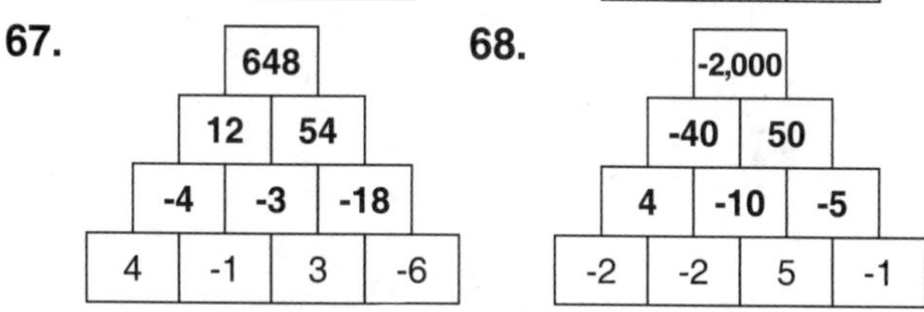

69. **70.**

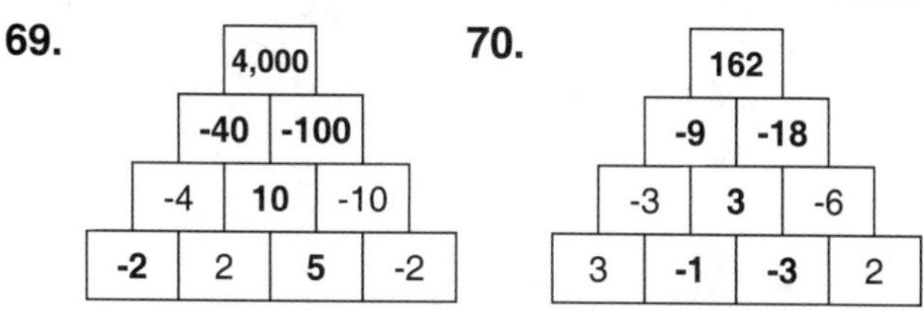

71. **72.**

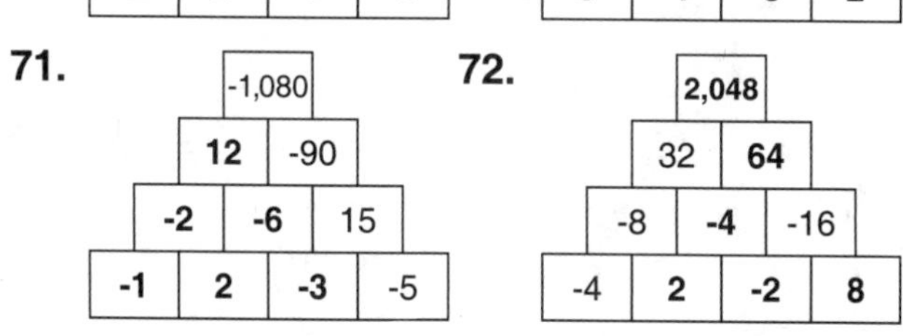

73. Since $1\times\boxed{-1}=-1$, we fill in the block in the row below the top as shown.

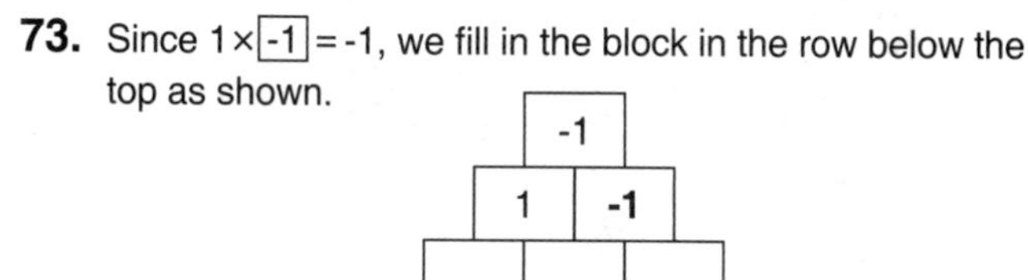

Then, we try to fill in the row that is third from the top. The two blocks below the 1 must have the same sign, and the two blocks below the -1 must have different signs. This gives us the following possibilities:

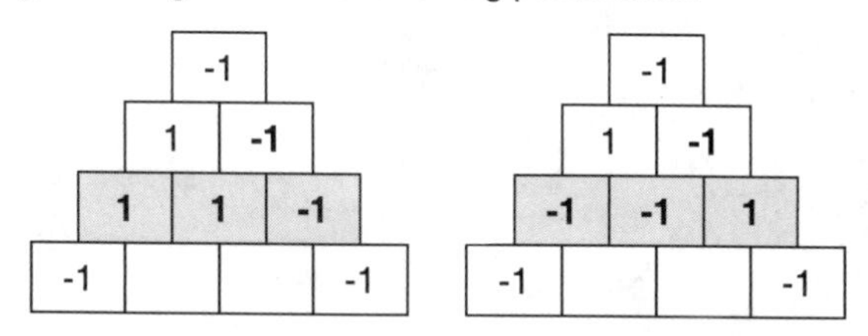

Finally, we try to fill in the bottom row. In the left diagram, the blank on the left must be filled with a -1. However, this makes it impossible to fill the blank on the right to satisfy both products above it.

In the right diagram, we fill the blanks as shown to complete the puzzle. This is the only possible solution.

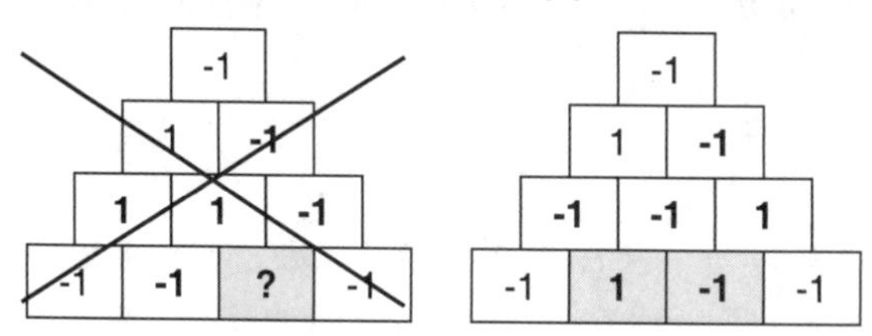

74. Since $2{,}160\div90=24$, we know $24\times90=2{,}160$.
So, $\boxed{-24}\times(-90)=2{,}160$.

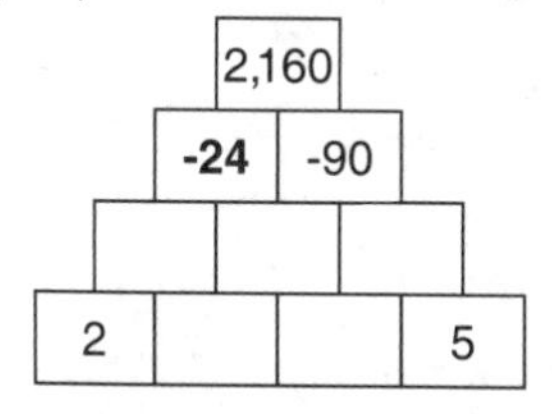

Then, we assign variables a and b to the two unlabeled blocks in the bottom row. This allows us to write expressions in terms of a and b as shown below:

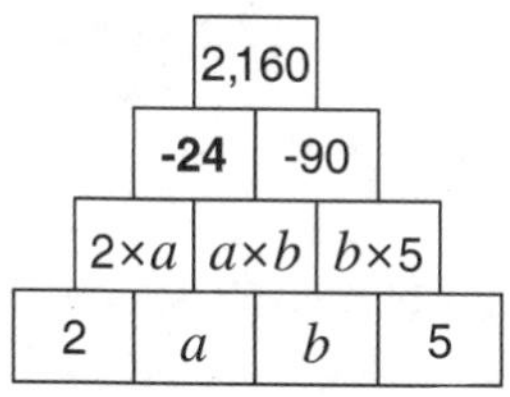

Using the blocks below -24, we write the equation

$$2\times a\times a\times b=-24.$$

Since $2\times\boxed{-12}=-24$, we have

$$a\times a\times b=-12.$$

The product $a\times a$ is a perfect square, and therefore positive. So, b must be negative. There are only two ways to get -12 by multiplying a perfect square by a negative number:

$$1\times(-12)\quad \textbf{or} \quad 4\times(-3).$$

So, b is either -12 or -3.

From the blocks below -90, we can write

$$a\times b\times b\times5=-90.$$

If b were -12, then $b\times b$ would be $(-12)\times(-12)=144$, and we cannot multiply 144 by an integer to get -90.

So, b must be -3, and our equation for the blocks below -90 becomes

$$a\times(-3)\times(-3)\times5=-90.$$

So, $a\times45=-90$. Since $\boxed{-2}\times45=-90$, we have $a=-2$.

We complete the block mountain as shown.

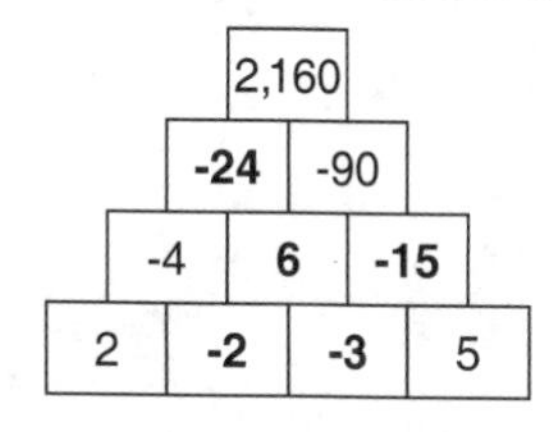

INTEGERS — Product Hives 50–51

Since a completed puzzle path must cross all hexagons, we ignore paths that cannot cross all hexagons.

75. The least product is 6×(-3) = -18. We connect 6 and -3, then continue the path of increasing products.

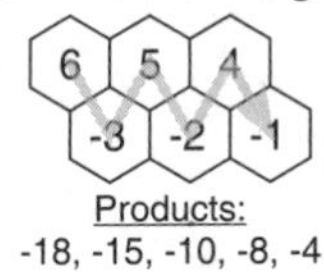

Products:
-18, -15, -10, -8, -4

76. We start by considering the least possible product in the hive: -7×4 = -28.

However, every path that begins with -7 and 4 results in a dead end.

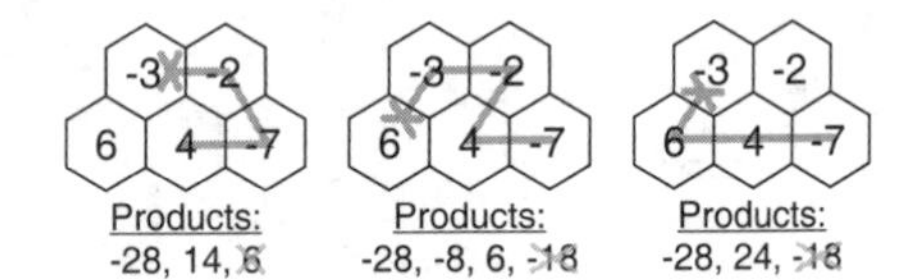

Products: -28, 14, 6 | Products: -28, -8, 6, -18 | Products: -28, 24, -18

So, we put a "wall" between 4 and -7. Then, the least remaining product is 6×(-3) = -18. We connect 6 and -3, and look for a path of increasing products. The following solution is the only possible path of increasing products:

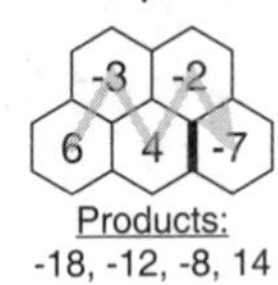

Products:
-18, -12, -8, 14

77. Since 5 is the only positive number in the hive, any pair of numbers that includes 5 creates a negative product, and any pair of numbers that does not include 5 creates a positive product.

So, the first product in our path must include the 5. Otherwise, we would start with a positive product and at some point reach a negative product.

We test the paths that begin with products that include 5 until we find a path of increasing products.

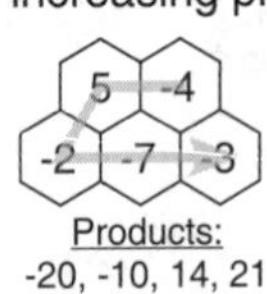

Products:
-20, -10, 14, 21

78. Similar to the previous problem, since 7 is the only positive number in the hive, 7 must be included in the first product.

We cannot pair the 7 with either the -5 or -4 first because doing so will force the path to cross some hexagon twice in order to reach every hexagon.

So, the only starting pairs we must consider are (7 and -3), and (7 and -2). We test paths that begin with these pairs until we find a path of increasing products.

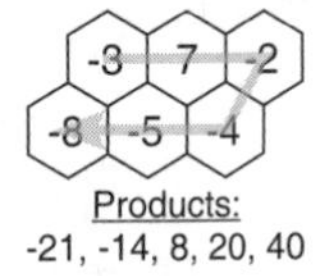

Products:
-21, -14, 8, 20, 40

To solve the remaining Product Hive puzzles, we use the strategies outlined in the previous solutions.

79.

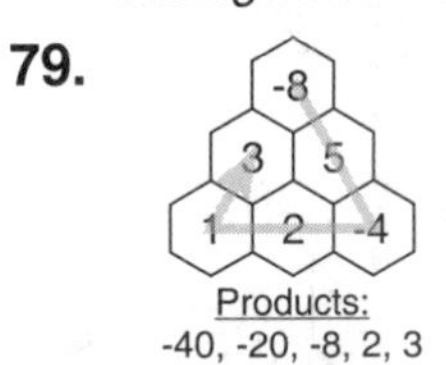

Products:
-40, -20, -8, 2, 3

80.

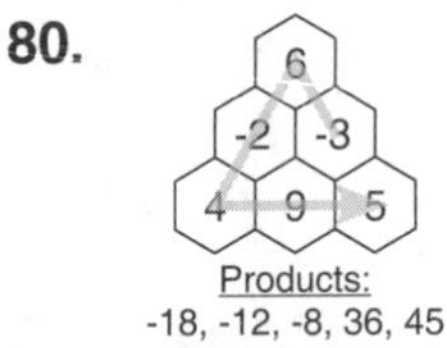

Products:
-18, -12, -8, 36, 45

81.

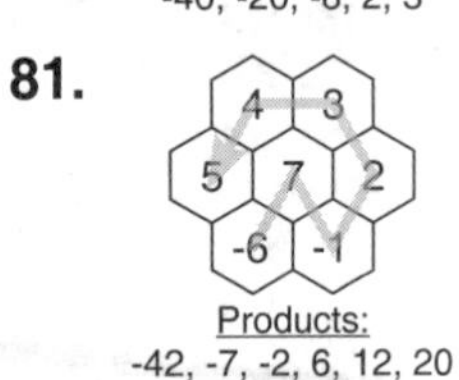

Products:
-42, -7, -2, 6, 12, 20

82.

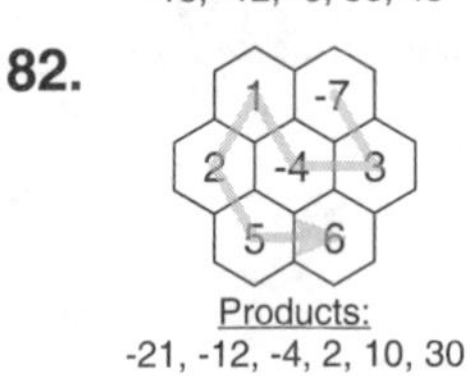

Products:
-21, -12, -4, 2, 10, 30

83.

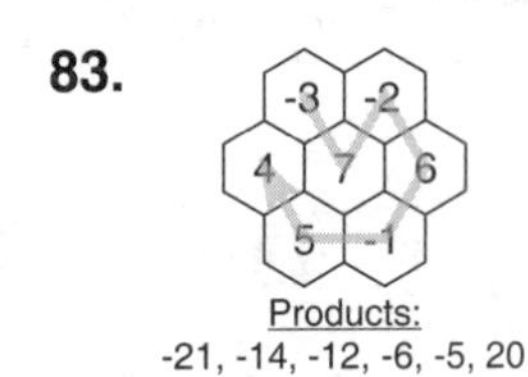

Products:
-21, -14, -12, -6, -5, 20

84.

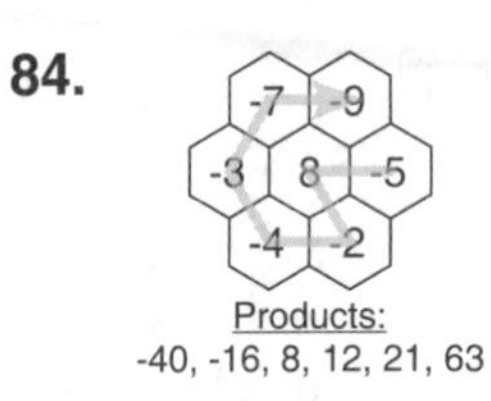

Products:
-40, -16, 8, 12, 21, 63

85.

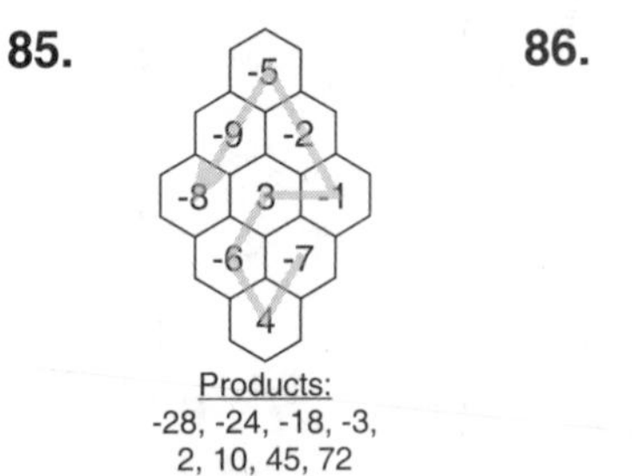

Products:
-28, -24, -18, -3, 2, 10, 45, 72

86.

Products:
-18, -15, -10, -8, -4, 8, 56, 63

INTEGERS — Longer Products 52–53

87. Multiplying from left to right, we have

$$5\times(-3)\times(-1) = -15\times(-1) = \mathbf{15}.$$

— or —

We multiply the numbers without considering the signs: 5×3×1 = 15. Then, since the number of negatives in the product is even, the final result is positive.

So, 5×(-3)×(-1) = **15**.

88. 8×4×2 = 64. Then, since the number of negatives in the product is odd, the final result is negative.

So, 8×(-4)×2 = **-64**.

89. -4×(-5)×(-9) = **-180**.

90. 8×(-6)×4×(-2) = **384**.

91. -1×(-3)×(-5)×(-7) = **105**.

92. -100×4×(-5)×(-6) = **-12,000**.

93. 3×(-2)×(-3)×(-2)×3 = **-108**.

94. 5×(-7)×10×(-7)×5 = **12,250**.

95. -6×(-5)×(-4)×(-3)×2×(-1) = **-720**.

96. -7×6×5×(-4)×3×(-2)×(-1) = **5,040**.

97. We evaluate each power, then add.

$(-2)^1 = -2.$
$(-2)^2 = (-2)\times(-2) = 4.$
$(-2)^3 = (-2)^2\times(-2) = (4)\times(-2) = -8.$
$(-2)^4 = (-2)^3\times(-2) = (-8)\times(-2) = 16.$

So, we have:

$$(-2)^1+(-2)^2+(-2)^3+(-2)^4=(-2)+4+(-8)+16=\mathbf{10}.$$

98. Raising -1 to an even power gives 1, and raising -1 to an odd power gives -1.

So, the sum

$$(-1)^1+(-1)^2+(-1)^3+\cdots+(-1)^{99}+(-1)^{100}$$

is equivalent to the sum

$$(-1)+1+(-1)+\cdots+(-1)+1.$$

The terms in this sum alternate between -1 and 1. Since every -1 is followed by 1, and $-1+1=0$, the entire expression is equal to **0**.

99. When adding two numbers, we will only get an even sum if both numbers are odd or if both are even. Since $a+b=76$, an even number, a and b must both be even or both odd.

Odd powers of -1 equal -1. Even powers of -1 equal 1.

If a and b are both even, then $(-1)^a\times(-1)^b=1\times1=1$.
If a and b are both odd, then $(-1)^a\times(-1)^b=(-1)\times(-1)=1$.
In both cases, we get a result of 1. Therefore, $(-1)^a\times(-1)^b=\mathbf{1}$.

— *or* —

$(-1)^a$ is the product of a copies of -1. Similarly, $(-1)^b$ is the product of b copies of -1. So, $(-1)^a\times(-1)^b$ is the product of $a+b$ copies of -1.

$a+b=76$, so $(-1)^a\times(-1)^b$ is the product of 76 copies of -1. We write this as $(-1)^a\times(-1)^b=(-1)^{76}$.

Raising -1 to an even power gives us 1. Therefore, $(-1)^a\times(-1)^b=(-1)^{76}=\mathbf{1}$.

100. The sign of the answer is determined by the number of negatives in the product. The negative numbers in the product are -2, -4, -6, -8, ..., -96, -98.

To count the number of negatives in the list above, we multiply each number in the list by -1, then divide each resulting number by 2. This gives us

1, 2, 3, 4, ..., 48, 49.

There are 49 numbers in this list, so there are 49 negative numbers in the product

$$99\times(-98)\times97\times(-96)\times\cdots\times(-4)\times3\times(-2)\times1.$$

Any product with an odd number of negatives (and no zeros) is **negative**.

101. The list of integers from -10 to 10 includes 0. Since 0 times anything is 0, the product is equal to **0**.

102. The integers with absolute value less than 10 are the integers from -9 to 9, inclusive.

To create the least possible product, Winnie must pick integers with the greatest absolute value. For example, if Winnie multiplies just *two* different integers, the least product she can make is $-9\times9=-81$. Similarly, the least product she can make with *three* different integers is $-9\times9\times8=-81\times8=-648$.

However, if Winnie picks -9, 9, -8, and 8 as her four integers, this will result in a positive product (which is greater than any negative product). To make a negative product, either 1 or 3 of Winnie's four numbers must be negative.

So, to create the least possible product, Winnie must choose either (-9, 9, 8, 7), or (-9, 9, -8, -7). $9\times9\times8\times7=81\times8\times7=648\times7=4{,}536$, so both choices have product **-4,536**.

INTEGERS

Sign Wars 54-55

103. Alex has two possible moves.

-1	1	-1
	1	1
-1	1	

If Alex places a -1 in the bottom-right square, then Grogg must place a 1 in the middle-left square, as shown below. In this case, Grogg wins by a score of 6 to 0.

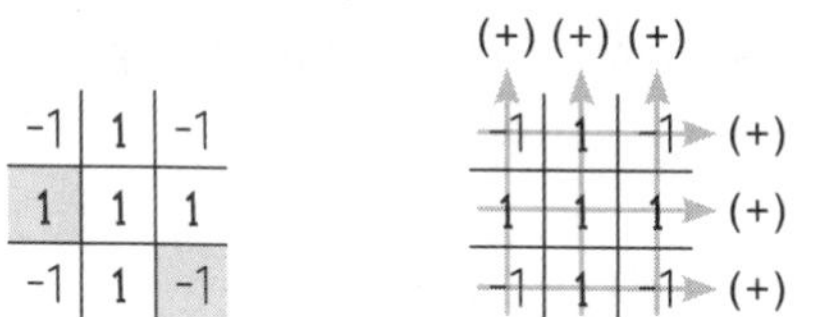
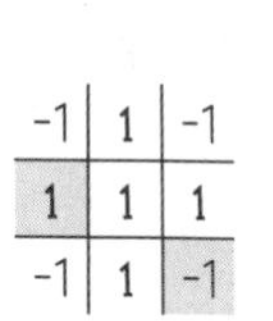

If Alex places a -1 in the middle-left square, then Grogg must place a 1 in the bottom-right square, as shown below. In this case, Alex wins by a score of 4 to 2.

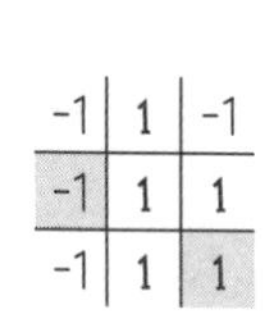
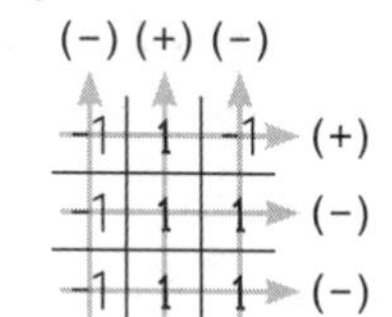

So, **Alex must place a -1 in the middle-left square.**

-1	1	-1
-1	1	1
-1	1	

104. Alex has two possible moves. If he places a -1 in the bottom-left square, then Grogg must place a 1 in the center square as shown. In this case, Grogg wins by a score of 4 to 2.

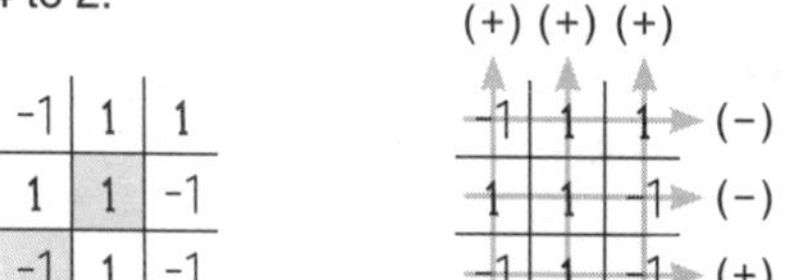
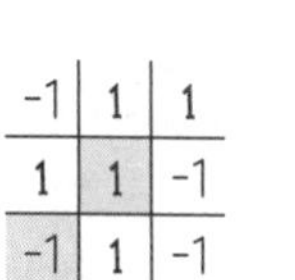

If Alex places a -1 in the center square, then Grogg must place a 1 in the bottom-left square as shown. In this case, Alex wins by a score of 4 to 2.

-1	1	1
1	-1	-1
1	1	-1

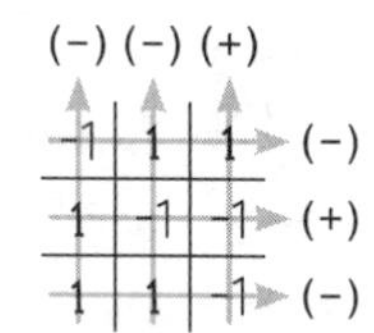

So, **Alex must place a -1 in the center square.**

-1	1	1
1	-1	-1
	1	-1

105. Alex has two possible moves. If he places a -1 in the middle-left square, then Grogg must place a 1 in the top-middle square as shown. In this case, Grogg wins by a score of 4 to 2.

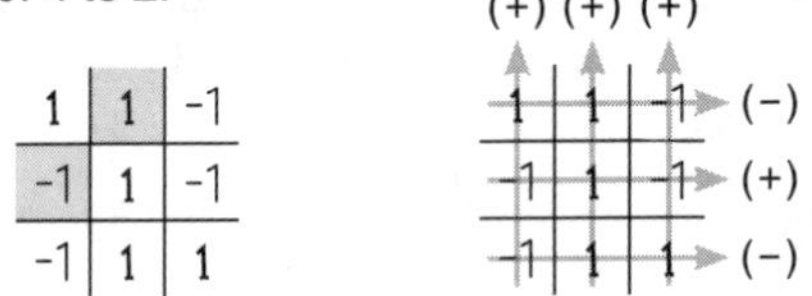

If Alex places a -1 in the top-middle square, then Grogg must place a 1 in the middle-left square as shown. In this case, Alex wins by a score of 4 to 2.

1	-1	-1
1	1	-1
-1	1	1

(-)	(-)	(+)	
1	-1	-1	(+)
1	1	-1	(-)
-1	1	1	(-)

So, **Alex must place a -1 in the top-middle square.**

1	-1	-1
	1	-1
-1	1	1

For Alex to win, he must score 4 or more points. There are just 3 row and 3 column products. So, if all row or all column products are positive, Alex cannot score more than 3 points and therefore cannot win.

106. Alex has four possible moves on this board.

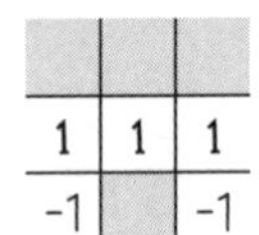

If Alex places a -1 in the bottom-middle square, then the three remaining numbers (1, -1, and 1) will be placed in the squares in the top row:

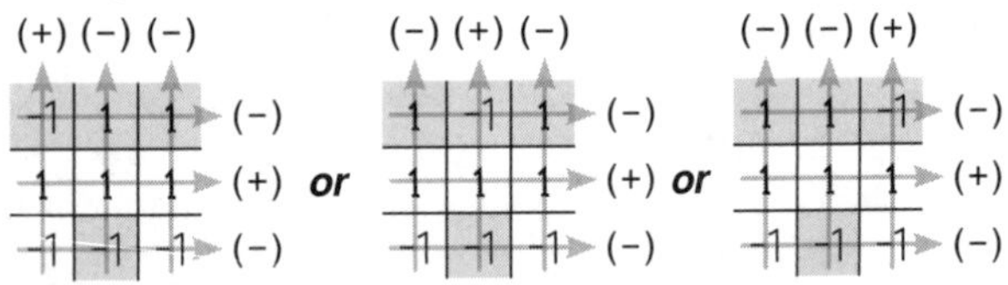

No matter how the three numbers are arranged in the top row, there are 4 negative and 2 positive products. Alex wins the game with a score of 4 to 2.

If Alex instead places -1 in any of the squares in the top row, then Grogg can place 1 in the bottom-middle square. This forces Alex to place another -1 into the top row, creating three positive row products.

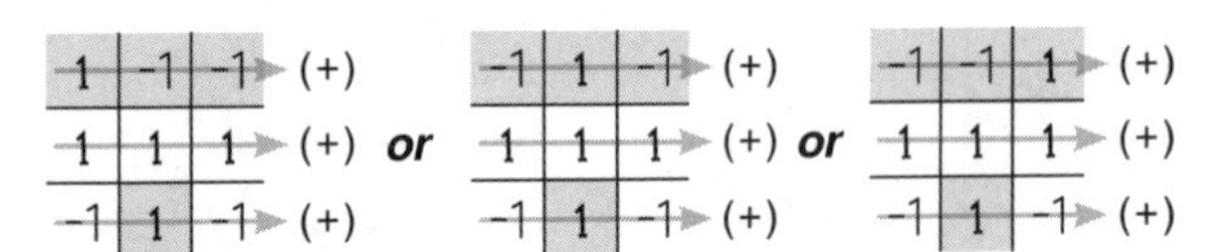

So, **to guarantee a win, Alex must place a -1 in the bottom-middle square.**

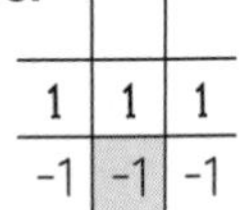

For the problems that follow, note that only -1's affect the sign of the product in a row or column. So, only Alex can affect the product of a row or column, and all product signs are determined once Alex places his fourth -1.

107. Alex has four possible moves. If he places a -1 in the bottom-middle square, then the three remaining numbers (1, -1, and 1) will be placed in the middle row:

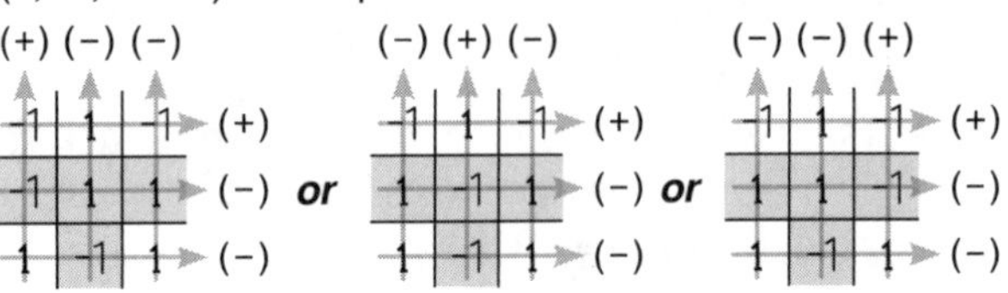

No matter how the three numbers are arranged in the middle row, Alex wins the game by a score of 4 to 2.

If Alex instead places -1 in any of the squares in the middle row, then Grogg can place 1 in the bottom-middle square. This forces Alex to place another -1 into the middle row, creating three positive row products.

So, **to guarantee a win, Alex must place a -1 in the bottom-middle square.**

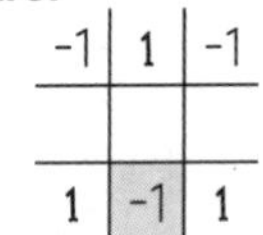

108. If Alex places a -1 in the top-middle square, he guarantees that he will place the remaining -1 in the left column, so the left and middle columns will both have a negative product.

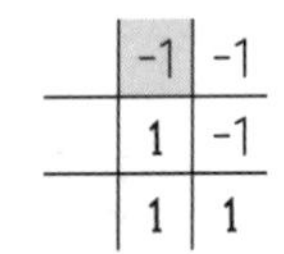

No matter where Grogg places a 1 in the left column, Alex can force a negative product in 2 rows by placing his final -1 in either the top-left or bottom-left square.

After Grogg fills in the remaining square with a 1, we have one of the two boards below:

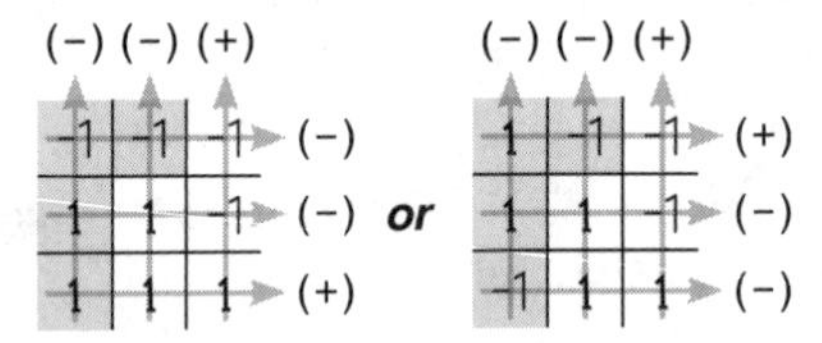

In both cases, we get 4 negative and 2 positive products. Alex wins the game by a score of 4 to 2.

If Alex instead places -1 in any of the squares in the left column, then Grogg can place 1 in the top-middle square. This forces Alex to place another -1 in the left column, creating three positive column products.

So, **to guarantee a win, Alex must place a -1 in the top-middle square.**

	-1	-1
	1	-1
	1	1

109. By placing a 1 in the top-right square, Grogg forces Alex to place a second -1 in the left column.

	-1	1
	-1	1
-1	1	1

Then, wherever Alex places his -1, all three columns will have a positive product. Alex's next -1 will also create a positive product in one of the top two rows, giving Grogg 4 points.

Therefore, Grogg wins by a score of 4 to 2.
No other move guarantees that Grogg will win.

110. Grogg has three possible moves. If Grogg places a 1 in the middle-left square, he forces Alex to place a second -1 in the bottom row.

-1	1	-1
1	1	1
	-1	

Wherever Alex places his next -1, all three rows will have a positive product. Alex's next -1 will also create a positive product in either the left or right column, giving Grogg 4 points.

Therefore, Grogg wins by a score of 4 to 2.
No other move guarantees that Grogg will win.

111. Grogg has three possible moves. By placing a 1 in the bottom-left square, Grogg guarantees that the products of the bottom row and middle column are both positive.

	-1	1
1	1	
1	-1	-1

This leaves Alex with 2 possible moves. Wherever Alex places his next -1, Grogg fills in the remaining square with 1. This gives one of the two boards below:

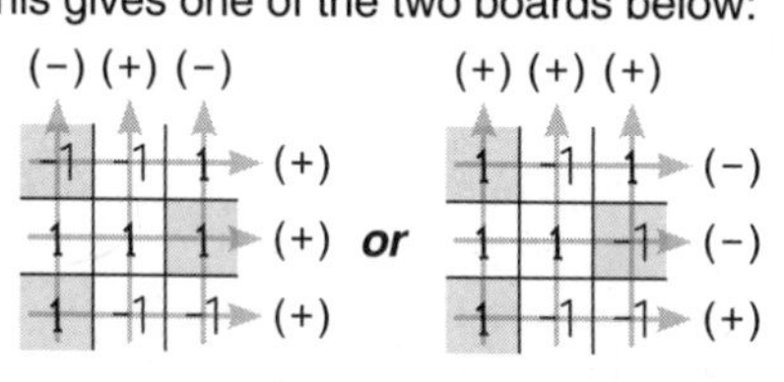

In both cases, we get 4 positive and 2 negative products. Grogg wins the game with a score of 4 to 2.

No other move guarantees that Grogg will win.

112. If Grogg places a 1 in the bottom-left square indicated below, he forces Alex to place a -1 in one of the two right columns.

1		-1
1	-1	
1		

Whichever column Alex places his next -1 in, Grogg can complete the column with a 1, making the product of this column positive. This forces Alex to play the last -1 in the remaining column, making its product positive. So, by placing a 1 in the bottom-left square, Grogg can guarantee a positive product in all three columns.

In the rows, we will have either three positive products or one positive and two negative products.

Using this strategy, Grogg wins by a score of 6 to 0 or a score of 4 to 2. No other move guarantees that Grogg will win.

113. If Grogg places a 1 in the top-right square, he forces Alex to play his next -1 in one of the two left columns.

-1	-1	1
		1
		1

Whichever column Alex places his next -1 in, Grogg can complete the column with a 1, making the product of this column positive. This forces Alex to play the last -1 in the remaining column, making its product positive.

So, by placing a 1 in the top-right square, Grogg can guarantee a positive product in all three columns.

Since the product in the top row is positive, Grogg will have at least 4 points and is guaranteed to win. No other move guarantees that Grogg will win.

114. If Grogg places a 1 in the middle-right square, he forces Alex to place a -1 in the top or bottom row.

		-1
1	1	1
-1		

Whichever row Alex places his -1 in, Grogg can complete the row with a 1. This forces Alex to play the last -1 in the remaining row. So, Grogg's move guarantees a positive product in all three rows.

In the columns, we will have either three positive products or one positive and two negative products.

Therefore, Grogg wins by a score of 6 to 0 or a score of 4 to 2. No other move guarantees that Grogg will win.

115. We consider the placement of the -1's on the game board, since the -1's determine the sign of the product. Consider a game in which Alex gets three -1's in the top row:

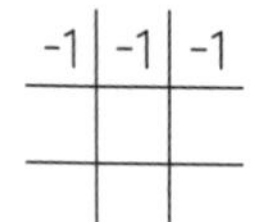

Row products: The top-row product is negative. The fourth -1 will be placed in one of the two bottom rows, so one of the products of the bottom two rows will be positive and the other will be negative.

Column products: The fourth -1 will be placed in one of the three columns. All three columns already contain one -1, so one of the column products will be positive, while the other two will be negative.

For example,

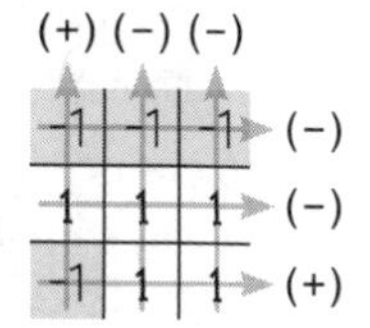

In all cases, Alex wins by a score of 4 to 2.

We can use the same logic to show that Alex will also win if his row of three is the middle or bottom row.

-1	-1	-1

-1	-1	-1

We can also use similar logic to show that Alex is guaranteed to win if he gets three -1's in any column.

-1		
-1		
-1		

	-1	
	-1	
	-1	

		-1
		-1
		-1

Finally, we consider the diagonals. Consider a game in which Alex gets three -1's in one diagonal.

-1		
	-1	
		-1

The fourth -1 will be placed in both a row and column that already contain a -1. So, the game will end with 1 positive and 2 negative row products, as well as 1 positive and 2 negative column products.

For example,

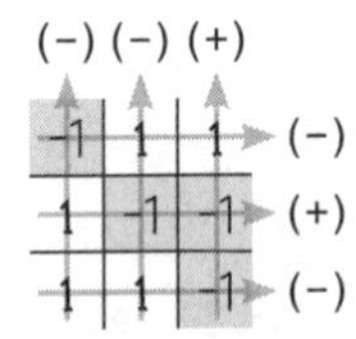

In all cases, Alex wins by a score of 4 to 2.

We can use similar logic to show that Alex is guaranteed to win if he gets three -1's in the other diagonal.

		-1
	-1	
-1		

Therefore, **Alex is correct. If he gets three -1's in a row, column, or diagonal, he is guaranteed to win with 4 points.**

116. For the player placing -1's to get 6 points, the product in every row and column must be negative. A row or column with no -1's has a positive product. So, for the product of every row and column to be negative, there must be at least one -1 in each row and column.

For example,

	-1	
-1		
		-1

However, in a game of Sign Wars, a total of *four* -1's are placed on the board. Wherever the fourth -1 is placed, it must be in a row and a column where another -1 has already been placed. The product in a row or column with exactly two -1's is positive.

So, all six products cannot be negative. The player placing -1's cannot win with 6 points.

117. To tie, each player must score exactly 3 points.

There are four -1's and five 1's on a final game board. Each of these nine integers appears in exactly one row and one column.

If we find the product of each row, then multiply these three products, we get the product of all nine integers on the board: $((-1)^4\times 1^5) = 1$. The product of three numbers is only positive if we have an *even* number of negatives, so either 0 or 2 row products are negative.

We can use the same logic to show that either 0 or 2 column products are negative.

All together, we can only have 0, 2, or 4 negative products on a game board. We can never have exactly 3 negative products. So, it is impossible to tie in a game of Sign Wars.

Something to think about: Do your answers to these last two questions change if Player 1 places -1's and Player 2 places 1's? If so, how? If not, why not?

118. Strategy 1: *Look at the corners first.* A number in a corner only touches two adjacent numbers, and it must be grouped with at least one adjacent number.

For example, the 5 in the top-left corner must be grouped with the 4 below it or the -4 to its right. Since $5\times(-4) = -20$, we pair the 5 with the -4 as shown.

5	-4	2
4	-5	-10
-2	-2	-5

Then, the 2 in the top-right corner must be paired with the -10 below it.

5	-4	2
4	-5	-10
-2	-2	-5

The remaining numbers are grouped into blobs as shown so that the product of the numbers in each blob is -20.

5	-4	2
4	-5	-10
-2	-2	-5

119. Strategy 2: *Use prime factorization.* Prime factorizations of both the target number and the composite numbers in the grid make it easier to see how to group the numbers.

$36 = 2\times 2\times 3\times 3$, so we need two 2's and two 3's in each blob. We must also have an even number of negative integers in each blob to make the product positive.

For example, we look at -9 in the center. $9 = 3\times 3$, so the blob with -9 needs two more 2's. We could group -9 with two -2's or one -4. However, $-9\times(-2)\times(-2)$ is negative, while $-9\times(-4)$ is positive. Our target product, 36, is positive. So, -9 must be grouped with the -4 to its right.

-2	-3	-3
-2	-9	-4
6	-2	-3

The remaining numbers are grouped into blobs as shown so that the product of the numbers in each blob is 36.

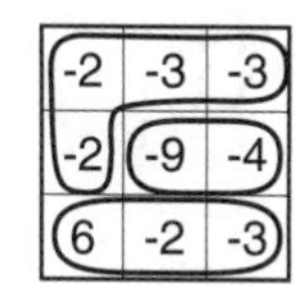

120. <u>Strategy 3:</u> *Start with numbers that have many factors in their prime factorizations.*

For example, $45 = 3\times3\times5$. Since $225 = 3\times3\times5\times5$, the blob with 45 requires exactly one more factor of 5. To get that factor of 5, we can only group 45 with the -5 above it. Then, since $-5\times45 = -225$, this blob is complete.

5	-5	-3
9	45	-1
-5	25	-3

The remaining numbers are grouped into blobs as shown so that the product of the numbers in each blob is -225.

5	-5	-3
9	45	-1
-5	25	-3

121. <u>Strategy 4:</u> *Separate numbers that cannot be in the same blob with "walls."*

For example, since $126 = 2\times3\times3\times7$, we cannot have two 7's in the same blob. So, we draw a wall between pairs of adjacent 7's or any other numbers that are multiples of 7.

3	2	9
-6	-7	7
-2	-9	7

Similarly, we cannot have two multiples of 2 in the same blob. So, we separate the -2 and -6.

3	2	9
-6	-7	7
-2	-9	7

The walls make it easier to see which numbers must be part of the same blob. We group the numbers without crossing walls so that the product of the numbers in each blob is 126.

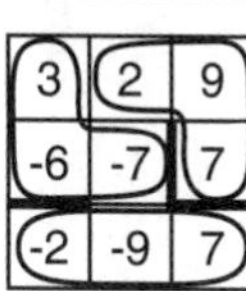

To solve the remaining Integer Blob puzzles, we use the strategies outlined in the previous solutions.

122. ***Target:*** -198

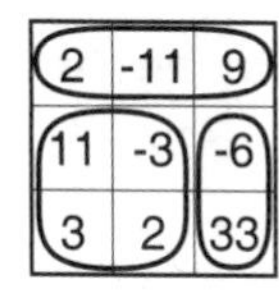

123. ***Target:*** -84

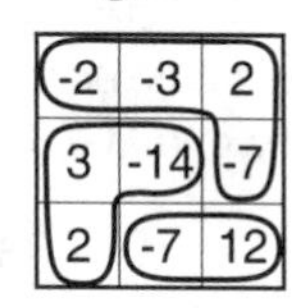

124. ***Target:*** 450

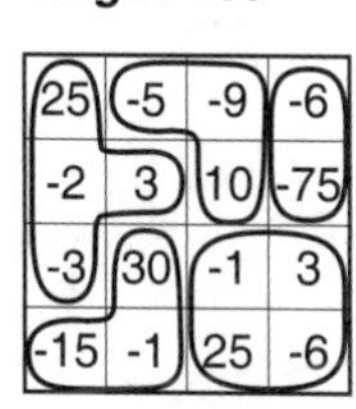

125. ***Target:*** -1,950

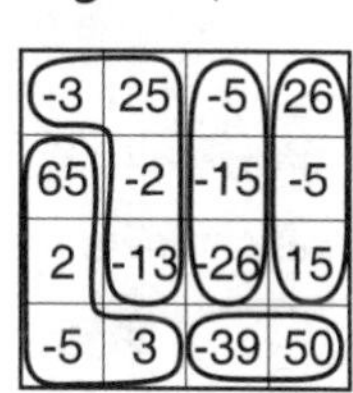

126. ***Target:*** -660

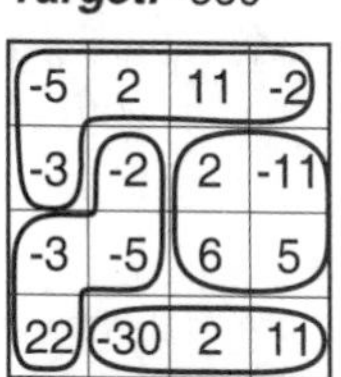

127. ***Target:*** 1,155

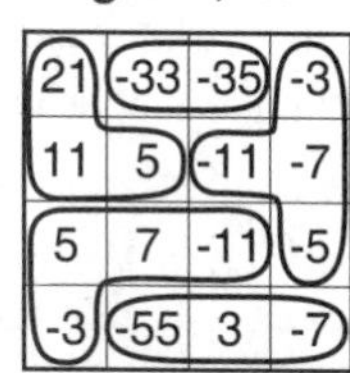

INTEGERS

Division 58-59

128. To divide $24\div(-3)$, we find the number that can be multiplied by -3 to get 24. We can use the multiplication fact $\boxed{-8}\times(-3) = 24$ to see that $24\div(-3) = \boxed{-8}$.

129. Since $\boxed{-12}\times(-2) = 24$, we have $24\div(-2) = \boxed{-12}$.

130. Since $\boxed{2}\times(-12) = -24$, we have $-24\div(-12) = \boxed{2}$.

131. Since $\boxed{11}\times(-3) = -33$, we have $-33\div(-3) = \boxed{11}$.

132. Since $\boxed{-4}\times9 = -36$, we have $-36\div9 = \boxed{-4}$.

133. Since $\boxed{3}\times(-11) = -33$, we have $-33\div(-11) = \boxed{3}$.

134. Since $\boxed{-9}\times5 = -45$, we have $-45\div5 = \boxed{-9}$.

135. Since $\boxed{-9}\times(-5) = 45$, we have $45\div(-5) = \boxed{-9}$.

The problems are matched as shown below:

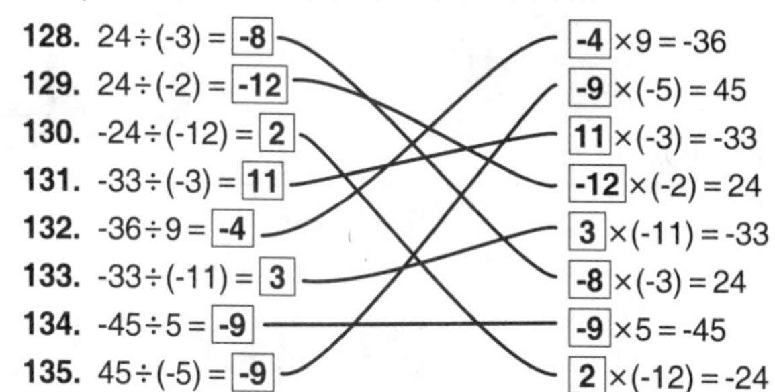

136. Two numbers with different signs have a negative quotient. $27\div9 = 3$, so $27\div(-9) = \mathbf{-3}$.

137. Two numbers with different signs have a negative quotient. $-24\div8 = \mathbf{-3}$.

138. Two numbers with different signs have a negative quotient. $-48\div6 = \mathbf{-8}$.

139. Two numbers with the same sign have a positive quotient. $-32\div(-8) = \mathbf{4}$.

140. Two numbers with different signs have a negative quotient. $45\div(-3) = \mathbf{-15}$.

141. Two numbers with different signs have a negative quotient. $90\div(-5) = \mathbf{-18}$.

142. Two numbers with the same sign have a positive quotient. $-420\div(-70) = \mathbf{6}$.

143. Two numbers with different signs have a negative quotient. $-9{,}600\div12 = \mathbf{-800}$.

144. $(-21\times10)\div15 = -210\div15$. Two numbers with different signs have a negative quotient, so $-210\div15 = \mathbf{-14}$.

145. Two numbers with different signs have a negative quotient, so $-120\div5 = -24$, and we have $(-120\div5)\div(-3) = -24\div(-3)$. Two numbers with the same sign have a positive quotient, so $-24\div(-3) = \mathbf{8}$.

146. Two numbers with different signs have a negative quotient, so 84÷(-2) = -42. Then, two numbers with the same sign have a positive quotient, so -42÷(-3) = **14**.

147. Lizzie: -24÷3 = -8, and -8+9 = 1.
Winnie: 9+(-24) = -15, and -15÷3 = -5.
Lizzie's result is 1−(-5) = 1+5 = **6** more than Winnie's.

INTEGERS Cross-Number Puzzles 60-61

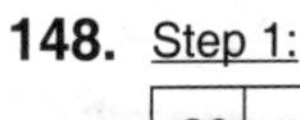

148. Step 1:

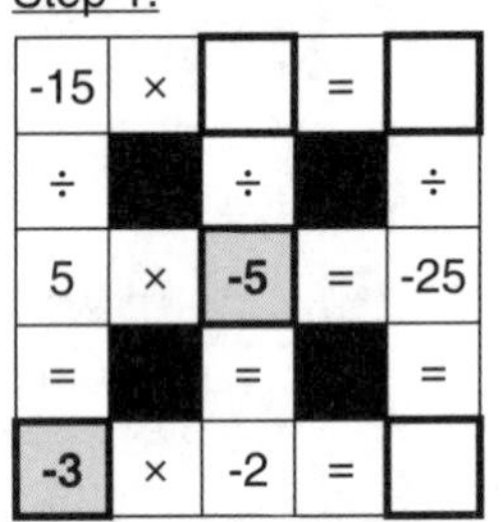

-80	÷	-4	=	**20**
÷		÷		÷
10	÷	-2	=	**-5**
=		=		=
-8	÷	**2**	=	

Final:

-80	÷	-4	=	**20**
÷		÷		÷
10	÷	-2	=	**-5**
=		=		=
-8	÷	**2**	=	**-4**

149. Step 1:

-15	×		=	
÷		÷		÷
5	×	**-5**	=	-25
=		=		=
-3	×	-2	=	

Step 2:

-15	×	**10**	=	
÷		÷		÷
5	×	**-5**	=	-25
=		=		=
-3	×	-2	=	**6**

Final:

-15	×	**10**	=	**-150**
÷		÷		÷
5	×	**-5**	=	-25
=		=		=
-3	×	-2	=	**6**

150.

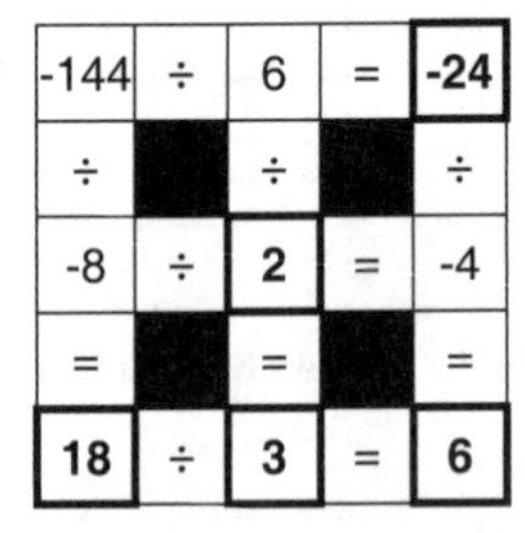

-144	÷	6	=	**-24**
÷		÷		÷
-8	÷	**2**	=	-4
=		=		=
18	÷	**3**	=	**6**

151.

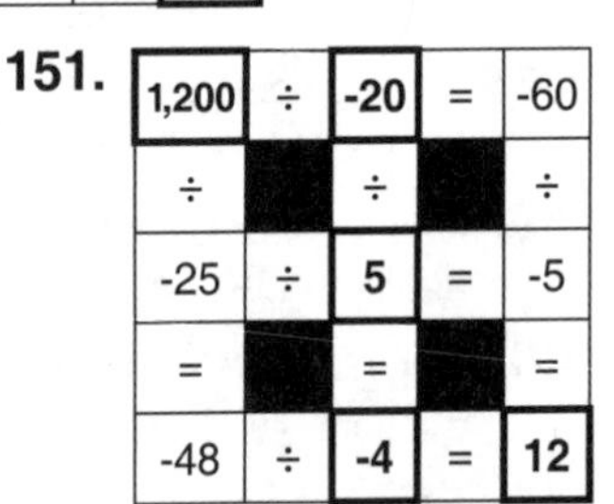

1,200	÷	**-20**	=	-60
÷		÷		÷
-25	÷	**5**	=	-5
=		=		=
-48	÷	**-4**	=	**12**

152.

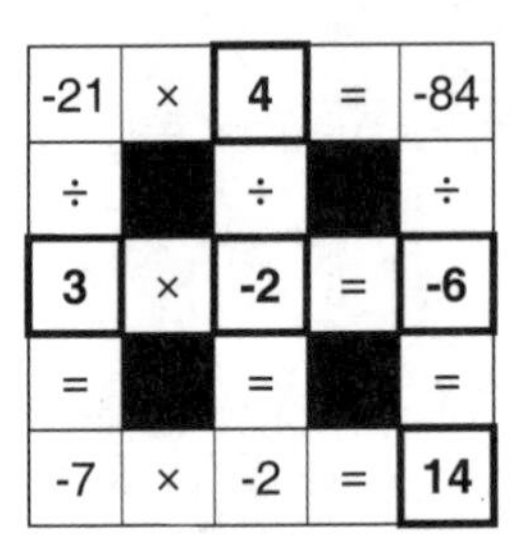

-21	×	**4**	=	-84
÷		÷		÷
3	×	**-2**	=	**-6**
=		=		=
-7	×	-2	=	**14**

153.

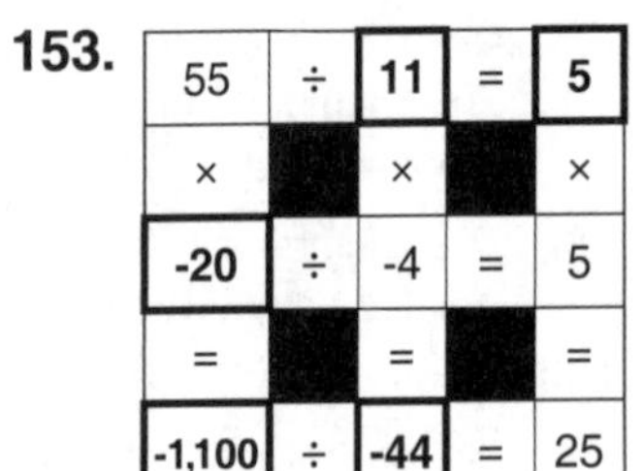

55	÷	**11**	=	**5**
×		×		×
-20	÷	-4	=	5
=		=		=
-1,100	÷	**-44**	=	25

154.

75	×	5	=	375
÷		×		÷
15	÷	**-1**	=	-15
=		=		=
5	×	**-5**	=	**-25**

155.

-24	÷	**8**	=	-3
×		÷		×
-21	×	**-4**	=	**84**
=		=		=
504	÷	**-2**	=	-252

INTEGERS Zig-Zags 62-63

156. Since -16 is not divisible by 5, we multiply: -16×5 = -80.

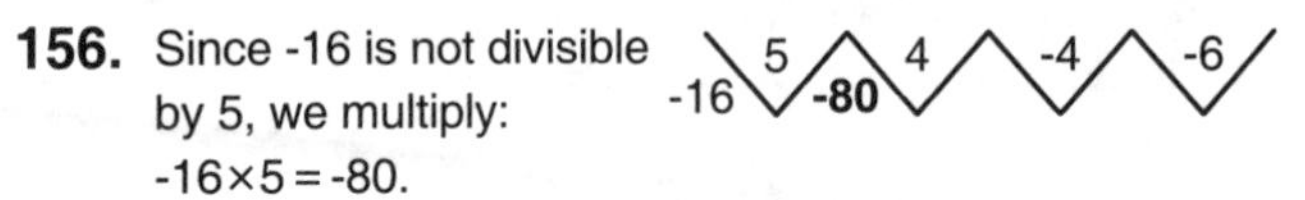

Bottom: -16, **-80**; Top: 5, 4, -4, -6

Since -80 is divisible by 4, we divide: -80÷4 = -20.

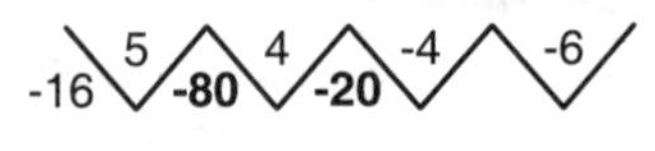

Bottom: -16, **-80**, **-20**; Top: 5, 4, -4, -6

Since -20 is divisible by -4, we divide: -20÷(-4) = 5.

Bottom: -16, **-80**, **-20**, **5**; Top: 5, 4, -4, -6

Since 5 is not divisible by -6, we multiply: 5×(-6) = -30.

Bottom: -16, **-80**, **-20**, **5**, **-30**; Top: 5, 4, -4, -6

157. Bottom: 240, **-120**, **40**, **-10**, **2**; Top: -2, -3, -4, -5

158. Bottom: 44, **4**, **-1**, **-25**, **75**; Top: 11, -4, 25, -3

159. Bottom: -320, **-20**, **4**, **-80**, **-5**; Top: 16, -5, -20, 16

160. Bottom: -10, **-5**, **15**, **3**, **-1**, **-2**; Top: 2, -3, 5, -3, 2

161. Bottom: 11, **-22**, **66**, **330**, **-110**, **55**; Top: -2, -3, 5, -3, -2

162. We begin by filling in the entries below.

Bottom: -5, **-30**, -15, 3, **24**, **2**, **-42**, 14, -1; Top: 6, 2, ( ), 8, 12, -21, ( ), ( )

The three remaining blank spaces are top numbers. We consider these one at a time.

Bottom: -15, 3; Top: ( )

This tells us that -15×☐ = 3 or -15÷☐ = 3.

Since -15÷[-5] = 3, the top number must be -5.

Bottom: -15, 3; Top: **-5**

Similarly, we can determine that -42÷[-3] = 14, and 14÷[-14] = -1.

Bottom: -5, **-30**, -15, 3, **24**, **2**, **-42**, 14, -1; Top: 6, 2, **-5**, 8, 12, -21, **-3**, **-14**

163. We fill in the following entries in the puzzle.

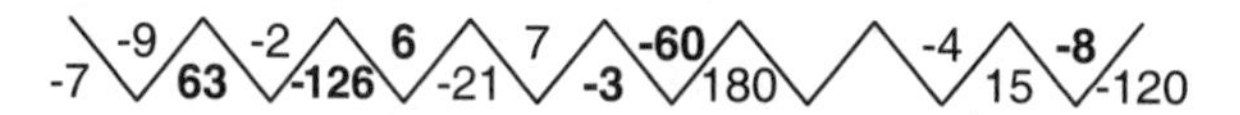

Looking at the remaining bottom number, we know that $\square \times (-4) = 15$ or $\square \div (-4) = 15$.

We cannot multiply -4 by an integer to get -15, but $\boxed{-60} \div (-4) = 15$. So, the remaining bottom number is -60.

Finally, since $180 \div \boxed{-3} = -60$, we complete the puzzle as shown below.

-7 -9 **63** -2 **-126** **6** -21 7 **-3** **-60** 180 **-3** **-60** -4 15 **-8** -120

164. We fill in the following entries in the puzzle.

35 **-7** -5 **2** -10 **-3** 30 -6 **-5** -4 **20** **6** 120 -7 28

Looking at the remaining bottom number, we know that $\square \div (-7) = 28$ or $\square \times (-7) = 28$.

Since $\boxed{-196} \div (-7) = 28$ and $\boxed{-4} \times (-7) = 28$, the remaining bottom number is either -196 or -4.

120 -196 -7 28   120 -4 -7 28

We cannot multiply or divide 120 by any integer to get -196, but we can divide $120 \div \boxed{-30}$ to get -4. So, we fill in these blanks as shown.

120 **-30** **-4** -7 28

The completed puzzle is shown below.

35 **-7** -5 **2** -10 **-3** 30 -6 **-5** -4 **20** **6** 120 **-30** **-4** -7 28

165. Looking at the first bottom number, we have either $\square \times 3 = 27$ or $\square \div 3 = 27$.

Since $\boxed{9} \times 3 = 27$ and $\boxed{81} \div 3 = 27$, the remaining bottom number is either 9 or 81.

However, 9 is divisible by 3. So, if the first number were 9, then we would follow the rule, "If the bottom number is divisible by the top number to its right, then divide." In that case, the next bottom number would be 3, not 27.

So, the first bottom number must be 81.

81 3 27

Then, we fill in the remaining entries in the puzzle.

81 3 27 -3 **-9** -4 **36** **2** **18** -9 -2 **5** -10 -6 **60** **-15** -4

166. Looking at the first bottom number, we have $\square \times 4 = -48$ or $\square \div 4 = -48$.

Since $\boxed{-12} \times 4 = -48$ and $\boxed{-192} \div 4 = -48$, the remaining bottom number is either -12 or -192.

However, -12 is divisible by 4. So, if the first number were -12, then we would follow the rule, "If the bottom number is divisible by the top number to its right, then divide." In that case, the next bottom number would be -3, not -48. So, the first bottom number must be -192.

-192 4 -48

We fill in more entries of the puzzle as shown:

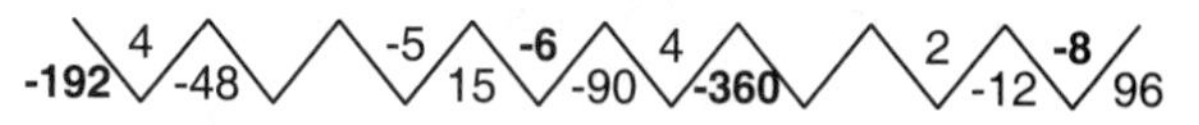

Then, we look at the following part of the zig-zag:

-48 -5 15

For the bottom number, we know that $\square \div (-5) = 15$ or $\square \times (-5) = 15$. Since $\boxed{-75} \div (-5) = 15$ and $\boxed{-3} \times (-5) = 15$, the bottom number is either -75 or -3.

-48 -75 -5 15   -48 -3 -5 15

We cannot multiply or divide -48 by an integer to get -75, but we can divide $-48 \div \boxed{16}$ to get -3. So, we fill in these blanks as shown.

-48 ? -75 -5 15   -48 **16** **-3** -5 15

Our puzzle is nearly complete, as shown:

-192 4 -48 **16** **-3** -5 15 **-6** -90 4 **-360** 2 -12 **-8** 96

Finally, looking at the remaining bottom number, we know that $\square \times 2 = -12$ or $\square \div 2 = -12$.

Since $\boxed{-6} \times 2 = -12$ and $\boxed{-24} \div 2 = -12$, the remaining bottom number is either -6 or -24.

However, if this number were -6, then we would follow the rule, "If the bottom number is divisible by the top number to its right, then divide." In that case, the next bottom number would be -3, not -12.

So, the remaining bottom number must be -24.

-24 2 -12

Finally, $-360 \div \boxed{15} = -24$, and our puzzle is complete.

-192 4 -48 **16** **-3** -5 15 **-6** -90 4 **-360** **15** **-24** 2 -12 **-8** 96

INTEGERS

Opposites 64

167. Writing a negative in front of an expression means to take its opposite. Since $6+4=10$, we have $-(6+4) = \mathbf{-10}$.

— or —

Taking the opposite of an expression is the same as multiplying by -1. So, we have

$$-(6+4) = -1 \times (6+4) = -1 \times 10 = \mathbf{-10}.$$

168. Since $4-8=-4$, we have $-(4-8) = -(-4) = \mathbf{4}$.

— or —

$-(4-8) = -1 \times (4-8) = -1 \times (-4) = \mathbf{4}$.

169. $-(-4+9-12) = -(5-12) = -(-7) = \mathbf{7}$.

170. $-(199+299-300) = -(498-300) = -(198) = \mathbf{-198}$.

171. $-(-8-7-13) = -(-15-13) = -(-28) = \mathbf{28}$.

172. $-(-23-34-45) = -(-57-45) = -(-102) = \mathbf{102}$.

173. Writing a negative in front of an expression means to take its opposite. So, "$-x$" means "the opposite of x." If x is 3, then the opposite of x is **-3**.

— or —

Taking the opposite of an expression is the same as multiplying by -1. So, when $x = 3$, we have $-x = -1\times3 = \mathbf{-3}$.

174. $-x = -(-5) = \mathbf{5}$.

— or —

$-x = -1\times(-5) = \mathbf{5}$.

175. When $x = -4$ and $y = -2$, this expression is equal to $-(-4)+(-2)$. The opposite of -4 is 4, so this expression is equal to $4+(-2) = \mathbf{2}$.

— or —

$-(-4)+(-2) = -1\times(-4)+(-2) = 4+(-2) = \mathbf{2}$.

176. $-(-4+(-2)) = -(-6) = \mathbf{6}$.

— or —

$-(-4+(-2)) = -1\times(-4+(-2)) = -1\times(-6) = \mathbf{6}$.

177. "$-n = 13$" tells us that the opposite of n is 13. The opposite of -13 is 13. Therefore, $n = \mathbf{-13}$.

— or —

"$-n = 13$" tells us that $-1\times n = 13$. Since $-1\times\boxed{-13} = 13$, we have $n = \mathbf{-13}$.

178. Since $\boxed{6}+4 = 10$, we know that $-m = 6$.

"$-m = 6$" tells us that the opposite of m is 6. The opposite of -6 is 6. Therefore, $m = \mathbf{-6}$.

— or —

"$-m = 6$" tells us that $-1\times m = 6$. Since $-1\times\boxed{-6} = 6$, we have $m = \mathbf{-6}$.

INTEGERS
Exponents 65-66

179. The expression -2^4 means "the opposite of 2^4," or $-(2^4)$. Since $2^4 = 16$, we have $-2^4 = \mathbf{-16}$.

180. To evaluate $(-2)^6$, we multiply six copies of -2. Multiplying an even number of negatives in a product gives a positive result, so $(-2)^6 = 2^6 = \mathbf{64}$.

181. -3^3 means "the opposite of 3^3." Since $3^3 = 27$, we have $-3^3 = \mathbf{-27}$.

182. To evaluate $(-3)^5$, we multiply five copies of -3. Multiplying an odd number of negatives in a product gives a negative result, so $(-3)^5 = \mathbf{-243}$.

183. $-(-6)^2 = -1\times((-6)\times(-6)) = -1\times36 = \mathbf{-36}$.

184. Following the order of operations, we first evaluate what is inside the parentheses: $-5^2 = -(5\times5) = -25$. So, $-(-5^2) = -(-25) = \mathbf{25}$.

185. We evaluate each expression:

$-4^2 = -(4\times4) = -16$. $\quad -(4^2) = -(4\times4) = -16$.
$(-4^2) = (-(4\times4)) = -16$. $\quad (-4)^2 = (-4)\times(-4) = 16$.
$-(-4^2) = -(-(4\times4)) = -(-16) = 16$.

So, the three circled expressions are equal to -16.

(-4^2) ($-(4^2)$) ((-4^2)) $\quad (-4)^2 \quad -(-4^2)$

186. We evaluate each expression:

$-5^3 = -(5\times5\times5) = -125$. $\quad -(5^3) = -(5\times5\times5) = -125$.
$(-5^3) = (-(5\times5\times5)) = -125$. $\quad (-5)^3 = (-5)\times(-5)\times(-5) = -125$.
$-(-5^3) = -(-(5\times5\times5)) = -(-125) = 125$.

So, the four circled expressions are equal to -125.

(-5^3) ($-(5^3)$) ((-5^3)) ($(-5)^3$) $\quad -(-5^3)$

187. $a^2 = a\times a = 3\times3 = \mathbf{9}$.
$(-a)^2 = (-a)\times(-a) = (-3)\times(-3) = \mathbf{9}$.
$-a^2 = -(a\times a) = -(3\times3) = \mathbf{-9}$.

188. $a^2 = a\times a = (-3)\times(-3) = \mathbf{9}$.
$(-a)^2 = (-(-3))^2 = (3)^2 = \mathbf{9}$.
$-a^2 = -(a\times a) = -((-3)\times(-3)) = -(9) = \mathbf{-9}$.

189. $b^3 = b\times b\times b = 6\times6\times6 = \mathbf{216}$.
$(-b)^3 = (-b)\times(-b)\times(-b) = (-6)\times(-6)\times(-6) = \mathbf{-216}$.
$-b^3 = -(b\times b\times b) = -(6\times6\times6) = -(216) = \mathbf{-216}$.

190. $b^3 = b\times b\times b = (-6)\times(-6)\times(-6) = \mathbf{-216}$.
$(-b)^3 = (-(-6))^3 = (6)^3 = \mathbf{216}$.
$-b^3 = -(b\times b\times b) = -((-6)\times(-6)\times(-6)) = -(-216) = \mathbf{216}$.

191. We consider each expression.

x^2: Multiplying any nonzero number by itself will always give a *positive* result, whether the number is positive or negative.
For example, $3^2 = 9$ and $(-3)^2 = 9$.

$-x^2$: Above, we determined that x^2 is positive for all nonzero values of x. So, its opposite, $-x^2$, is *negative* for all nonzero values of x.
For example, $-3^2 = -9$ and $-(-3)^2 = -9$.

$-(x^2)$: Since $-(x^2) = -x^2$, we know $-(x^2)$ is *negative* for all nonzero values of x, as we found above.

$(-x)^2$: Multiplying any nonzero number by itself will always give a *positive* result, whether the number is positive or negative.
For example, $(-3)^2 = 9$ and $3^2 = 9$.

$-(-x)^2$: Above, we determined that $(-x)^2$ is positive for all nonzero values of x. So, its opposite, $-(-x)^2$ is *negative* for all nonzero values of x.
For example, $-(-3)^2 = -9$ and $-(3)^2 = -9$.

So, only the two expressions circled below are positive for any nonzero value of x.

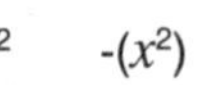
(x^2) $\quad -x^2 \quad -(x^2)$

($(-x)^2$) $\quad -(-x)^2$

192. We consider each expression.

y^3: If y is positive, then y^3 is positive. For example, if $y = 6$, then we have $6^3 = 216$.

$-y^3$: If y is negative, then y^3 is negative. So, its opposite, $-y^3$, is positive. For example, if $y = -6$, then we have $-((-6)^3) = 216$.

$-(y^3)$: Since $-(y^3) = -y^3$, we know this expression is positive when y is negative, as we found above.

$(-y)^3$: If y is negative, then $-y$ is positive and $(-y)^3$ is therefore positive. For example, if $y = -6$, then we have $(-(-6))^3 = 6^3 = 216$.

$-(-y)^3$: If y is positive, then $-y$ is negative, $(-y)^3$ is negative, and $-(-y)^3$ is positive. For example, if $y = 6$, then we have $-(-6)^3 = -(-216) = 216$.

Since each expression above can be made positive for some nonzero value of y, **none of these** expressions are negative for all nonzero values of y.

193. If a is nonzero, then a and $-a$ have opposite signs. **Two numbers with opposite signs always have a negative product. Therefore, $-a\times a$ is always negative for nonzero values of a.**

194. Since $12^{14}\times\boxed{12}=12^{15}$, we have $12^{15}\div12^{14}=\mathbf{12}$. You may have also written this as 12^1, although the exponent is not necessary for the first power of a number.

195. Since 24 is even, $(-8)^{24}$ is positive and equal to 8^{24}.

So, $(-8)^{24}\div8^{23}=8^{24}\div8^{23}$.

Then, since $8^{23}\times\boxed{8}=8^{24}$, we have $(-8)^{24}\div8^{23}=8^{24}\div8^{23}=\mathbf{8}$.

For these problems, it can help to begin by disregarding the signs of the numbers at first. Then, at the end, we determine whether the quotient is negative or positive.

196. $25^{44}\times\boxed{25}=25^{45}$, so $25^{45}\div25^{44}=25$.

Then, 25^{45} is positive, while -25^{44} is negative. Since 25^{45} and -25^{44} have opposite signs, their quotient is negative. Therefore, $25^{45}\div(-25^{44})=\mathbf{-25}$.

197. $19^{30}\div19^{30}=1$.

Then, -19^{30} is negative, while $(-19)^{30}$ is positive. Since -19^{30} and $(-19)^{30}$ have opposite signs, their quotient is negative. Therefore, $-19^{30}\div(-19)^{30}=-19^{30}\div19^{30}=\mathbf{-1}$.

198. $2\times\boxed{2^{99}}=2^{100}$, so $2^{100}\div2=2^{99}$.

Then, $(-2)^{100}$ is positive, while (-2) is negative. Since $(-2)^{100}$ and (-2) have opposite signs, their quotient is negative. Therefore, $(-2)^{100}\div(-2)=\mathbf{-2^{99}}$.

— or —

$(-2)\times\boxed{(-2)^{99}}=(-2)^{100}$. Therefore, $(-2)^{100}\div(-2)=\mathbf{-2^{99}}$.

199. $3^{35}\times\boxed{3}=3^{36}$, so $3^{36}\div3^{35}=3$.

Then, $-(-3)^{36}$ is negative, and $(-3)^{35}$ is also negative. Since $-(-3)^{36}$ and $(-3)^{35}$ have the same sign, their quotient is positive. Therefore, $-(-3)^{36}\div(-3)^{35}=\mathbf{3}$.

200. First, we simplify the expression in parentheses: $-10\times10^{29}=-1\times10\times10^{29}=-1\times10^{30}=-10^{30}$.

So, $(-10\times10^{29})\div10^{30}=-10^{30}\div10^{30}$.

Since -10^{30} is negative and 10^{30} is positive, $-10^{30}\div10^{30}=\mathbf{-1}$.

201. Since 30 is even, $(-4)^{30}$ is positive and $(-4)^{30}\div2^{59}=4^{30}\div2^{59}$.

These two expressions have different bases. Since $4=2\times2$, we have $4^{30}=(2\times2)^{30}$. Multiplying 30 copies of 2×2 is the same as multiplying 60 copies of 2, so $4^{30}=(2\times2)^{30}=2^{60}$.

So, we have $(-4)^{30}\div2^{59}=4^{30}\div2^{59}=2^{60}\div2^{59}$.

Since $2^{59}\times\boxed{2}=2^{60}$, we have $2^{60}\div2^{59}=\mathbf{2}$.

202. We begin with the expression in parentheses on the left.

Since 10 is even, $(-99)^{10}$ is positive and equal to 99^{10}. -99^{10} and 99^{10} have opposite signs, so we have $-99^{10}\div(-99)^{10}=-99^{10}\div99^{10}=-1$.

Next, we consider the expression in parentheses on the right. Since 9 is odd, $(-99)^9=-99^9$. So, $(-99)^9\div(-99^9)=1$.

Therefore, the original expression simplifies to

$$\begin{aligned}&(-99^{10}\div(-99)^{10})+((-99)^9\div(-99^9))\\=&(-99^{10}\div99^{10})+((-99^9)\div(-99^9))\\=&-1+1\\=&\mathbf{0}.\end{aligned}$$

203. To evaluate Winnie's sum, we can pair each negative number from Grogg's sum with a positive number from Lizzie's sum so that the sum of each pair is 1. For example, $-1+2=1$ and $-99+100=1$.

$$\begin{array}{r}(-1)+(-3)+(-5)+\cdots+(-95)+(-97)+(-99)\\+\ 2\ +\ 4\ +\ 6\ +\cdots+\ 96\ +\ 98\ +\ 100\\\hline 1\ +\ 1\ +\ 1\ +\cdots+\ 1\ +\ 1\ +\ 1\end{array}$$

Every number from each original sum is paired. There are 50 numbers in each list, so we have 50 pairs. Each pair sums to 1, so when Winnie adds Grogg's and Lizzie's sums, her result is $50\times1=\mathbf{50}$.

204. First, we consider the factor pairs of 36. $36=1\times36=2\times18=3\times12=4\times9=6\times6$.

To get a product of -36, one of the two numbers in the product must be negative.

For each of the first 4 factor pairs above, we have 2 choices: either the smaller or larger number will be the negative. This gives $4\times2=8$ possible pairs whose product is -36.

In the last pair, we see that (-6, 6) is the same pair as (6,-6). Therefore, there are $8+1=\mathbf{9}$ different pairs of integers whose product is -36:
(-1, 36), (1, -36), (-2, 18), (2, -18), (-3, 12), (3, -12), (-4, 9), (4, -9), and (-6, 6).

205. Each of Angelica's numbers is either positive or negative, **So, at least two of Angelica's three integers have the same sign. Two numbers with the same sign always have a positive product.** So, at least one pair of Angelica's numbers has a positive product.

206. Whether Alex's original number is positive or negative, raising it to an even power gives a positive result. Then, raising this result to an odd power will give him a positive final result. For example, $(2^4)^3=16^3=4{,}096$ and $((-2)^4)^3=16^3=4{,}096$.

If Grogg's original number is positive, then raising it to an odd power gives a positive result.
If Grogg's original number is negative, then raising it to an odd power is the same as multiplying an odd number of negatives, which gives a negative result.

Whether Grogg's first result is positive or negative, raising this result to an even power gives him a positive final result. For example, $(2^3)^4=8^4=4{,}096$ and $((-2)^3)^4=(-8)^4=4{,}096$.

In all these cases, Alex's and Grogg's final numbers have **the same sign**: positive.

207. a. Belinda flips her coin three times. Each time, it lands heads or tails, so there are $2^3 = 8$ possible outcomes: HHH, HHT, HTH, HTT, THH, THT, TTH, and TTT.

An odd number of negatives in a product gives a negative result, while an even number of negatives gives a positive result.

So, Belinda's product will only be negative if exactly one or three of her three integers are negative (exactly one or three tails are flipped).

There are three ways for Belinda to flip 1 tails (HHT, HTH, and THH) and one way for Belinda to flip 3 tails (TTT). So, there are 4 ways for Belinda's product to be negative.

So, the probability that her product is negative is:

$$\frac{\text{\# of ways product is negative}}{\text{total \# of possible outcomes}} = \frac{3+1}{8} = \frac{4}{8} = \mathbf{\frac{1}{2}}.$$

b. Belinda flips her coin four times. Each time, it lands heads or tails, so there are $2^4 = 16$ possible outcomes.

An odd number of negatives in a product gives a negative result, while an even number of negatives gives a positive result. So, Belinda's product will be odd if exactly one or three of her four integers are negative (exactly one or three tails are flipped).

One negative: There are 4 ways to pick one of the four numbers to be the only negative (HHHT, HHTH, HTHH, and THHH).

Three negatives: Choosing three of the four numbers to be the only negative is the same as choosing one of the four numbers to be the only positive. There are 4 ways to pick one of the numbers to be the only positive (HTTT, THTT, TTHT, and TTTH).

So, the probability that her product is negative is

$$\frac{\text{\# of ways product is negative}}{\text{total \# of possible outcomes}} = \frac{4+4}{16} = \frac{8}{16} = \mathbf{\frac{1}{2}}.$$

— or —

If the product of the first three numbers is negative, then the four-number product is only negative when the fourth number is positive. The probability that the fourth number is positive is the same as the probability that a coin flip lands heads: $\frac{1}{2}$.

If the product of the first three numbers is positive, then the four-number product is only negative when the fourth number is negative. The probability that the fourth number is negative is the same as the probability that a coin flip lands tails: $\frac{1}{2}$.

So, no matter what the sign of the product of the first three numbers is, the probability that Belinda's four-number product is negative is $\mathbf{\frac{1}{2}}$.

c. If the product of the first 999 numbers is negative, then the thousand-number product is only negative when the thousandth number is positive, which happens with probability $\frac{1}{2}$.

If the product of the first 999 numbers is positive, then the thousand-number product is only negative when the thousandth number is negative, which also happens with probability $\frac{1}{2}$.

So, no matter what the sign of the product of the first 999 numbers is, the probability that Belinda's thousand-number product is negative is $\mathbf{\frac{1}{2}}$.

Multiplication Notation Page 71

1. $3 \cdot 7$ means 3×7. So, $3 \cdot 7 = \mathbf{21}$.

2. $-4 \cdot 11$ means -4×11. So, $-4 \cdot 11 = \mathbf{-44}$.

3. $5(9)$ means 5×9. So, $5(9) = \mathbf{45}$.

4. $(18)(3)$ means 18×3. So, $(18)(3) = \mathbf{54}$.

5. $-7 \cdot (-11)$ means $-7\times(-11)$. So, $-7 \cdot (-11) = \mathbf{77}$.

6. $2(3)(4)$ means $2\times3\times4$. So, $2(3)(4) = 6(4) = \mathbf{24}$.

7. $6(8+9)$ means $6\times(8+9)$. So, $6(8+9) = 6\times(17) = \mathbf{102}$.

8. $(4-6) \cdot 13$ means $(4-6)\times13$. So, $(4-6) \cdot 13 = -2 \cdot 13 = \mathbf{-26}$.

9. $(5 \cdot 6)+(7 \cdot 8)$ means $(5\times6)+(7\times8)$. So, $(5 \cdot 6)+(7 \cdot 8) = 30+56 = \mathbf{86}$.

10. $(6-7)(7-8)(8-9)$ means $(6-7)\times(7-8)\times(8-9)$. So, $(6-7)(7-8)(8-9) = (-1)\times(-1)\times(-1) = \mathbf{-1}$.

Multiplication with Variables 72

11. $10a$ means $10 \cdot a$. So, when $a=3$, we have $10a = 10 \cdot 3 = \mathbf{30}$.

12. $5b$ means $5 \cdot b$. So, when $b=-4$, we have $5b = 5 \cdot (-4) = \mathbf{-20}$.

13. $-7c$ means $-7 \cdot c$. So, when $c=5$, we have $-7c = -7 \cdot 5 = \mathbf{-35}$.

14. ac means $a \cdot c$. So, when $a=3$ and $c=5$, we have $ac = 3 \cdot 5 = \mathbf{15}$.

15. $2ab$ means $2 \cdot a \cdot b$. So, when $a=3$ and $b=-4$, we have $2ab = 2 \cdot 3 \cdot (-4) = 6 \cdot (-4) = \mathbf{-24}$.

16. $-6abc$ means $-6 \cdot a \cdot b \cdot c$. So, when $a=3$, $b=-4$, and $c=5$, we have $-6abc = -6 \cdot 3 \cdot (-4) \cdot 5 = \mathbf{360}$.

17. $2(3b)$ means $2 \cdot (3 \cdot b)$. When $b=-4$, we have $2 \cdot (3 \cdot (-4)) = 2 \cdot (-12) = \mathbf{-24}$.

— or —

Since multiplication is associative, we can write $2(3b)$ as $(2 \cdot 3)b = 6b$. When $b=-4$, we have $6b = 6 \cdot (-4) = \mathbf{-24}$.

18. $4(2ac)$ means $4 \cdot (2 \cdot a \cdot c)$. When $a=3$ and $c=5$, we have $4 \cdot (2 \cdot a \cdot c) = 4 \cdot (2 \cdot 3 \cdot 5) = 4 \cdot 30 = \mathbf{120}$.

— or —

Since multiplication is associative, we can write $4(2ac)$ as $(4 \cdot 2)(ac) = 8ac$. When $a=3$ and $c=5$, we have $8ac = 8 \cdot 3 \cdot 5 = \mathbf{120}$.

19. When $x=8$ and $y=6$, we have $2x+y = 2(8)+6 = 16+6 = \mathbf{22}$.

20. When $x=8$ and $y=6$, we have $5y-2x = 5(6)-2(8) = 30-16 = \mathbf{14}$.

21. When $x=8$ and $y=6$, we have $6-2xy = 6-2(8)(6) = 6-96 = \mathbf{-90}$.

22. When $x=8$ and $y=6$, we have

$$\begin{aligned}5xy+4x+3y+2 &= 5(8)(6)+4(8)+3(6)+2\\ &= 240+32+18+2\\ &= \mathbf{292}.\end{aligned}$$

Division as a Fraction 73-75

23. We evaluate the numerator of the fraction first, then divide. So, $\frac{4+14}{3} = \frac{18}{3} = \mathbf{6}$.

24. We evaluate the denominator of the fraction first, then divide. So, $\frac{24}{15-23} = \frac{24}{-8} = \mathbf{-3}$.

25. We evaluate the numerator and denominator of the fraction first, then divide. So, $\frac{27+8}{3+2} = \frac{35}{5} = \mathbf{7}$.

26. $\frac{20}{4(3)} = \frac{20}{12} = \mathbf{\frac{5}{3}}$ ***or*** $\mathbf{1\frac{2}{3}}$.

27. $\frac{6(-8)}{4(-6)} = \frac{-48}{-24} = \mathbf{2}$.

28. $\frac{3(4+5)}{2-11} = \frac{3(9)}{-9} = \frac{27}{-9} = \mathbf{-3}$.

29. When $a=4$, $b=6$, and $c=24$, we have $\frac{ab}{c} = \frac{4(6)}{24} = \frac{24}{24} = \mathbf{1}$.

30. When $a=4$, $b=6$, and $c=24$, we have $\frac{b+c}{a+5} = \frac{6+24}{4+5} = \frac{30}{9} = \mathbf{\frac{10}{3}}$ ***or*** $\mathbf{3\frac{1}{3}}$.

31. When $a=4$, $b=6$, and $c=24$, we have $\frac{c-a-b}{2a-6} = \frac{24-4-6}{2(4)-6} = \frac{14}{2} = \mathbf{7}$.

32. When $r=-2$, $s=3$, and $t=-7$, we have $\frac{s+t}{r} = \frac{3+(-7)}{-2} = \frac{-4}{-2} = \mathbf{2}$.

33. When $r=-2$, $s=3$, and $t=-7$, we have $\frac{12r+s}{-t} = \frac{12(-2)+3}{-(-7)} = \frac{-21}{7} = \mathbf{-3}$.

34. When $r=-2$, $s=3$, and $t=-7$, we have $\frac{-16r}{3s-t} = \frac{-16(-2)}{3(3)-(-7)} = \frac{32}{9+7} = \frac{32}{16} = \mathbf{2}$.

35. The grouped quantity $(30-20)$ is divided by the grouped quantity $(5-3)$. So, $(30-20)\div(5-3) = \boxed{\frac{30-20}{5-3}}$.

We evaluate the expression:

$$\frac{30-20}{5-3} = \frac{10}{2} = \mathbf{5}.$$

36. The grouped quantity $(30-20)$ is divided only by 5. So, $(30-20)\div5-3 = \boxed{\frac{30-20}{5}-3}$.

We evaluate the expression:

$$\frac{30-20}{5}-3 = \frac{10}{5}-3 = 2-3 = \mathbf{-1}.$$

37. Since division comes before subtraction in the order of operations, only 20 is divided by the grouped quantity $(5-3)$. So, $30-20\div(5-3) = \boxed{30-\frac{20}{5-3}}$.

We evaluate the expression:

$$30-\frac{20}{5-3} = 30-\frac{20}{2} = 30-10 = \mathbf{20}.$$

38. Since division comes before subtraction in the order of operations, only 20 is divided by 5.
So, $30-20\div5-3=\boxed{30-\frac{20}{5}-3}$.
We evaluate the expression:

$$30-\frac{20}{5}-3=30-4-3=\mathbf{23}.$$

Below is the correct matching for these problems:

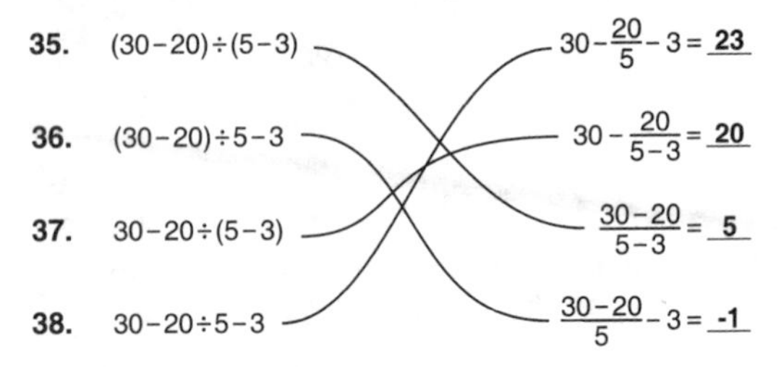

39. We evaluate the numerator of the fraction first, then divide, then add.

$$\frac{6+3}{3}+2=\frac{9}{3}+2=3+2=\mathbf{5}.$$

40. We evaluate the denominator of the fraction first, then divide, then subtract.

$$5-\frac{8}{6(4)}=5-\frac{8}{24}=5-\frac{1}{3}=\mathbf{4\tfrac{2}{3}}\ \textit{or}\ \mathbf{\tfrac{14}{3}}.$$

41. Evaluating the numerator of the fraction, we have:

$$3\cdot\frac{7+9}{2}=3\cdot\frac{16}{2}$$

$\frac{16}{2}$ simplifies to 8. So, $3\cdot\frac{7+9}{2}=3\cdot\frac{16}{2}=3\cdot8=\mathbf{24}$.

42. $\frac{3\cdot7+9}{2}=\frac{21+9}{2}=\frac{30}{2}=\mathbf{15}$.

43. $\frac{-3(4)}{(6-4)^2}=\frac{-12}{(2)^2}=\frac{-12}{4}=\mathbf{-3}$.

44. We simplify each fraction first, then add.

$$\begin{aligned}\frac{20^2}{2}+\frac{20}{2^2}+\left(\frac{20}{2}\right)^2&=\frac{400}{2}+\frac{20}{4}+(10)^2\\&=200+5+100\\&=\mathbf{305}.\end{aligned}$$

45. We simplify the fraction first, then multiply, then subtract.

$$\begin{aligned}17-2\left(\frac{1+11}{2\cdot3}\right)&=17-2\left(\frac{12}{6}\right)\\&=17-2(2)\\&=17-4\\&=\mathbf{13}.\end{aligned}$$

46. We simplify the fractions first, then multiply.

$$\begin{aligned}\frac{6^2}{7+5}\cdot\frac{7-6-5}{2}&=\frac{36}{12}\cdot\frac{-4}{2}\\&=3\cdot(-2)\\&=\mathbf{-6}.\end{aligned}$$

47. $\frac{8(7-3)^2}{-(3-7)^3}=\frac{8(4)^2}{-(-4)^3}=\frac{8(16)}{-(-64)}=\frac{128}{64}=\mathbf{2}$.

48. $\left(\frac{5+7+9}{2^5-5^2}\right)^3=\left(\frac{21}{32-25}\right)^3=\left(\frac{21}{7}\right)^3=(3)^3=\mathbf{27}$.

EXPRESSIONS & EQUATIONS
Like Terms Link 76

49.

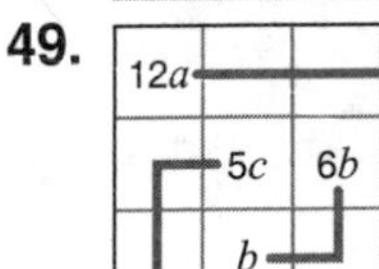

50.

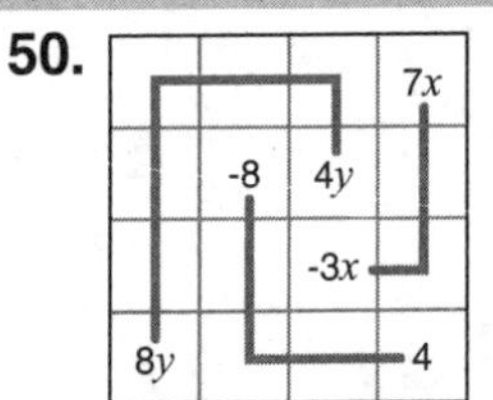

51.

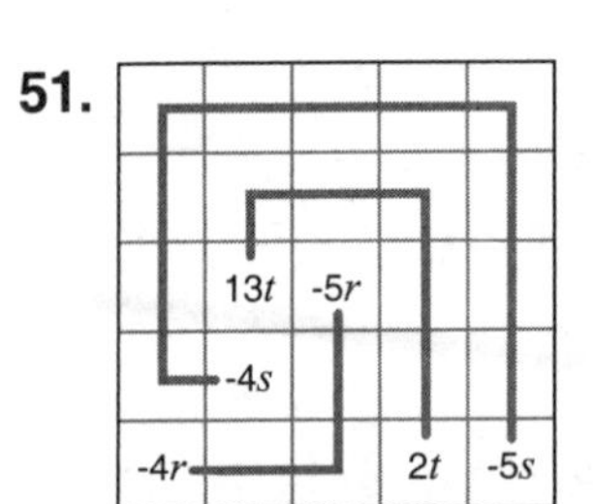

52.

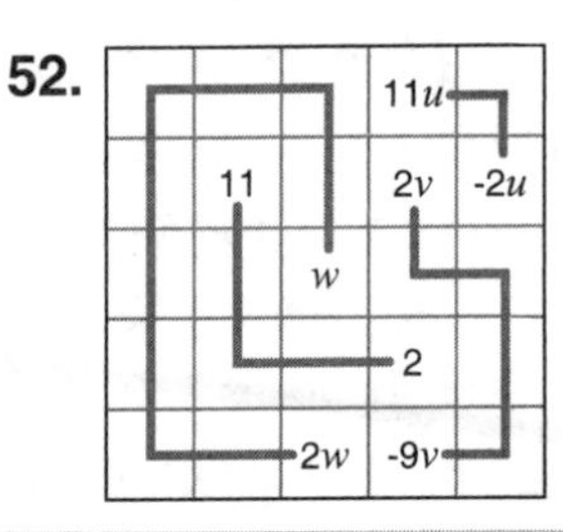

EXPRESSIONS & EQUATIONS
Combining Like Terms 77

53. $5x=x+x+x+x+x$, and $4x=x+x+x+x$. So,

$$\begin{aligned}5x+4x&=(x+x+x+x+x)+(x+x+x+x)\\&=x+x+x+x+x+x+x+x+x\\&=\mathbf{9x}.\end{aligned}$$

— *or* —

We factor x from each term: $5x+4x=(5+4)x=\mathbf{9x}$.

54. We factor y from each term: $10y-3y=(10-3)y=\mathbf{7y}$.

55. $3d+4d+5d=(3+4+5)d=\mathbf{12d}$.

56. We note that s can be written as $1s$. So,
$s+3s+15s=1s+3s+15s=(1+3+15)s=\mathbf{19s}$.

57. $-3w+12w=(-3+12)w=\mathbf{9w}$.

58. We note that $-p$ can be written as $-1p$.
So, $6p+(-p)=(6+(-1))p=\mathbf{5p}$.

59. $8c-14c+3c=(8-14+3)c=(-3)c=\mathbf{-3c}$.

60. $-22g+36g-12g=(-22+36-12)g=\mathbf{2g}$.

61. $12n-7n-5n=(12-7-5)n=0n=\mathbf{0}$.

62. We rewrite subtraction as addition, then rearrange terms to make our computation easier:

$$\begin{aligned}93k+47k-92k&=93k+47k+(-92k)\\&=93k+(-92k)+47k\\&=k+47k\\&=\mathbf{48k}.\end{aligned}$$

63. A square with side length s has perimeter $s+s+s+s=\mathbf{4s}$.

64. The rectangle has two sides of length x and two sides of length $3x$.

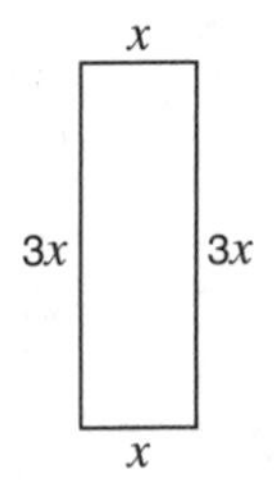

So, the perimeter of the rectangle is $x+x+3x+3x=\mathbf{8x}$.

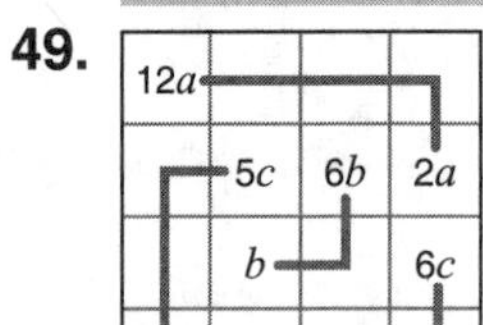

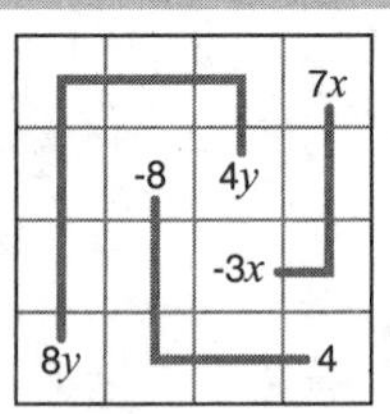

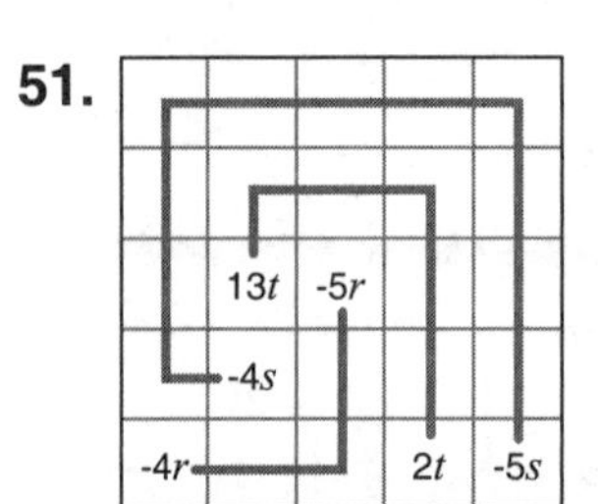

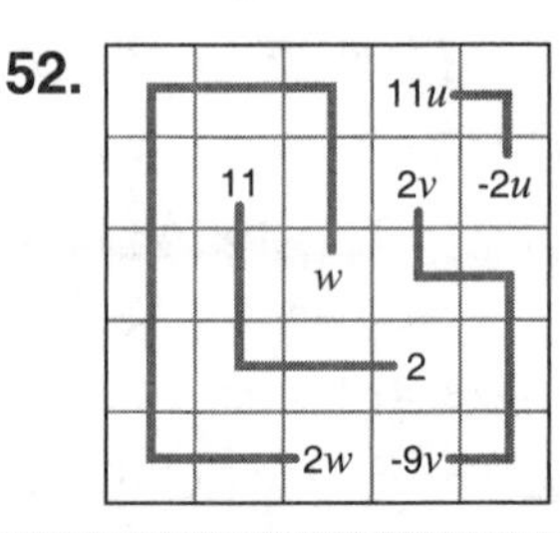

Areas for Like Terms Corrals are given in square units.

65. The like terms a and $5a$ have sum $a+5a=6a$. The coefficient of $6a$ is 6, so these terms must be fenced into a corral with area 6.

The like terms $7b$ and $3b$ have sum $7b+3b=10b$. The coefficient of $10b$ is 10, so these terms must be fenced into a corral with area 10.

There is only one way to fence like terms a and $5a$ into a corral with area 6 without splitting up the b-corral.

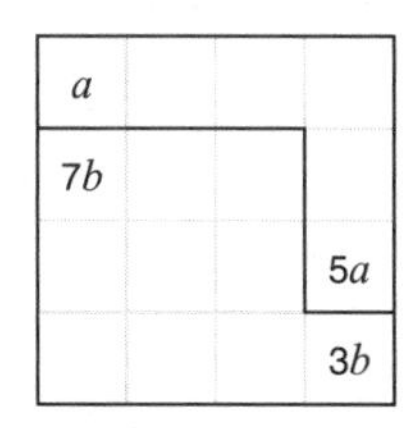

The remaining corral containing like terms $7b$ and $3b$ has area 10, as desired.

66. $12c+(-4c)=\underline{8}c$. So, we fence these like terms into a corral with area 8.

$2d+6d=\underline{8}d$. So, we fence these like terms into a corral with area 8.

There is only one way to fence like terms $12c$ and $-4c$ into a corral with area 8 without splitting up the d-corral.

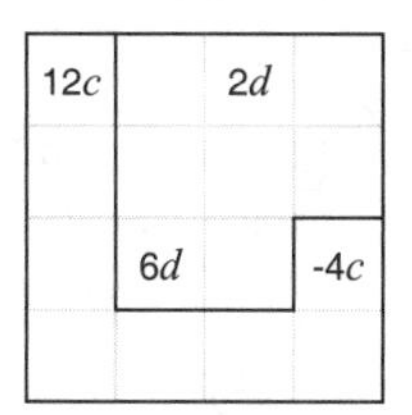

The remaining corral that contains like terms $2d$ and $6d$ has area 8, as desired.

67. $2r+2r+r+8r=\underline{13}r$, so the r-corral has area 13.

$6s+3s+(-2s)=\underline{7}s$, so the s-corral has area 7.

A corral that contains all terms with variable r without splitting up the s-corral must include the squares in gray below.

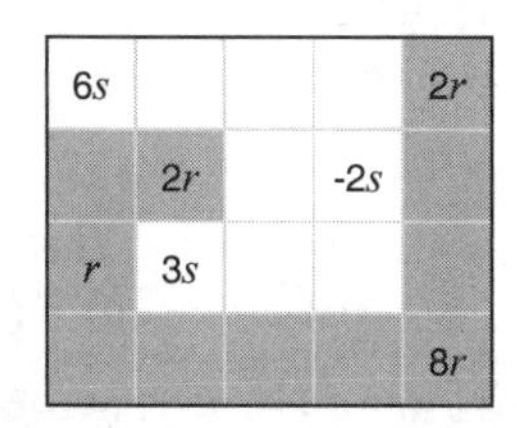

These gray squares have combined area 11. So, we need $13-11=2$ more squares to complete the r-corral. We shade the only two squares that can be added without splitting up the s-corral.

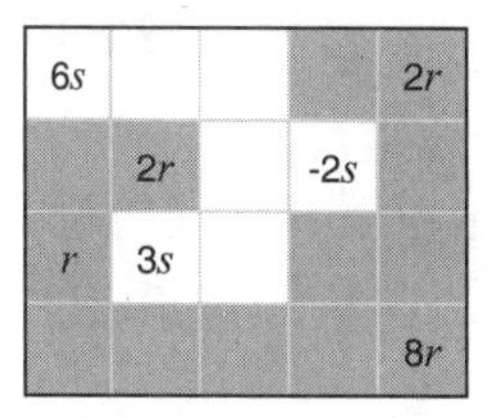

The two regions have areas 7 and 13, as desired. So, we draw fences as shown below.

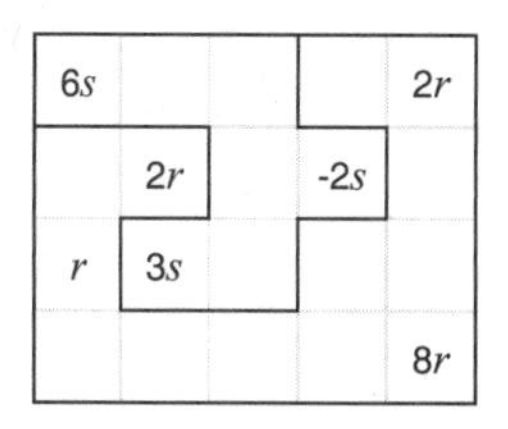

68. $u+3u=\underline{4}u$, so the u-corral has area 4.

$2v+7v=\underline{9}v$, so the v-corral has area 9.

$-3w+10w=\underline{7}w$, so the w-corral has area 7.

There are three ways we can group the terms with variable u into a corral with area 4.

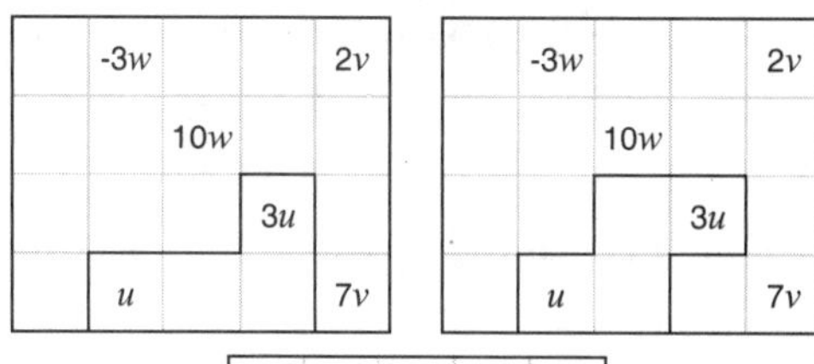

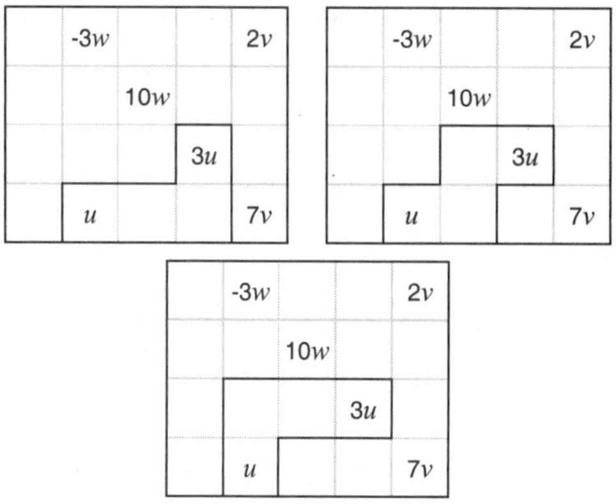

However, only the last way makes it possible to create a v-corral with area 9. So, we draw fences to complete the remaining corrals as shown below.

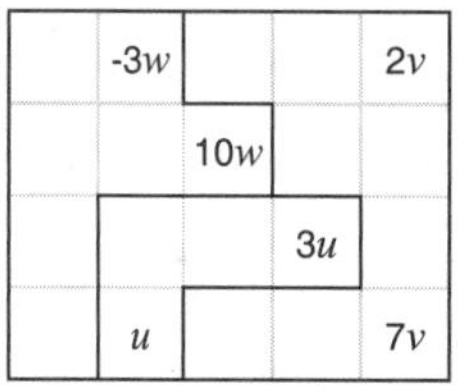

For Problems 69-74, we use strategies that are similar to those discussed in the previous problems.

69.

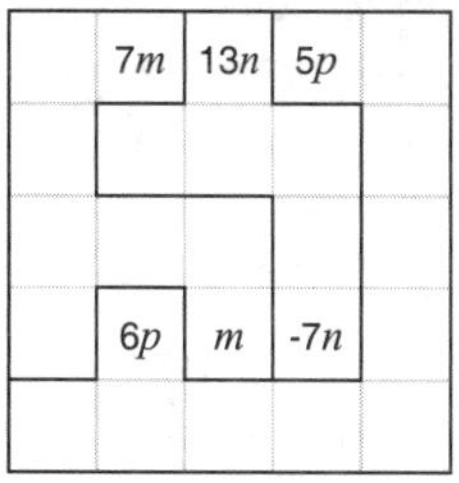

70.

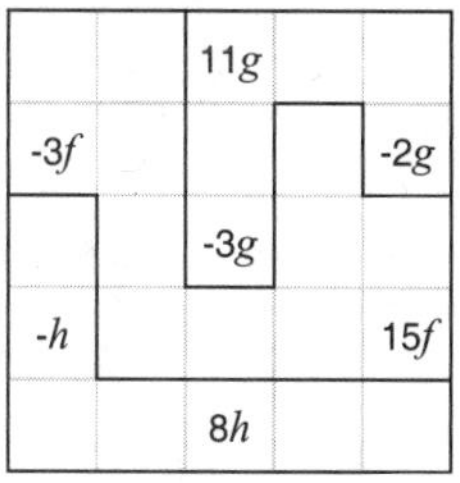

71.

$4x$ $15y$ $20z$ $-11z$ $6x$ $-9y$

72.

$11i$ $-7j$ $-2i$ $9i$ $20j$ $-6i$

73.

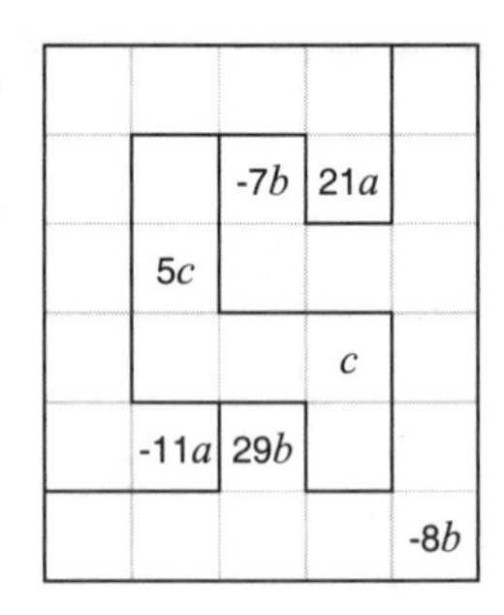

74.

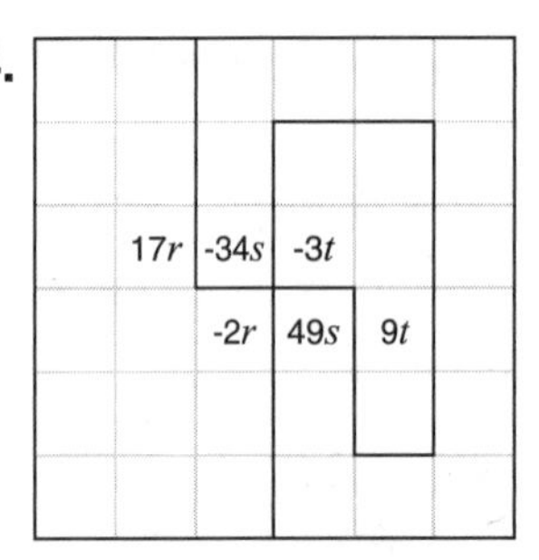

EXPRESSIONS & EQUATIONS

Simplifying Expressions 80-81

75. $6a$ and $9a$ are like terms, and 4 and 12 are like terms. Combining these like terms, we have

$$\begin{aligned}6a+9a+4+12&=(6a+9a)+(4+12)\\&=\mathbf{15a+16}.\end{aligned}$$

Since $15a$ and 16 are not like terms, we cannot simplify any further.

76. We rearrange the expression so that like terms are together. Then, we combine like terms.

$$\begin{aligned}15+3b+5+18b&=(3b+18b)+(15+5)\\&=\mathbf{21b+20}.\end{aligned}$$

77. We rewrite all subtraction as addition, then rearrange the expression so that like terms are together. Then, we combine like terms.

$$\begin{aligned}22-4c-3+6c&=22+(-4c)+(-3)+6c\\&=(-4c+6c)+(22+(-3))\\&=\mathbf{2c+19}.\end{aligned}$$

78. We combine like terms as shown below:

$$\begin{aligned}8d-4+15-12d&=8d+(-4)+15+(-12d)\\&=(8d+(-12d))+((-4)+15)\\&=\mathbf{-4d+11}.\end{aligned}$$

You may have also written this answer as $\mathbf{11-4d}$.

79. We combine like terms as shown below:

$$\begin{aligned}19e+21f+9e+11f&=(19e+9e)+(21f+11f)\\&=\mathbf{28e+32f}.\end{aligned}$$

80. We combine like terms as shown below:

$$\begin{aligned}-4g+10+10g-6h+3h&=-4g+10+10g+(-6h)+3h\\&=(-4g+10g)+((-6h)+3h)+10\\&=6g+(-3h)+10\\&=\mathbf{6g-3h+10}.\end{aligned}$$

81. No two terms in the expression have the same variable. So, there are no like terms, and the expression **cannot be simplified**. We circle the expression:

$$12m-5n+11p+7.$$

82. We combine like terms as shown below:

$$\begin{aligned}14r+8q-3r-16q+10r&=14r+8q+(-3r)+(-16q)+10r\\&=(-16q+8q)+(14r+(-3r)+10r)\\&=\mathbf{-8q+21r}.\end{aligned}$$

You may have also written this answer as $\mathbf{21r-8q}$.

83. We combine like terms as shown below:

$$\begin{aligned}3u+7v-2u-v-u&=3u+7v+(-2u)+(-v)+(-u)\\&=(3u+(-2u)+(-u))+(7v+(-v))\\&=0u+6v\\&=\mathbf{6v}.\end{aligned}$$

84. We combine like terms as shown below:

$$\begin{aligned}7-8s+10t-3s+11t-12&=7+(-8s)+10t+(-3s)+11t+(-12)\\&=(-8s+(-3s))+(10t+11t)+(7+(-12))\\&=-11s+21t+(-5)\\&=\mathbf{-11s+21t-5}.\end{aligned}$$

You may have also written this answer as $\mathbf{21t-11s-5}$.

85. When $a=7$, we have
$12a+8a=12(7)+8(7)=84+56=\mathbf{140}$.

— or —

We simplify the expression first: $12a+8a=20a$.

When $a=7$, we have $20a=20(7)=\mathbf{140}$.

86. We simplify the expression: $5b-4b=b$.

When $b=-17$, we have $5b-4b=b=\mathbf{-17}$.

87. We simplify the expression: $11c-2c+11c=20c$.

When $c=29$, we have $20c=20(29)=\mathbf{580}$.

88. We simplify the expression:
$3a-c+2a=(3a+2a)+(-c)=5a-c$.

When $a=7$ and $c=29$, we have $5a-c=5(7)-29=\mathbf{6}$.

89. We simplify the expression:

$$\begin{aligned}r+2t+3r+4t&=(r+3r)+(2t+4t)\\&=4r+6t.\end{aligned}$$

When $r=11$ and $t=-5$, we have
$4r+6t=4(11)+6(-5)=44+(-30)=\mathbf{14}$.

90. We simplify the expression:

$$\begin{aligned}3s-11+t+7s+9&=(3s+7s)+t+(-11+9)\\&=10s+t-2.\end{aligned}$$

When $s=13$ and $t=-5$, we have
$10s+t-2=10(13)+(-5)-2=130+(-5)-2=\mathbf{123}$.

91. We simplify the expression:

$$\begin{aligned}2s-6r-5s+19r-8s&=(-6r+19r)+(2s+(-5s)+(-8s))\\&=13r-11s.\end{aligned}$$

When $r=11$ and $s=13$, we have
$13r-11s=13(11)-11(13)=\mathbf{0}$.

92. To simplify the expression, we first distribute the 14:

$$\begin{aligned}14(5+s)-13s+17&=14(5)+14(s)-13s+17\\&=70+14s-13s+17.\end{aligned}$$

Then, we combine like terms:

$$\begin{aligned}70+14s-13s+17&=(14s+(-13s))+(70+17)\\&=s+87.\end{aligned}$$

When $s=13$, we have $s+87=13+87=$ **100**.

EXPRESSIONS & EQUATIONS
Short Circuit Puzzles 82-83

93. We use what we know about addition and subtraction to rewrite the expressions on the left.

$\underline{b-(-a)}$: Subtracting a number is the same as adding its opposite. So, we have

$$b-(-a)=b+a=a+b.$$

$\underline{-a+b}$: Using the commutative property of addition to rearrange terms, we have

$$-a+b=b+(-a)=b-a.$$

$\underline{-b-a}$: Changing the subtraction to addition, then using the commutative property of addition, we have

$$-b-a=-b+(-a)=-a+(-b)=-a-b.$$

We connect these pairs without crossing wires as shown:

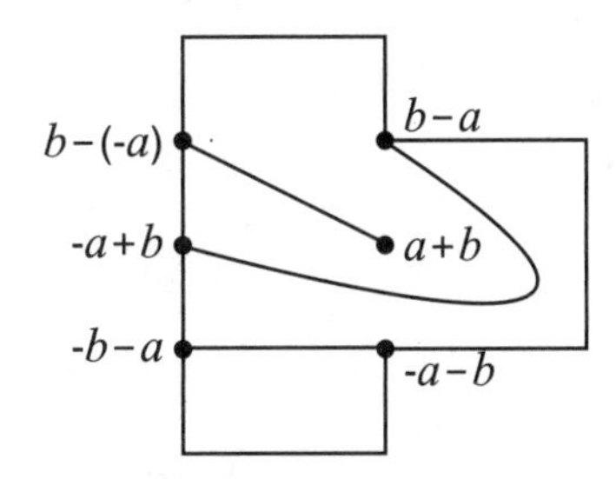

94. We first simplify the expressions on the left. When helpful, we rewrite subtraction as addition to group like terms. Then, we combine like terms.

$$\begin{aligned}4+x+6x+1&=(x+6x)+(4+1)\\&=7x+5.\end{aligned}$$

$$\begin{aligned}3x-2+4x&=3x+(-2)+4x\\&=(3x+4x)+(-2)\\&=7x-2.\end{aligned}$$

$$\begin{aligned}9x-3x-2-1&=9x-3x+(-2)+(-1)\\&=6x+(-3)\\&=6x-3.\end{aligned}$$

We connect these pairs without crossing wires as shown:

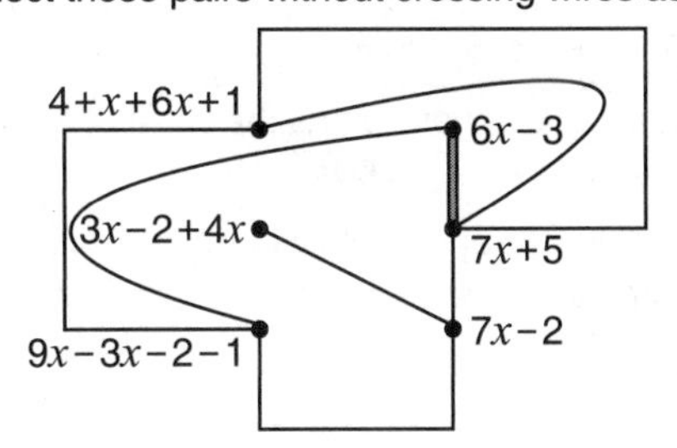

95. We simplify the expressions on the left. When helpful, we rewrite subtraction as addition to group like terms. Then, we combine like terms.

$$\begin{aligned}y+1-2y&=y+1+(-2y)\\&=(y+(-2y))+1\\&=-y+1\\&=1-y.\end{aligned}$$

$$\begin{aligned}y+y-1-y+1+1&=y+y+(-1)+(-y)+1+1\\&=(y+y+(-y))+(-1+1+1)\\&=y+1.\end{aligned}$$

$$\begin{aligned}1-y-1+y+y-1&=1+(-y)+(-1)+y+y+(-1)\\&=(-y+y+y)+(1+(-1)+(-1))\\&=y+(-1)\\&=y-1.\end{aligned}$$

We connect these pairs without crossing wires as shown:

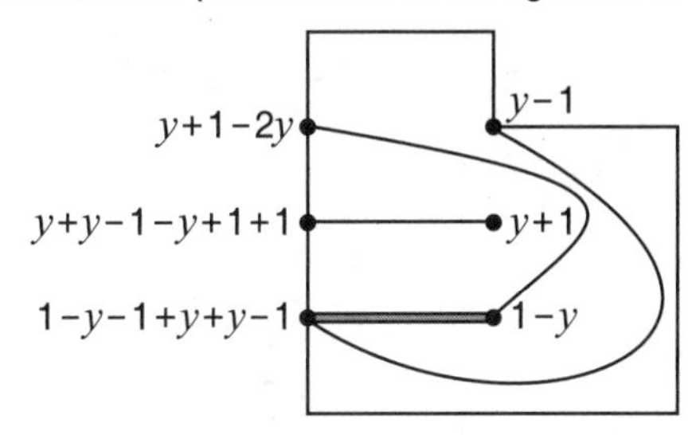

96. We simplify the expressions on the left:

$$\begin{aligned}3d+1-2d&=3d+(-2d)+1\\&=d+1.\end{aligned}$$

$$d+3d-2d=2d.$$

$$\begin{aligned}d+1-d-1-d&=(d+(-d)+(-d))+(1+(-1))\\&=-d+0\\&=-d.\end{aligned}$$

We connect these pairs without crossing wires as shown:

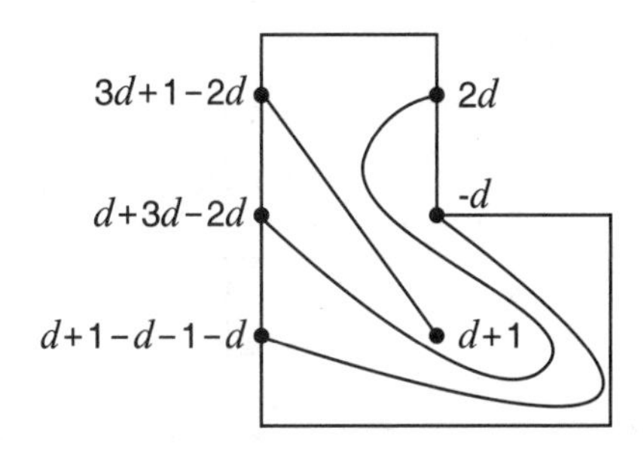

97. We simplify the expressions on the left:

$$\begin{aligned}u+2v-v-2u&=(u+(-2u))+(2v+(-v))\\&=-u+v\\&=v-u.\end{aligned}$$

$$\begin{aligned}v-4u+v+5u&=(-4u+5u)+(v+v)\\&=u+2v.\end{aligned}$$

$$\begin{aligned}u+3u-v-2u&=(u+3u+(-2u))+(-v)\\&=2u-v.\end{aligned}$$

We connect these pairs without crossing wires as shown:

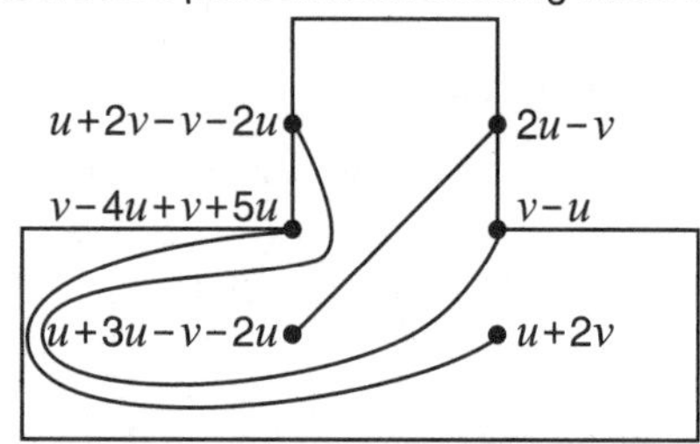

98. We simplify the expressions on the left:

$$\begin{aligned}3s+4r-2r+2s &= (4r+(\text{-}2r))+(3s+2s)\\ &= 2r+5s.\end{aligned}$$

$$\begin{aligned}r-s+2r-3s &= (r+2r)+(\text{-}s+(\text{-}3s))\\ &= 3r-4s.\end{aligned}$$

$$\begin{aligned}s-3s+4r+5s &= 4r+(s+(\text{-}3s)+5s)\\ &= 4r+3s.\end{aligned}$$

We connect these pairs without crossing wires as shown:

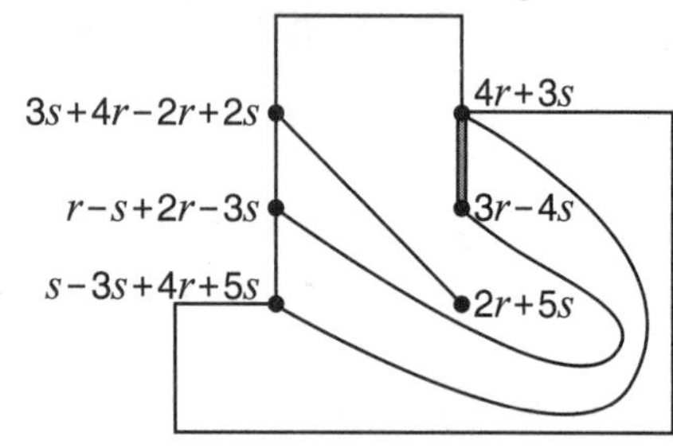

99. We simplify the expressions on the left:

$$\begin{aligned}\text{-}8k+3-k+14k-7 &= (\text{-}8k+(\text{-}k)+14k)+(3+(\text{-}7))\\ &= 5k-4.\end{aligned}$$

$$\begin{aligned}4+2k-3-7k-1 &= (2k+(\text{-}7k))+(4+(\text{-}3)+(\text{-}1))\\ &= \text{-}5k+0\\ &= \text{-}5k.\end{aligned}$$

$$\text{-}k-2k-3k = \text{-}6k.$$

We connect these pairs without crossing wires as shown:

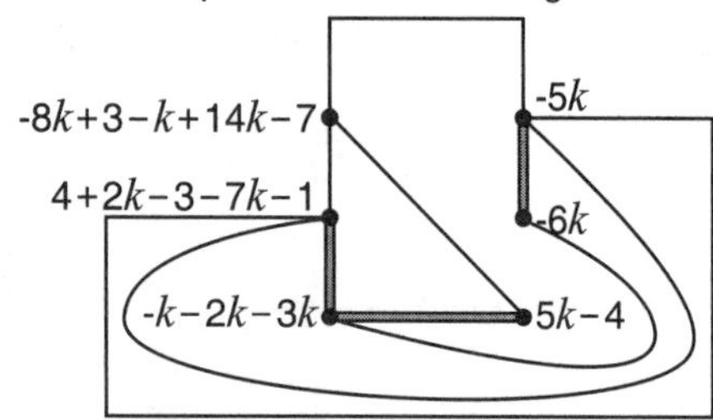

100. We simplify the expressions on the left:

$$\begin{aligned}4-m+2n+4m-8 &= (\text{-}m+4m)+2n+(4+(\text{-}8))\\ &= 3m+2n-4.\end{aligned}$$

$$\begin{aligned}3m-n+1+4n-5n+3 &= 3m+(\text{-}n+4n+(\text{-}5n))+(1+3)\\ &= 3m-2n+4.\end{aligned}$$

$$\begin{aligned}4-m+n-2m+n &= (\text{-}m+(\text{-}2m))+(n+n)+4\\ &= \text{-}3m+2n+4\\ &= 2n-3m+4.\end{aligned}$$

We connect these pairs without crossing wires as shown:

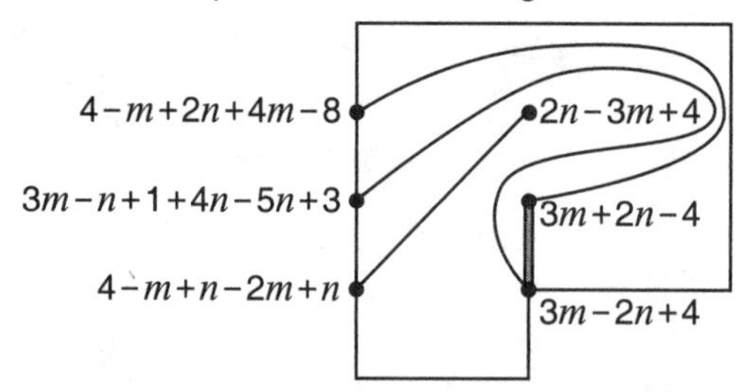

EXPRESSIONS & EQUATIONS Review 84

101. There are no like terms to combine, so we replace n with 8, then evaluate the expression.

$$\begin{aligned}6+\frac{3n-12}{4} &= 6+\frac{3(8)-12}{4}\\ &= 6+\frac{24-12}{4}\\ &= 6+\frac{12}{4}\\ &= 6+3\\ &= \mathbf{9}.\end{aligned}$$

102. First, we simplify the expression.

$$\begin{aligned}31r-24s+11s+19r+13s &= (31r+19r)+(\text{-}24s+11s+13s)\\ &= 50r+0s\\ &= 50r.\end{aligned}$$

When $r=13$, we have $50r = 50(13) = \mathbf{650}$.

Notice that since all the terms with s were eliminated, we did not need to use the value of s!

103. First, we simplify each expression.

$4+a-1+a-3 = (a+a)+(4+(\text{-}1)+(\text{-}3)) = 2a+0 = 2a.$
$3a+4a-8a = 7a-8a = \text{-}a.$
$5a-2-3a+7 = (5a+(\text{-}3a))+(\text{-}2+7) = 2a+5.$

Then, we order the expressions from least to greatest.

Since a is positive, $2a$ is also positive. Adding 5 to any positive number gives a greater positive number, so $2a+5$ is both positive and greater than $2a$.

The opposite of a positive number is always negative, so $\text{-}a$ is negative. Since $\text{-}a$ is the only expression with a negative value, it is the least of the three expressions.

So, the order of the expressions from least to greatest is $\mathbf{\text{-}a, 2a, 2a+5}$.

104. We simplify each expression.

$2b-3b-6+8b = (2b+(\text{-}3b)+8b)+(\text{-}6) = 7b-6.$
$\text{-}6b+2+3b+4 = (\text{-}6b+3b)+(2+4) = \text{-}3b+6.$
$2b+7-5b-7 = (2b+(\text{-}5b))+(7+(\text{-}7)) = \text{-}3b+0 = \text{-}3b.$

Then, we order the expressions from least to greatest.

Since b is negative, and the product of two negatives is positive, $\text{-}3b$ is positive.

Adding 6 to any positive number gives a greater positive number, so $\text{-}3b+6$ is both positive and greater than $\text{-}3b$.

92. To simplify the expression, we first distribute the 14:

$$14(5+s)-13s+17 = 14(5)+14(s)-13s+17$$
$$= 70+14s-13s+17.$$

Then, we combine like terms:

$$70+14s-13s+17 = (14s+(-13s))+(70+17)$$
$$= s+87.$$

When $s=13$, we have $s+87 = 13+87 = \mathbf{100}$.

EXPRESSIONS & EQUATIONS
Short Circuit Puzzles 82–83

93. We use what we know about addition and subtraction to rewrite the expressions on the left.

$\underline{b-(-a)}$: Subtracting a number is the same as adding its opposite. So, we have

$$b-(-a) = b+a = a+b.$$

$\underline{-a+b}$: Using the commutative property of addition to rearrange terms, we have

$$-a+b = b+(-a) = b-a.$$

$\underline{-b-a}$: Changing the subtraction to addition, then using the commutative property of addition, we have

$$-b-a = -b+(-a) = -a+(-b) = -a-b.$$

We connect these pairs without crossing wires as shown:

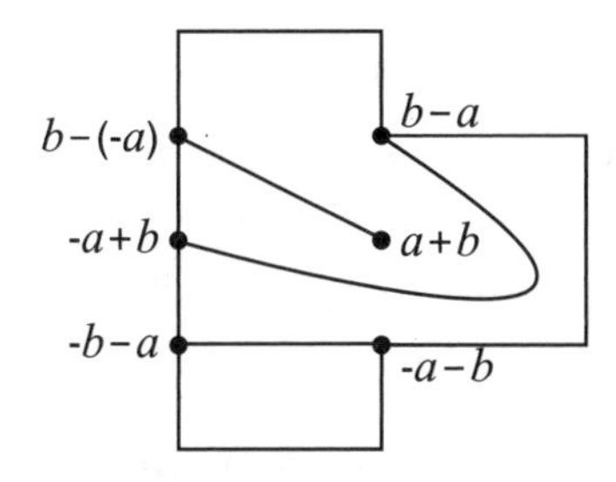

94. We first simplify the expressions on the left. When helpful, we rewrite subtraction as addition to group like terms. Then, we combine like terms.

$$4+x+6x+1 = (x+6x)+(4+1)$$
$$= 7x+5.$$

$$3x-2+4x = 3x+(-2)+4x$$
$$= (3x+4x)+(-2)$$
$$= 7x-2.$$

$$9x-3x-2-1 = 9x-3x+(-2)+(-1)$$
$$= 6x+(-3)$$
$$= 6x-3.$$

We connect these pairs without crossing wires as shown:

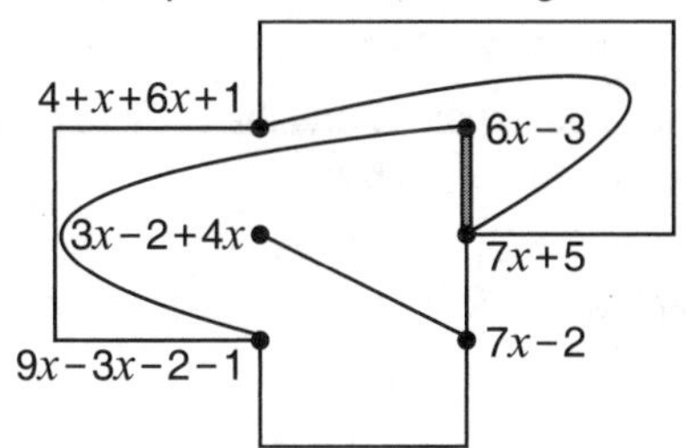

95. We simplify the expressions on the left. When helpful, we rewrite subtraction as addition to group like terms. Then, we combine like terms.

$$y+1-2y = y+1+(-2y)$$
$$= (y+(-2y))+1$$
$$= -y+1$$
$$= 1-y.$$

$$y+y-1-y+1+1 = y+y+(-1)+(-y)+1+1$$
$$= (y+y+(-y))+(-1+1+1)$$
$$= y+1.$$

$$1-y-1+y+y-1 = 1+(-y)+(-1)+y+y+(-1)$$
$$= (-y+y+y)+(1+(-1)+(-1))$$
$$= y+(-1)$$
$$= y-1.$$

We connect these pairs without crossing wires as shown:

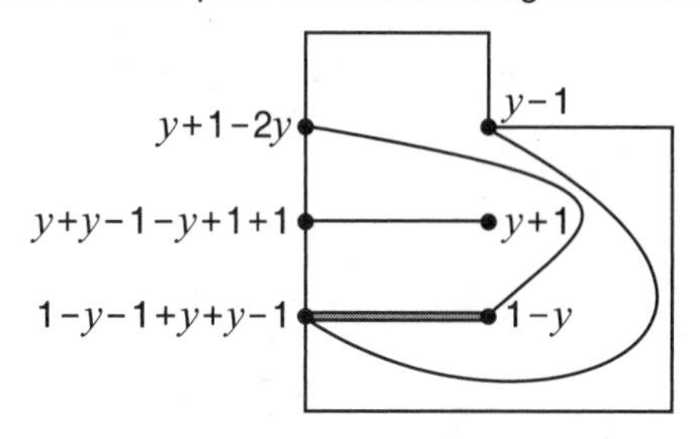

96. We simplify the expressions on the left:

$$3d+1-2d = 3d+(-2d)+1$$
$$= d+1.$$

$$d+3d-2d = 2d.$$

$$d+1-d-1-d = (d+(-d)+(-d))+(1+(-1))$$
$$= -d+0$$
$$= -d.$$

We connect these pairs without crossing wires as shown:

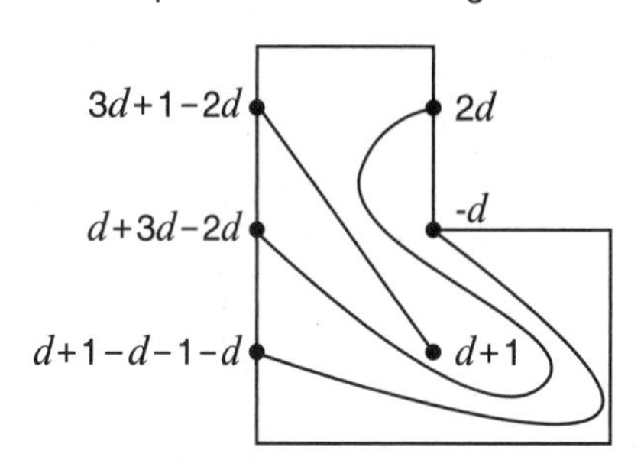

97. We simplify the expressions on the left:

$$u+2v-v-2u = (u+(-2u))+(2v+(-v))$$
$$= -u+v$$
$$= v-u.$$

$$v-4u+v+5u = (-4u+5u)+(v+v)$$
$$= u+2v.$$

$$u+3u-v-2u = (u+3u+(-2u))+(-v)$$
$$= 2u-v.$$

We connect these pairs without crossing wires as shown:

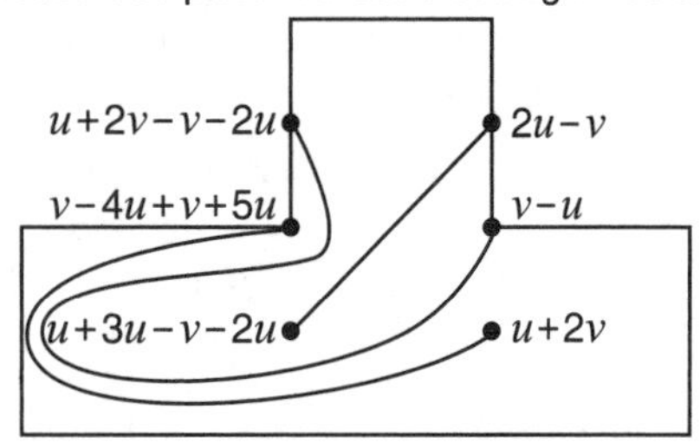

98. We simplify the expressions on the left:

$$\begin{aligned}3s+4r-2r+2s &= (4r+(-2r))+(3s+2s)\\ &= 2r+5s.\end{aligned}$$

$$\begin{aligned}r-s+2r-3s &= (r+2r)+(-s+(-3s))\\ &= 3r-4s.\end{aligned}$$

$$\begin{aligned}s-3s+4r+5s &= 4r+(s+(-3s)+5s)\\ &= 4r+3s.\end{aligned}$$

We connect these pairs without crossing wires as shown:

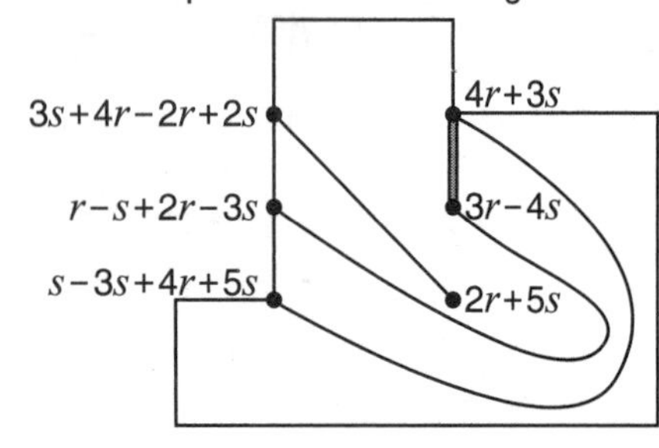

99. We simplify the expressions on the left:

$$\begin{aligned}-8k+3-k+14k-7 &= (-8k+(-k)+14k)+(3+(-7))\\ &= 5k-4.\end{aligned}$$

$$\begin{aligned}4+2k-3-7k-1 &= (2k+(-7k))+(4+(-3)+(-1))\\ &= -5k+0\\ &= -5k.\end{aligned}$$

$$-k-2k-3k = -6k.$$

We connect these pairs without crossing wires as shown:

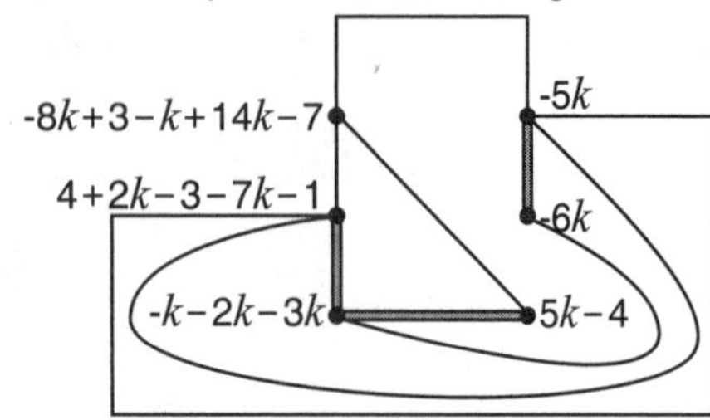

100. We simplify the expressions on the left:

$$\begin{aligned}4-m+2n+4m-8 &= (-m+4m)+2n+(4+(-8))\\ &= 3m+2n-4.\end{aligned}$$

$$\begin{aligned}3m-n+1+4n-5n+3 &= 3m+(-n+4n+(-5n))+(1+3)\\ &= 3m-2n+4.\end{aligned}$$

$$\begin{aligned}4-m+n-2m+n &= (-m+(-2m))+(n+n)+4\\ &= -3m+2n+4\\ &= 2n-3m+4.\end{aligned}$$

We connect these pairs without crossing wires as shown:

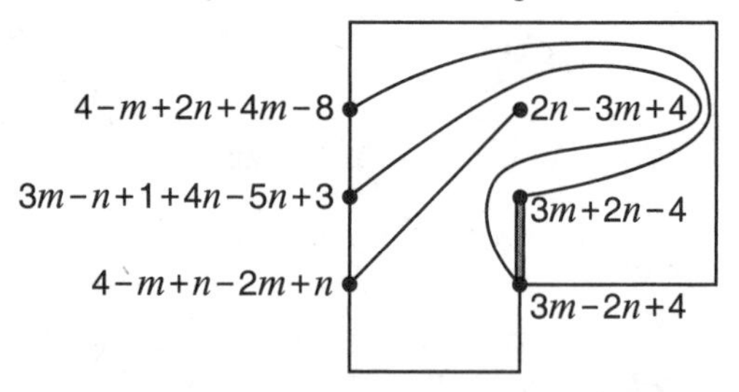

EXPRESSIONS & EQUATIONS Review 84

101. There are no like terms to combine, so we replace n with 8, then evaluate the expression.

$$\begin{aligned}6+\frac{3n-12}{4} &= 6+\frac{3(8)-12}{4}\\ &= 6+\frac{24-12}{4}\\ &= 6+\frac{12}{4}\\ &= 6+3\\ &= \mathbf{9}.\end{aligned}$$

102. First, we simplify the expression.

$$\begin{aligned}31r-24s+11s+19r+13s &= (31r+19r)+(-24s+11s+13s)\\ &= 50r+0s\\ &= 50r.\end{aligned}$$

When $r=13$, we have $50r=50(13)=\mathbf{650}$.

Notice that since all the terms with s were eliminated, we did not need to use the value of s!

103. First, we simplify each expression.

$4+a-1+a-3=(a+a)+(4+(-1)+(-3))=2a+0=2a$.
$3a+4a-8a=7a-8a=-a$.
$5a-2-3a+7=(5a+(-3a))+(-2+7)=2a+5$.

Then, we order the expressions from least to greatest.

Since a is positive, $2a$ is also positive. Adding 5 to any positive number gives a greater positive number, so $2a+5$ is both positive and greater than $2a$.

The opposite of a positive number is always negative, so $-a$ is negative. Since $-a$ is the only expression with a negative value, it is the least of the three expressions.

So, the order of the expressions from least to greatest is $\boldsymbol{-a, 2a, 2a+5}$.

104. We simplify each expression.

$2b-3b-6+8b=(2b+(-3b)+8b)+(-6)=7b-6$.
$-6b+2+3b+4=(-6b+3b)+(2+4)=-3b+6$.
$2b+7-5b-7=(2b+(-5b))+(7+(-7))=-3b+0=-3b$.

Then, we order the expressions from least to greatest.

Since b is negative, and the product of two negatives is positive, $-3b$ is positive.

Adding 6 to any positive number gives a greater positive number, so $-3b+6$ is both positive and greater than $-3b$.

Since b is negative, and the product of a positive and a negative is negative, $7b$ is negative. Subtracting 6 from any negative number gives a negative number, so $7b-6$ is negative. Since $7b-6$ is the only expression with a negative value, it is the least of the three expressions.

So, the order of the expressions from least to greatest is $\mathbf{7b-6, -3b, -3b+6}$.

105. We begin by drawing a diagram of two congruent rectangles attached along their widths.

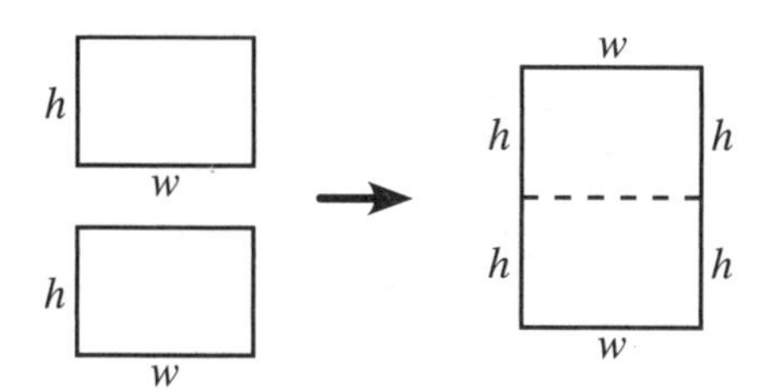

The resulting rectangle has width w, and height $h+h=2h$.

So, the perimeter of the new rectangle is $2h+2h+w+w$. Combining like terms, this simplifies to $\mathbf{4h+2w}$.

106. We start with the innermost parentheses and work our way out.

$$2(x+2(x+2(\underline{x+2})))$$

Since x and 2 are not like terms, we cannot simplify $x+2$. So, we move to the second pair of parentheses.

$$2(x+2(\underline{x+2(x+2)}))$$

Distributing the 2, then combining like terms, we have

$$\begin{aligned}2(x+2(\underline{x+2(x+2)})) &= 2(x+2(\underline{x+2(x)+2(2)}))\\ &= 2(x+2(\underline{3x+4})).\end{aligned}$$

We continue to work our way from the inside out.

$$\begin{aligned}2(x+2(3x+4)) &= 2(x+2(3x)+2(4))\\ &= 2(x+6x+8)\\ &= 2(7x+8)\\ &= 2(7x)+2(8)\\ &= \mathbf{14x+16}.\end{aligned}$$

There are no like terms to combine, so our answer is fully simplified.

EXPRESSIONS & EQUATIONS

Inverse Operations 85

107. a. When $a=44$, we have

$$a-24+24=44-24+24=20+24=\mathbf{44}.$$

— or —

Subtracting 24 then adding 24 is the same as doing nothing. So, $a-24+24=a$. Therefore, when $a=44$, the expression equals **44**.

b. When $a=44$, we have

$$\frac{a}{11}\cdot 11=\frac{44}{11}\cdot 11=4\cdot 11=\mathbf{44}.$$

— or —

Dividing by 11 then multiplying by 11 is the same as doing nothing. So, $\frac{a}{11}\cdot 11=a$. Therefore, when $a=44$, the expression equals **44**.

c. When $a=44$, we have

$$\frac{71a}{71}=\frac{71(44)}{71}=\frac{3{,}124}{71}=\mathbf{44}.$$

— or —

Multiplying by 71 then dividing by 71 is the same as doing nothing. So, $\frac{71a}{71}=a$. Therefore, when $a=44$, the expression equals **44**.

108. a. When $b=34$, we have

$$(b+4)+17-17=(34+4)+17-17=38+17-17=\mathbf{38}.$$

— or —

Adding 17 then subtracting 17 is the same as doing nothing. So, $(b+4)+17-17=b+4$. Therefore, when $b=34$, the expression equals $34+4=\mathbf{38}$.

b. When $b=34$, we have

$$\frac{9(b-7)}{9}=\frac{9(34-7)}{9}=\frac{9(27)}{9}=\frac{243}{9}=\mathbf{27}.$$

— or —

Multiplying by 9 then dividing by 9 is the same as doing nothing. So, $\frac{9(b-7)}{9}=b-7$. Therefore, when $b=34$, the expression equals $34-7=\mathbf{27}$.

c. When $b=34$, we have

$$18-2b-18=18-2(34)-18=18-68-18=\mathbf{-68}.$$

— or —

Rearranging terms, we see $18-2b-18=-2b+18-18$. Adding 18 then subtracting 18 is the same as doing nothing. So, $-2b+18-18=-2b$. Therefore, when $b=34$, the expression equals $-2(34)=\mathbf{-68}$.

109. We consider each expression separately.

$87+46-46$: Since adding 46 then subtracting 46 is the same as doing nothing, $87+46-46=87$.

$\frac{87(13)}{13}$: Since multiplying by 13 then dividing by 13 is the same as doing nothing, $\frac{87(13)}{13}=87$.

$34-87+34$: Evaluating from left to right, we have $34-87+34=-53+34=-19$. So, this expression is not equal to 87.

$\frac{-5(87)}{5}$: Evaluating the numerator, then dividing, we have:

$$\frac{-5(87)}{5}=\frac{-435}{5}=-87.$$

So, this expression is not equal to 87.

Careful! Multiplying by 5 and dividing by 5 are inverse operations, but multiplying by -5 and dividing by 5 are not.

$9\cdot\frac{87}{9}$: Since multiplying by 9 and dividing by 9 is the same as doing nothing, $9\cdot\frac{87}{9}=87$.

We circle the expressions that are equal to 87:

$\boxed{87+46-46}$ $\quad\boxed{\frac{87(13)}{13}}$ $\quad 34-87+34$ $\quad\frac{-5(87)}{5}$ $\quad\boxed{9\cdot\frac{87}{9}}$

110. We consider each expression separately.

$\frac{6x}{x}$: Multiplying by x then dividing by x is the same as doing nothing (unless $x=0$). So, $\frac{6x}{x}$ is equal to 6, not x.

$14+x-14$: Rearranging terms, we see that $14+x-14$ is equal to $x+14-14=x$.

$\frac{-11x}{-11}$: Multiplying by -11 then dividing by -11 is the same as doing nothing. So, $\frac{-11x}{-11}$ is equal to x.

$-6+x-6$: Rearranging terms, we see that $-6+x-6$ is equal to $x-6-6=x-12$. So, this expression is not equal to x.

$\frac{x}{3}\cdot 3$: Dividing by 3 then multiplying by 3 is the same as doing nothing. So, $\frac{x}{3}\cdot 3$ is equal to x.

We circle the expressions that are equal to x:

$\frac{6x}{x}$ 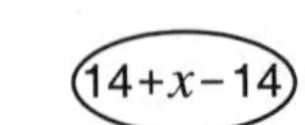 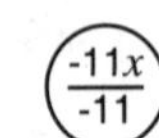 $-6+x-6$ 

EXPRESSIONS & EQUATIONS
Mystery Numbers 86

111. We work backwards. To find the number Worlag *added* 7 to in order to get 55, we *subtract* 7 from 55 to get $55-7=48$.

Then, to find the number Worlag *multiplied* by 8 to get 48, we *divide* 48 by 8 to get $48\div 8=6$.

So, Worlag's original number was **6**.

We check our answer: $6\cdot 8=48$, and $48+7=55$. ✓

112. We work backwards. To find the number Hubert *subtracted* 10 from to get 17, we *add* 10 to 17 to get $10+17=27$.

Then, to find the number Hubert *divided* by 4 to get 27, we *multiply* 27 by 4 to get $27\cdot 4=108$.

So, Hubert's favorite number is **108**.

We check our answer: $108\div 4=27$, and $27-10=17$. ✓

113. We work backwards. To find the age Chloe *divided* by 8 to get 7, we *multiply* 7 by 8 to get $7\cdot 8=56$.

Then, to find the age Chloe *added* 5 to in order to get 56, we *subtract* 5 from 56 to get $56-5=51$.

So, Chloe is **51** years old.

We check our answer: $51+5=56$, and $56\div 8=7$. ✓

114. We work backwards. To find the length Barbara *tripled* to get 57, we *divide* 57 by 3 to get $57\div 3=19$.

Then, to find the length Barbara *cut* 11 inches from to get 19, we *add* 11 to 19 to get $11+19=30$.

So, Barbara's ponytail was **30 inches** long before she trimmed it.

We check our answer: If Barbara cuts 11 inches from a 30-inch ponytail, its will be $30-11=19$ inches long. Then, after it triples in length, it will be $19\cdot 3=57$ inches long. ✓

EXPRESSIONS & EQUATIONS
Multi-Step Equations 87-89

115. In this equation, we multiply w by 8, then add 7 to the result to get 55. To isolate w, we undo these steps in reverse order.

To undo adding 7, we subtract 7 from both sides of the equation. Since $8w+7-7$ simplifies to $8w$, we have:

$$\begin{array}{r} 8w+7=55 \\ -7\ \ -7 \\ \hline 8w=48 \end{array}$$

Then, to undo multiplying by 8, we divide both sides of the equation by 8. Since $\frac{8w}{8}$ simplifies to w, we have:

$$\frac{8w}{8}=\frac{48}{8}$$
$$w=6$$

So, $w=$ **6**.

We check our answer by replacing w with 6 in the original equation: $8w+7=8(6)+7=48+7=55$. ✓

116. In this equation, we divide h by 4, then subtract 10 from the result to get 17. To isolate h, we undo these steps in reverse order.

To undo subtracting 10, we add 10 to both sides of the equation. Since $\frac{h}{4}-10+10$ simplifies to $\frac{h}{4}$, we have:

$$\begin{array}{r} \frac{h}{4}-10=17 \\ +10\ \ +10 \\ \hline \frac{h}{4}=27 \end{array}$$

Then, to undo dividing by 4, we multiply both sides of the equation by 4. Since $\frac{h}{4}\cdot 4$ simplifies to h, we have:

$$\frac{h}{4}\cdot 4=27\cdot 4$$
$$h=108$$

So, $h=$ **108**.

We check our answer by replacing h with 108 in the original equation: $\frac{h}{4}-10=\frac{108}{4}-10=27-10=17$. ✓

117. In this equation, we add 5 to c, then divide the result by 8 to get 7. To isolate c, we undo these steps in reverse order.

To undo dividing by 8, we multiply both sides of the equation by 8. Since $\frac{c+5}{8}\cdot 8$ simplifies to $c+5$, we have:

$$\frac{c+5}{8}\cdot 8=7\cdot 8$$
$$c+5=56$$

Then, to undo adding 5, we subtract 5 from both sides of the equation. Since $c+5-5$ simplifies to c, we have:

$$\begin{array}{r} c+5=56 \\ -5\ \ -5 \\ \hline c=51 \end{array}$$

So, $c=$ **51**.

We check our answer by replacing c with 51 in the original equation: $\frac{c+5}{8}=\frac{51+5}{8}=\frac{56}{8}=7$. ✓

118. In this equation, we subtract 11 from b, then multiply the result by 3 to get 57. To isolate b, we undo these steps in reverse order.

To undo multiplying by 3, we divide both sides of the equation by 3. Since $\frac{3(b-11)}{3}$ simplifies to $b-11$, we have:

$$\frac{3(b-11)}{3}=\frac{57}{3}$$
$$b-11=19$$

Then, to undo subtracting 11, we add 11 to both sides of the equation. Since $b-11+11$ simplifies to b, we have:

$$\begin{array}{r} b-11=19 \\ +11\ \ +11 \\ \hline b=30 \end{array}$$

So, $b=\mathbf{30}$.

We check our answer by replacing b with 30 in the original equation: $3(b-11)=3(30-11)=3(19)=57$. ✓

Notice that the equations in Problems 115-118 can be used to solve the word problems from Problems 111-114!

119. In this equation, we divide x by 2, then add 5 to the result to get -3. We isolate x by undoing these steps in reverse order.

Subtract 5	Multiply by 2
$5+\frac{x}{2}=-3$ $-5\quad -5$ $\frac{x}{2}=-8$	$\frac{x}{2}\cdot 2=-8\cdot 2$ $x=-16$

So, $x=\mathbf{-16}$.

We check our answer: $5+\frac{x}{2}=5+\frac{-16}{2}=5+(-8)=-3$. ✓

120. In this equation, we subtract 4 from b, then multiply the result by 7 to get 84. We isolate b by undoing these steps in reverse order.

Divide by 7	Add 4
$\frac{7(b-4)}{7}=\frac{84}{7}$ $b-4=12$	$b-4=12$ $+4\quad +4$ $b=16$

So, $b=\mathbf{16}$.

We check our answer: $7(b-4)=7(16-4)=7(12)=84$. ✓

121. In this equation, we add 3 to m, then divide the result by -6 to get -11. We isolate m by undoing these steps in reverse order.

Multiply by -6	Subtract 3
$\frac{m+3}{-6}\cdot(-6)=-11\cdot(-6)$ $m+3=66$	$m+3=66$ $-3\quad -3$ $m=63$

So, $m=\mathbf{63}$.

We check our answer: $\frac{m+3}{-6}=\frac{63+3}{-6}=\frac{66}{-6}=-11$. ✓

122. In this equation, we multiply n by 13, then add -7 to the result to get 58. We isolate n by undoing these steps in reverse order.

To undo adding -7, we subtract -7. Subtracting -7 is the same as adding 7. So, to undo adding -7, we add 7.

Add 7	Divide by 13
$-7+13n=58$ $+7\quad\quad +7$ $13n=65$	$\frac{13n}{13}=\frac{65}{13}$ $n=5$

So, $n=\mathbf{5}$.

We check our answer: $-7+13n=-7+13(5)=-7+65=58$. ✓

123. In this equation, we add 5 to r, then multiply the result by 4 to get 72. We isolate r by undoing these steps in reverse order.

Divide by 4	Subtract 5
$\frac{72}{4}=\frac{4(r+5)}{4}$ $18=r+5$	$18=r+5$ $-5\quad -5$ $13=r$

So, $r=\mathbf{13}$.

We check our answer: $4(r+5)=4(13+5)=4(18)=72$. ✓

124. In this equation, we divide v by 15, then add 19 to the result to get 49. We isolate v by undoing these steps in reverse order.

Subtract 19	Multiply by 15
$19+\frac{v}{15}=49$ $-19\quad -19$ $\frac{v}{15}=30$	$\frac{v}{15}\cdot 15=30\cdot 15$ $v=450$

So, $v=\mathbf{450}$.

We check our answer: $19+\frac{v}{15}=19+\frac{450}{15}=19+30=49$. ✓

125. In this equation, we multiply z by -3, then add 8 to the result to get 2. We isolate z by undoing these steps in reverse order.

Subtract 8	Divide by -3
$-3z+8=2$ $-8\quad -8$ $-3z=-6$	$\frac{-3z}{-3}=\frac{-6}{-3}$ $z=2$

So, $z=\mathbf{2}$.

We check our answer: $-3z+8=-3(2)+8=-6+8=2$. ✓

126. In this equation, we subtract 9 from k, then divide the result by -3 to get 13. We isolate k by undoing these steps in reverse order.

Multiply by -3	Add 9
$\frac{k-9}{-3}\cdot(-3)=13\cdot(-3)$ $k-9=-39$	$k-9=-39$ $+9\quad +9$ $k=-30$

So, $k=\mathbf{-30}$.

We check our answer: $\frac{k-9}{-3}=\frac{-30-9}{-3}=\frac{-39}{-3}=13$. ✓

127. We first simplify the right side of the equation. Since $2x-2+4x$ simplifies to $6x-2$, the equation simplifies to

$$16=6x-2.$$

Then, adding 2 to both sides of the equation gives $18=6x$. Dividing both sides by 6, we have $\mathbf{3}=x$.

We check our answer:
$2x-2+4x=2(3)-2+4(3)=6-2+12=16$. ✓

For each problem below, we can check our work by plugging our answer into the original equation.

128. We first simplify the left side of the equation. Since $3-4y+5+11y$ simplifies to $7y+8$, the equation simplifies to

$$7y+8=85.$$

Subtracting 8 from both sides gives $7y=77$. Dividing both sides by 7, we have $y=\mathbf{11}$.

129. We first simplify the left side of the equation. Distributing, we have $2n+3(n+6)=2n+3(n)+3(6)=5n+18$. So, the equation simplifies to

$$5n+18=43.$$

Subtracting 18 from both sides gives $5n=25$. Dividing both sides by 5, we have $n=\mathbf{5}$.

130. We first get all terms with p on one side of the equation. We can eliminate $6p$ from the right side of the equation by subtracting $6p$ from both sides.

$$\begin{array}{r} 9p = 6p+21 \\ -6p \quad -6p \\ \hline 3p = 21 \end{array}$$

Then, dividing both sides of $3p=21$ by 3, we have $p=\mathbf{7}$.

131. We first get all terms with r on one side of the equation. We can undo subtracting $4r$ on the left side of the equation by adding $4r$ to both sides.

$$\begin{array}{r} 99-4r = 7r \\ +4r \quad +4r \\ \hline 99 = 11r \end{array}$$

Then, dividing both sides of $99=11r$ by 11, we have $\mathbf{9}=r$.

132. We first get all terms with t on one side of the equation. We can eliminate $8t$ on the left side of the equation by subtracting $8t$ from both sides.

$$\begin{array}{r} 8t+13=20t-35 \\ -8t \qquad -8t \qquad \\ \hline 13=12t-35 \end{array}$$

Then, adding 35 to both sides of $13=12t-35$ gives $48=12t$. Dividing both sides by 12, we have $\mathbf{4}=t$.

133. We first get all terms with a on one side of the equation. We can eliminate $24a$ on the left side of the equation by subtracting $24a$ from both sides. Since $24a-31-24a$ simplifies to -31, we have:

$$\begin{array}{r} 24a-31 = 30a+11 \\ -24a \qquad -24a \qquad \\ \hline -31 = 6a+11 \end{array}$$

Then, subtracting 11 from both sides of $-31=6a+11$ gives $-42=6a$. Dividing both sides by 6, we have $\mathbf{-7}=a$.

— or —

We can eliminate $30a$ on the right side of the equation by subtracting $30a$ from both sides.

$$\begin{array}{r} 24a-31 = 30a+11 \\ -30a \qquad -30a \qquad \\ \hline -6a-31 = 11 \end{array}$$

Then, adding 31 to both sides of $-6a-31=11$ gives $-6a=42$. Dividing both sides by -6, we have $a=\mathbf{-7}$.

134. We first simplify the left side of the equation. Distributing, we have $6(c+5)-2c=6(c)+6(5)-2c=4c+30$. So, the equation simplifies to

$$4c+30=56-9c.$$

We can undo subtracting $9c$ on the right side of the equation by adding $9c$ to both sides.

$$\begin{array}{r} 4c+30=56-9c \\ +9c \qquad\quad +9c \\ \hline 13c+30=56 \quad \end{array}$$

Then, subtracting 30 from both sides of $13c+30=56$ gives $13c=26$. Dividing both sides by 13, we have $c=\mathbf{2}$.

135. The top circle is the sum of the two circles below it. So,

$$a+16=84.$$

Subtracting 16 from both sides of this equation gives $a=\mathbf{68}$.

136. The top circle is the sum of the two circles below it. So,

$$2b+b=81.$$

Combining like terms on the left side gives

$$3b=81.$$

Dividing both sides of this equation by 3, we have $b=\mathbf{27}$.

137. The blank circle is the sum of the two circles below it. So, we label it $g+1$.

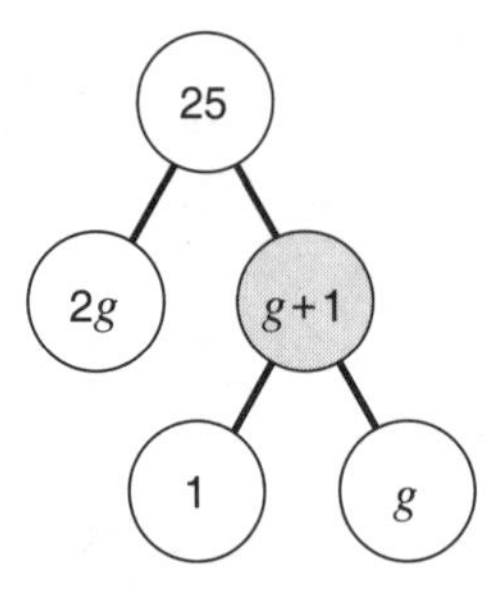

Then, we use the top three circles to write an equation:

$$2g+(g+1)=25.$$

Combining like terms, the equation simplifies to

$$3g+1=25.$$

Subtracting 1 from both sides gives $3g=24$. Dividing both sides by 3, we have $g=\mathbf{8}$.

138. The blank circle is the sum of the two circles below it. So, we label it $x+(-6)=x-6$.

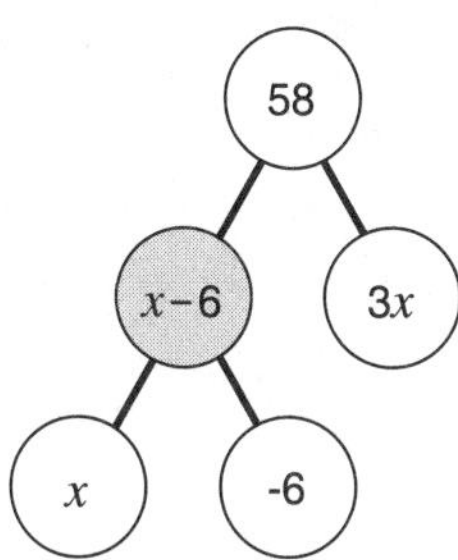

Then, we use the top three circles to write an equation:

$$(x-6)+3x=58.$$

Combining like terms, the equation simplifies to

$$4x-6=58.$$

Adding 6 to both sides gives $4x=64$. Dividing both sides by 4, we have $x=\mathbf{16}$.

139. Each circle is labeled with the sum of the numbers below it. So, we label the left blank circle $s+s=2s$ and the right blank circle $s+(-2)=s-2$.

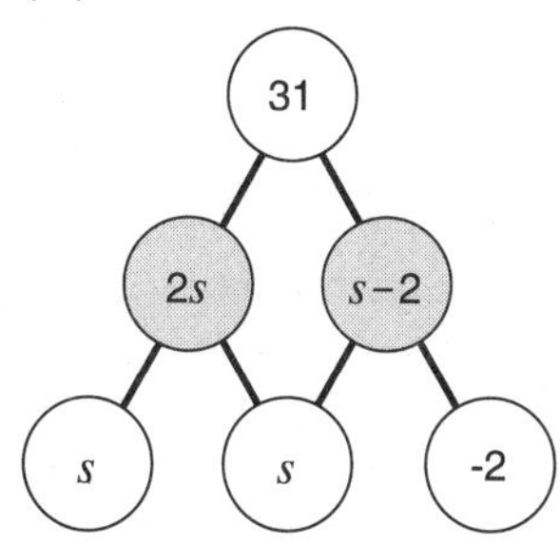

Then, we use the top three circles to write an equation:

$$2s+(s-2)=31.$$

Combining like terms, the equation simplifies to

$$3s-2=31.$$

Adding 2 to both sides gives $3s=33$. Dividing both sides by 3, we have $s=\mathbf{11}$.

140. We label the left blank circle $4d+6$ and the right blank circle $-2d+6$.

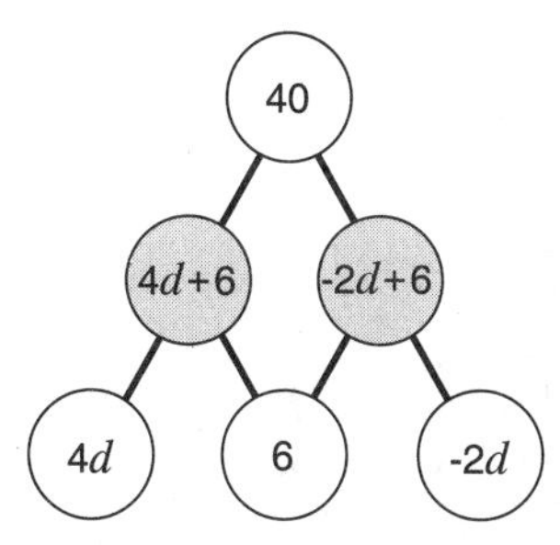

Then, we use the top three circles to write an equation:

$$(4d+6)+(-2d+6)=40.$$

Combining like terms, the equation simplifies to

$$2d+12=40.$$

Subtracting 12 from both sides gives $2d=28$. Dividing both sides by 2, we have $d=\mathbf{14}$.

141. We label the left blank circle $2n+4$ and the right blank circle $5n+4$.

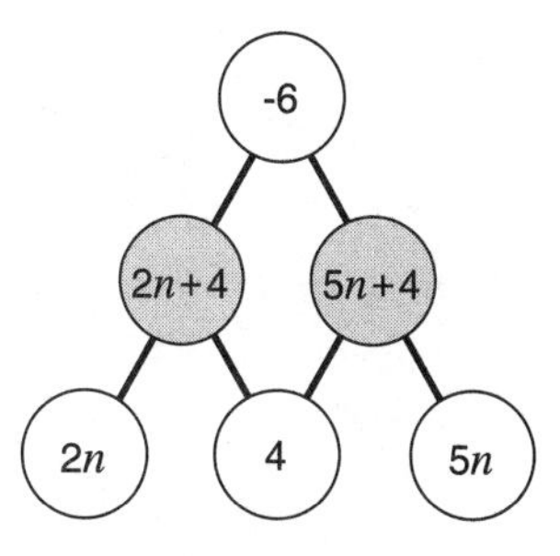

Then, we use the top three circles to write an equation:

$$(2n+4)+(5n+4)=-6.$$

Combining like terms, the equation simplifies to

$$7n+8=-6.$$

Subtracting 8 from both sides gives $7n=-14$. Dividing both sides by 7, we have $n=\mathbf{-2}$.

142. We label the left blank circle $2p-8$ and the right blank circle $3p+3$.

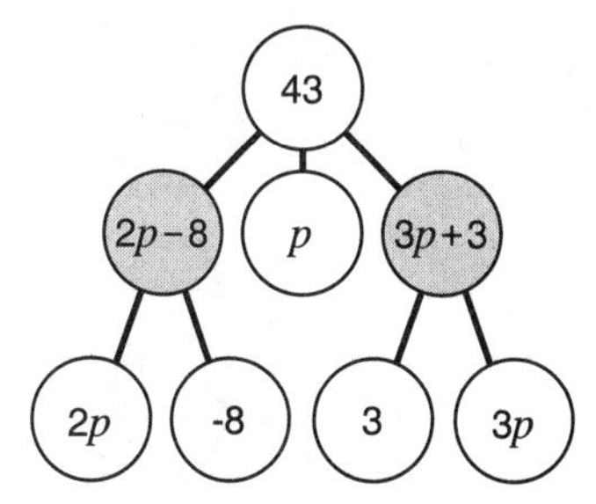

Then, we use the top four circles to write an equation:

$$(2p-8)+p+(3p+3)=43.$$

Combining like terms, the equation simplifies to

$$6p-5=43.$$

Adding 5 to both sides gives $6p=48$. Dividing both sides by 6, we have $p=\mathbf{8}$.

143. We label the left blank circle $k+5$, the middle blank circle $4k$, and the right blank circle $3k+7$.

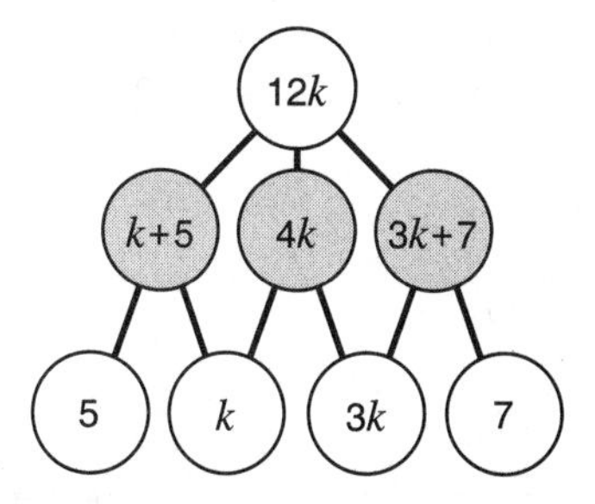

Then, we use the top four circles to write an equation:

$$(k+5)+4k+(3k+7)=12k.$$

Combining like terms, the equation simplifies to

$$8k+12=12k.$$

We can eliminate $8k$ on the left side of the equation by subtracting $8k$ from both sides.

$$\begin{array}{r} 8k+12=12k \\ -8k \qquad -8k \\ \hline 12=4k \end{array}$$

Then, dividing both sides of $12=4k$ by 4, we have $\mathbf{3}=k$.

144. We first label the blank circle whose two circles beneath it are already filled in.

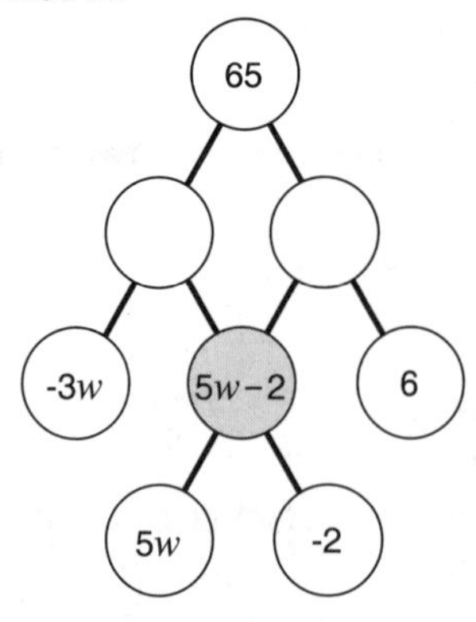

Next, we label the remaining blank circles as shown.

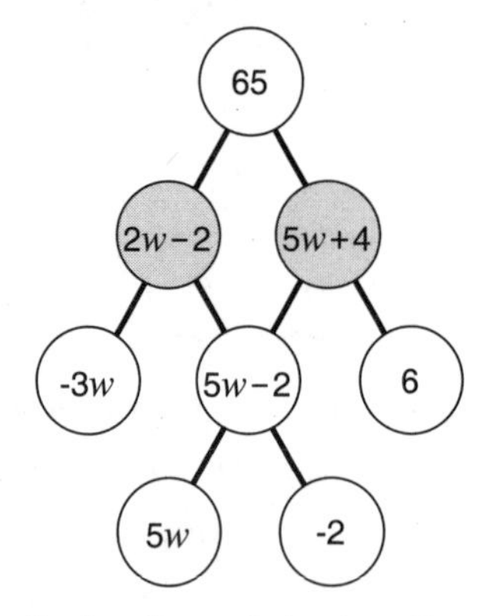

Then, we use the top three circles to write an equation:

$$(2w-2)+(5w+4)=65.$$

Combining like terms, the equation simplifies to

$$7w+2=65.$$

Subtracting 2 from both sides gives $7w=63$. Dividing both sides by 7, we have $w=\mathbf{9}$.

145. Step 1: Step 2:

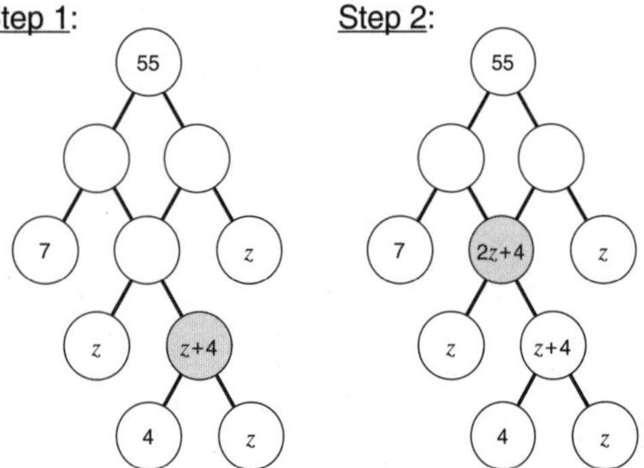

Step 3:

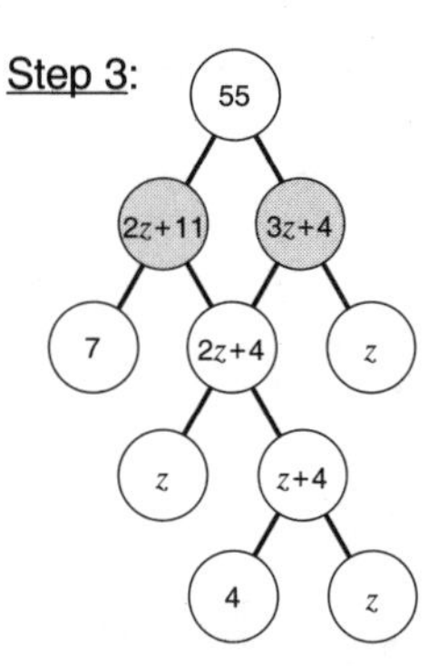

We use the top three circles to write an equation:

$$(2z+11)+(3z+4)=55.$$

Combining like terms, the equation simplifies to

$$5z+15=55.$$

Subtracting 15 from both sides gives $5z=40$. Dividing both sides by 5, we have $z=\mathbf{8}$.

146. Step 1: Step 2:

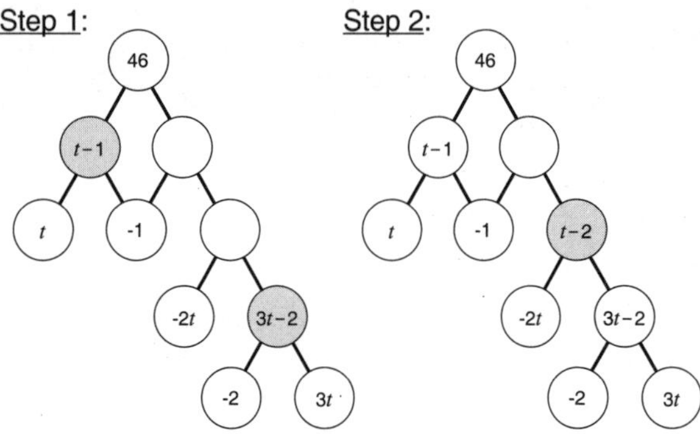

Step 3:

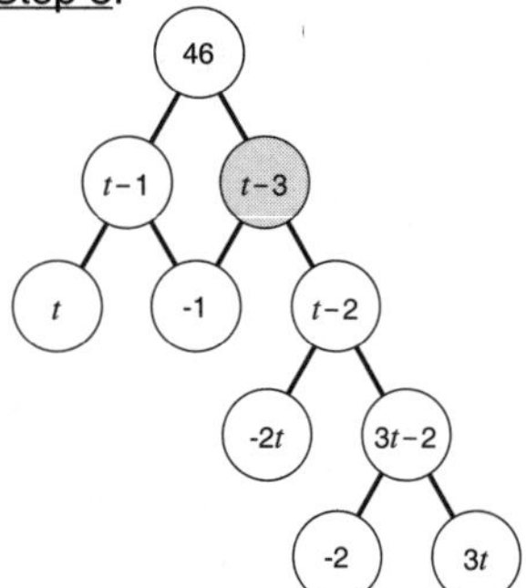

We use the top three circles to write an equation:

$$(t-1)+(t-3)=46.$$

Combining like terms, the equation simplifies to

$$2t-4=46.$$

Adding 4 to both sides gives $2t=50$. Dividing both sides by 2, we have $t=\mathbf{25}$.

147. Step 1: Step 2:

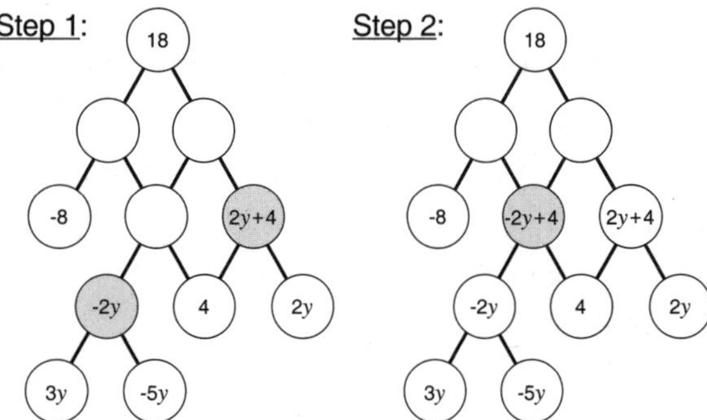

Step 3:

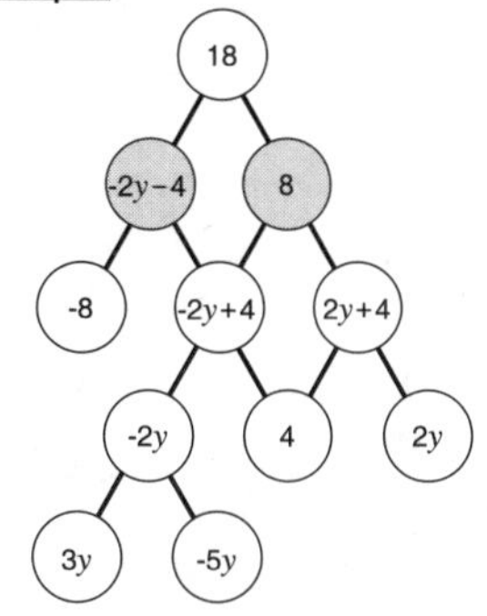

We use the top three circles to write an equation:

$$(-2y-4)+8=18.$$

Combining like terms, the equation simplifies to

$$-2y+4=18.$$

Subtracting 4 from both sides gives $-2y=14$. Dividing both sides by -2, we have $y=\mathbf{-7}$.

148. Step 1: Step 2:

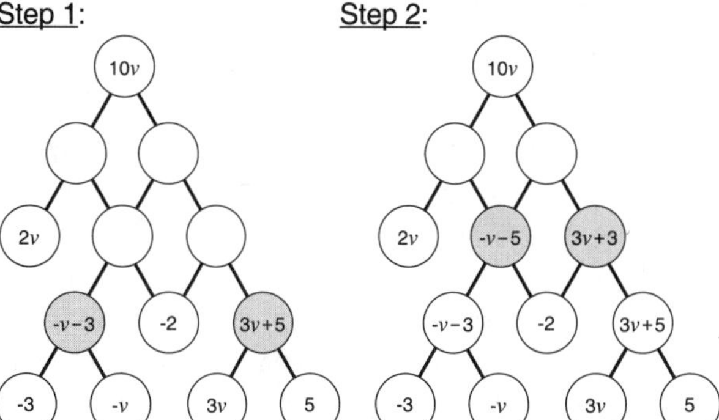

Step 3:

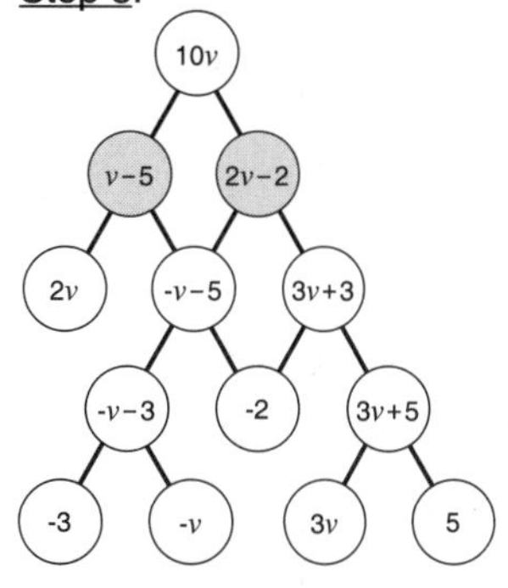

We use the top three circles to write an equation:

$$(v-5)+(2v-2)=10v.$$

Combining like terms, the equation simplifies to

$$3v-7=10v.$$

We can eliminate $3v$ on the left side of the equation by subtracting $3v$ from both sides. Since $3v-7-3v$ simplifies to -7, we have:

$$\begin{array}{r} 3v-7=10v \\ -3v \quad\;\; -3v \\ \hline -7=7v \end{array}$$

Then, dividing both sides of $-7=7v$ by 7, we have $v=\mathbf{-1}$.

149. The given rectangle has width y inches and height 12 inches. So, its perimeter is $y+y+12+12=2y+24$. We are given that the perimeter is 36 inches, so we write an equation:

$$2y+24=36.$$

Subtracting 24 from both sides gives $2y=12$. Dividing both sides by 2, we have $y=\mathbf{6}$ in.

We check our work: $6+6+12+12=36$. ✓

150. The sum of the side lengths of the triangle is $(x-7)+(x-7)+x=3x-14$ inches. We are given that the perimeter is 52 inches, so we write an equation:

$$3x-14=52.$$

Adding 14 to both sides gives $3x=66$. Dividing both sides by 3, we have $x=\mathbf{22}$ inches.

We check our work:
$(22-7)+(22-7)+22=15+15+22=52$. ✓

151. The height of the rectangle is 7 inches greater than its width. So, if the width is w, the height is $w+7$. The perimeter of a rectangle with width w and height $w+7$ is

$$w+w+(w+7)+(w+7)=4w+14.$$

We are given that the perimeter of the rectangle is 46 inches, so we write an equation:

$$4w+14=46.$$

Subtracting 14 from both sides gives $4w=32$. Dividing both sides by 4, we have $w=\mathbf{8}$ inches.

We check our work: If the rectangle is 8 inches wide, then its height is $8+7=15$ inches, and its perimeter is $8+8+15+15=46$ inches. ✓

152. The perimeter of a rectangle with width $k+5$ inches and height $k+2$ inches is $(k+5)+(k+5)+(k+2)+(k+2)$ inches. Combining like terms, this simplifies to $4k+14$ inches.

We are given that the perimeter is $6k$ inches, so we write an equation:

$$4k+14=6k.$$

We can eliminate $4k$ on the left side of the equation by subtracting $4k$ from both sides.

$$\begin{array}{r} 4k+14=6k \\ -4k \qquad -4k \\ \hline 14=2k \end{array}$$

Then, dividing both sides of $14=2k$ by 2, we have $7=k$.

So, the rectangle has width $k+5=7+5=12$ inches, and height $k+2=7+2=9$ inches.

Therefore, the area of the rectangle is $12\cdot 9=\mathbf{108}$ **sq in**.

153. We draw triangle ABC. Since we are looking for the length of side AB, we label this length x.

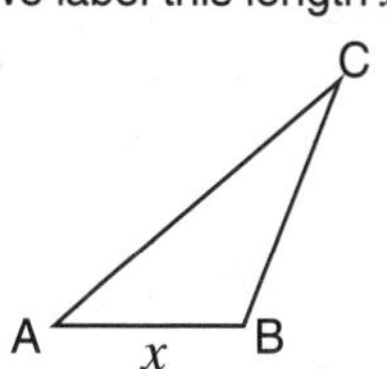

We are told that side BC is 3 inches longer than side AB, so side BC has length $x+3$.

We are also told that side AC is 5 inches longer than side BC. So, side AC has length $(x+3)+5=x+8$.

We label these lengths on our diagram.

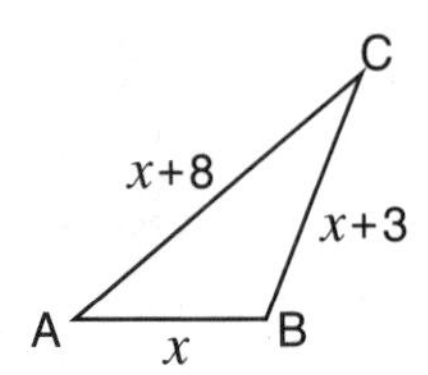

The perimeter of this triangle is 35 inches. So,

$$x+(x+3)+(x+8)=35.$$

Combining like terms, this simplifies to

$$3x+11=35.$$

Subtracting 11 from both sides gives $3x=24$. Dividing both sides by 3, we have $x=8$.

Since x represents the length of side AB, the length of side AB is **8 inches**.

We check our work: If side AB is 8 inches long, then side BC is $8+3=11$ inches long and side AC is $11+5=16$ inches long. $8+11+16=35$. ✓

154. We draw a diagram. When the octagons are joined along 4-foot sides, the resulting shape has two sides of length 4 and twelve sides of length s.

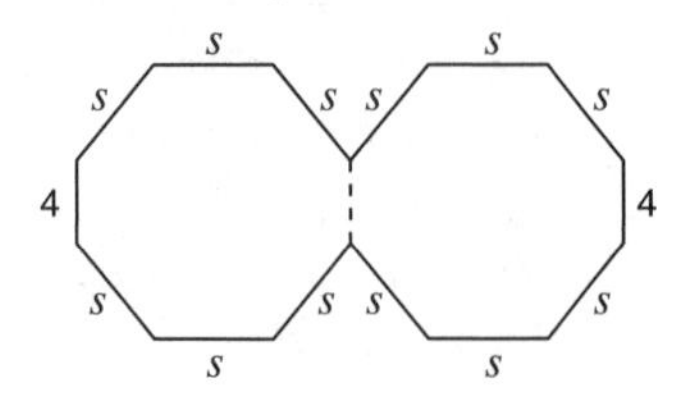

So, the perimeter of the resulting shape is $12s+2(4)=12s+8$. We are given that this perimeter is 80 ft, so we write an equation:

$$12s+8=80.$$

Subtracting 8 from both sides gives $12s=72$. Dividing both sides by 12, we have $s=\mathbf{6}$.

155. We use n to represent the unknown number. The product of the number and seven is $7n$. Three more than this product is $7n+3$. This expression is equal to 59, so we have

$$7n+3=59.$$

Subtracting 3 from both sides gives $7n=56$. Dividing both sides by 7, we have $n=8$. So, the unknown number is **8**.

We check our work: $7(8)+3=56+3=59$. ✓

156. We use n to represent the unknown number. Six times the number is $6n$, so five less than 6 times the number is $6n-5$. This expression is equal to 37, so we have

$$6n-5=37.$$

Adding 5 to both sides gives $6n=42$. Dividing both sides by 6, we have $n=7$. So, the unknown number is **7**.

We check our work: $6(7)-5=42-5=37$. ✓

157. We use m to represent the age of Borg's monkaroo. Subtracting 7 from the age of the monkaroo gives $m-7$. Multiplying this by 5 gives $5(m-7)$. This expression is equal to 20, so we have

$$5(m-7)=20.$$

Dividing both sides by 5 gives $m-7=4$. Adding 7 to both sides, we have $m=11$. So, Borg's monkaroo is **11** years old.

We check our work: $5(11-7)=5(4)=20$. ✓

158. We use c to represent the number of candies Grogg originally pulled from his pocket. After Grogg eats 11 pieces, there are $c-11$ candies remaining. Grogg divides these remaining candies among 7 friends, so each friend gets $\frac{c-11}{7}$ candies.

We are told that each friend received 9 candies. So, we write an equation:

$$\frac{c-11}{7}=9.$$

Multiplying both sides by 7 gives $c-11=63$. Adding 11 to both sides, we have $c=74$. So, Grogg originally had **74** candies in his pocket.

We check our work: If Grogg has 74 candies in his pocket, then eats 11 of them, he is left with $74-11=63$ candies. When he divides these among 7 friends, each friend gets $63\div7=9$ candies. ✓

159. We let j represent Jorple's age in years. Since Swillard's age is twice Jorple's age, Swillard's age is $2j$.

The sum of the two ages is 42. So, we have

$$j+2j=42.$$

Combining like terms, the equation simplifies to

$$3j=42.$$

Dividing both sides by 3, we have $j=14$.

Since j represents Jorple's age, Jorple is **14** years old.

We check our work: If Jorple is 14, then Swillard is $2(14)=28$. The sum of their ages is then $14+28=42$. ✓

160. We let g represent the cost of a grapefruit in dollars. Since a slingshot costs 2 dollars more than 6 times the cost of a grapefruit, the slingshot costs $6g+2$ dollars.

The combined cost of a slingshot and a grapefruit is 23 dollars. So, we write an equation:

$$g+(6g+2)=23.$$

Combining like terms, the equation simplifies to

$$7g+2=23.$$

Subtracting 2 from both sides gives $7g=21$. Dividing both sides by 7, we have $g=3$.

Since g represents the cost of a grapefruit, the grapefruit costs **3** dollars.

We check our work: If a grapefruit is 3 dollars, then a slingshot is is $6(3)+2=20$ dollars. The combined cost of a grapefruit and a slingshot is then $3+20=23$ dollars. ✓

161. We let t represent Todd's weight in pounds. Rod weighs 7 pounds less than 3 times Todd's weight, so Rod weighs $3t-7$ pounds.

The combined weight of both brothers is 213 pounds. So, we write an equation:

$$t+(3t-7)=213.$$

Combining like terms, the equation simplifies to

$$4t-7=213.$$

Adding 7 to both sides gives $4t=220$. Dividing both sides by 4, we have $t=55$.

Since t represents Todd's weight, Todd weighs 55 pounds. Therefore, Rod weighs $3(55)-7=165-7=$ **158** pounds.

We check our work: $55+158=213$. ✓

162. Since we are given Fleet's jogging distance in terms of Wheezy's, we let w represent the number of miles that Wheezy jogged and write an expression for Fleet's distance. Fleet jogged 5 less than 4 times as many miles as Wheezy. So, Fleet jogged $4w-5$ miles.

We are also told that Fleet jogged 22 more miles than Wheezy. So, Fleet jogged $w+22$ miles.

We now have two expressions representing the number of miles Fleet jogged: $w+22$ and $4w-5$. Since each expression represents the same quantity, they must be equal. So, we write an equation:

$$w+22=4w-5.$$

We can eliminate w on the left side of the equation by subtracting w from both sides.

$$\begin{array}{r} w+22=4w-5 \\ -w \qquad -w \\ \hline 22=3w-5 \end{array}$$

Then, adding 5 to both sides of $22=3w-5$ gives $27=3w$. Dividing both sides by 3, we have $w=9$.

So, Wheezy jogged $w=9$ miles. To find the number of miles Fleet jogged, we can plug $w=9$ into either of $w+22$ or $4w-5$:

$w+22=9+22=31$.
$4w-5=4(9)-5=36-5=31$.

So, Fleet jogged **31** miles.

— or —

We let w represent the number of miles that Wheezy jogged. Fleet jogged 5 less than 4 times the number of miles Wheezy jogged, so Fleet jogged $4w-5$ miles.

We are told that Fleet jogged 22 more miles than Wheezy. So,

(# of miles Fleet jogged) − (# of miles Wheezy jogged) = 22.

So, we write an equation:

$$(4w-5)-w=22.$$

Combining like terms, this equation simplifies to

$$3w-5=22.$$

Adding 5 to both sides gives $3w=27$. Dividing both sides by 3, we have $w=9$. We plug this into our expression for the number of miles Fleet jogged:

$4w-5=4(9)-5=36-5=31$.

So, Fleet jogged **31** miles.

We check our work: If Wheezy jogged 9 miles, then Fleet jogged $4(9)-5=31$ miles. So, Fleet jogged $31-9=22$ more miles than Wheezy. ✓

EXPRESSIONS & EQUATIONS
Consecutive Numbers 97

163. We let n represent the smallest of the five integers.

Since the integers are consecutive, the second integer is one more than n which is $n+1$. Then, the third integer is one more than the second: $(n+1)+1=n+2$.

Similarly, the fourth integer is $(n+2)+1=n+3$, and the fifth integer is $(n+3)+1=n+4$.

The sum of all five integers is 45. So,

$$n+(n+1)+(n+2)+(n+3)+(n+4)=45.$$

Rearranging terms, we have

$$(n+n+n+n+n)+(1+2+3+4)=45.$$

This equation simplifies to

$$5n+10=45.$$

Subtracting 10 from both sides gives $5n=35$. Dividing both sides by 5, we have $n=7$.

Since n represents the smallest of the five integers, the smallest integer is **7**.

We check our work: $7+8+9+10+11=45$. ✓

164. We let n represent the smallest of the nine integers in Norbert's list. So, the next eight integers are $n+1$, $n+2$, $n+3$, $n+4$, $n+5$, $n+6$, $n+7$, and $n+8$.

The sum of all nine integers is 108. So, we have

$$n+(n+1)+(n+2)+\cdots+(n+7)+(n+8)=108.$$

Combining like terms, this equation simplifies to

$$9n+36=108.$$

Subtracting 36 from both sides gives $9n=72$. Dividing both sides by 9, we have $n=8$.

So, the smallest of the nine integers is 8. Our expression for the fifth integer is $n+4$, so the fifth integer in Norbert's list is $8+4=$ **12**.

— or —

We let n represent the *fifth* integer. Our expressions for the nine consecutive integers are then

$$n-4,\ n-3,\ n-2,\ n-1,\ n,\ n+1,\ n+2,\ n+3,\ n+4.$$

The sum of these integers is 108. So, we have the equation

$$(n-4)+(n-3)+\cdots+(n+3)+(n+4)=108.$$

When we combine like terms, the number terms can be

grouped into pairs with sum zero: $-4+4=0$, $-3+3=0$, $-2+2=0$, and $-1+1=0$. So, all of the number terms sum to 0, and the equation simplifies to

$$9n=108.$$

Dividing both sides by 9 gives $n=12$. Since n represents the fifth integer, the fifth integer in Norbert's list is **12**.

We check our work: If the fifth integer is 12, then the nine numbers are the integers from 8 to 16, inclusive. The sum of these integers is 108. ✓

165. The first number Grogg says is x. Since Grogg is skip-counting by 7, the next number Grogg says is $x+7$, followed by $(x+7)+7=x+14$, and finally $(x+14)+7=x+21$.

The sum of these four numbers is 58. So,

$$x+(x+7)+(x+14)+(x+21)=58.$$

Combining like terms, we have

$$4x+42=58.$$

Subtracting 42 from both sides gives $4x=16$. Dividing both sides by 4, we have $x=\mathbf{4}$.

We check our work: If Grogg starts at 4 and skip-counts by 7, the first four numbers he says will be 4, 11, 18, and 25. The sum of these four numbers is $4+11+18+25=58$. ✓

166. The first number Alex says is -10. Since Alex is skip-counting by y, the second number he says is $-10+y$, and the third number he says is $(-10+y)+y=-10+2y$.

The sum of these numbers is 3. So,

$$-10+(-10+y)+(-10+2y)=3.$$

Combining like terms, we have:

$$3y-30=3.$$

Adding 30 to both sides gives $3y=33$. Dividing both sides by 3, we have $y=\mathbf{11}$.

We check our work: If Alex starts at -10 and skip-counts by 11, the first three numbers he says will be -10, 1, and 12. The sum of these three numbers is $-10+1+12=3$. ✓

EXPRESSIONS & EQUATIONS — Think About It 98

167. To isolate x, we divide both sides of $7x=3{,}122$ by 7. This gives $x=446$.

So, $21x$ is $21(446)=\mathbf{9{,}366}$.

— or —

Rather than solving for x and then multiplying by 21, we notice that $3\cdot 7x=21x$.

Since $7x=3{,}122$, to find $21x$ we can multiply 3,122 by 3:

$$21x=3\cdot 7x=3\cdot 3{,}122=\mathbf{9{,}366}.$$

168. We note that $-17=17(-1)$. Therefore, if $-17=17x^{101}$, then $x^{101}=-1$.

We know that -1 raised to any odd power is -1. Since 101 is odd, $(-1)^{101}=-1$. Therefore, $x=\mathbf{-1}$.

We check our work: $17(-1)^{101}=17(-1)=-17$. ✓

169. We notice that $4b$ is a term on both sides of the equation. If we subtract $4b$ from both sides, we eliminate all terms with variable b from the equation.

$$\begin{array}{r}3a+4b+5c+6=6a+5c+4b+3\\ -4b \qquad\qquad\qquad -4b\\ \hline 3a+5c+6=6a+5c+3\end{array}$$

Similarly, we notice that $5c$ is a term on both sides of the equation. So, subtracting $5c$ from both sides eliminates all terms with variable c from the equation.

$$\begin{array}{r}3a+5c+6=6a+5c+3\\ -5c \qquad\quad -5c\\ \hline 3a+6=6a+3\end{array}$$

We are left with an equation that contains only the variable a. We can eliminate $3a$ from the left side of the equation by subtracting $3a$ from both sides.

$$\begin{array}{r}3a+6=6a+3\\ -3a \quad\; -3a\\ \hline 6=3a+3\end{array}$$

Subtracting 3 from both sides of $6=3a+3$ gives $3=3a$. Dividing both sides by 3, we have $\mathbf{1}=a$.

170. Since $3(x+y)=3x+3y$, we can write $3x+3y-5$ as $3(x+y)-5$. Since $x+y=7$, we replace $x+y$ with 7 in the expression:

$$3(x+y)-5=3(7)-5=21-5=\mathbf{16}.$$

171. Rather than solving for x and dealing with complicated multiplication and division, we look for a clever approach.

We notice that the difference between $29x$ and $17x$ is $29x-17x=12x$. Since $29x=6{,}844$ and $17x=4{,}012$, we have $12x=29x-17x=6{,}844-4{,}012=\mathbf{2{,}832}$.

EXPRESSIONS & EQUATIONS — Challenge Problems 99-101

172. Factoring n from each term on the left side of the equation, we have

$$(3+6+9+12+15+18+21+24+27)n=270.$$

The sum of the numbers within the parentheses is

$$\begin{aligned}&(3+27)+(6+24)+(9+21)+(12+18)+15\\ &=30+30+30+30+15\\ &=135.\end{aligned}$$

So, the equation simplifies to

$$135n=270.$$

Since $135(2)=270$, we have $n=\mathbf{2}$.

— or —

Factoring $3n$ from each term on the left side of the

equation, we have

$$3n(1+2+3+4+5+6+7+8+9)=270.$$

We divide both sides of the equation by 3 to get:

$$n(1+2+3+4+5+6+7+8+9)=90.$$

Since $1+2+3+4+5+6+7+8+9=45$, we have $45n=90$. Since $45(2)=90$, we have $n=\mathbf{2}$.

173. The two numbers with absolute value 12 are 12 and -12. Therefore, if $|6c-30|=12$, then $6c-30=12$ *or* $6c-30=-12$. We solve both equations.

$\underline{6c-30=12}$: Adding 30 to both sides of this equation gives $6c=42$. Dividing both sides by 6, we have $c=7$.

$\underline{6c-30=-12}$: Adding 30 to both sides of this equation gives $6c=18$. Dividing both sides by 6, we have $c=3$.

So, the values of c that make this equation true are **3 and 7**.

We check our work:
$|6(3)-30|=|18-30|=|-12|=12.$ ✓
$|6(7)-30|=|42-30|=|12|=12.$ ✓

174. The two numbers we can square to get 49 are 7 and -7.

So, the equation $(x-3)^2=49$ is true if $x-3$ is equal to either 7 or -7. We consider both cases:

$\underline{x-3=7}$: In this case, we add 3 to both sides to get $x=10$. This is Grogg's solution. We check that it works: $(10-3)^2=7^2=49.$ ✓

$\underline{x-3=-7}$: In this case, we add 3 to both sides to get $x=-4$. We check that this solution works: $(-4-3)^2=(-7)^2=49.$ ✓

So, the solution Winnie found is $\boldsymbol{x=-4}$.

175. Since triangle XYZ is isosceles, two of its sides have equal length. There are two ways we can draw triangle XYZ: one where side YZ is congruent to the shorter side XZ, and one where side YZ is congruent to the longer side XY.

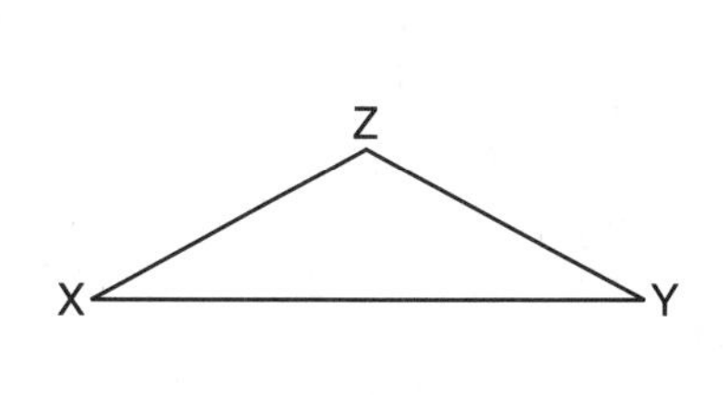

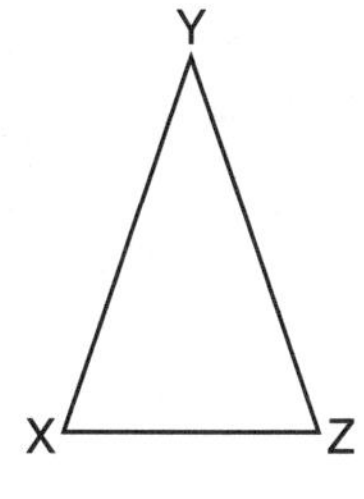

We consider each case.

<u>Side YZ is short</u>:
We let s represent the length of short sides XZ and YZ. Then, $s+6$ represents the length of side XY.

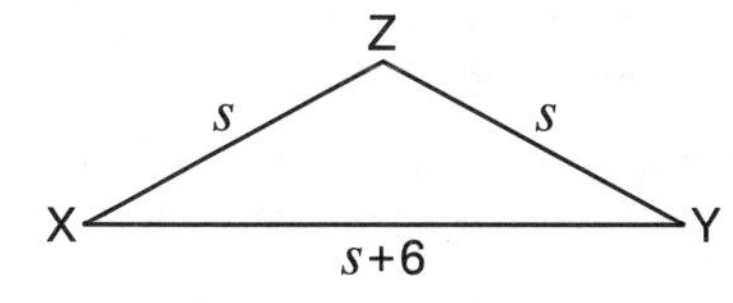

The perimeter of the triangle is 21 cm. Therefore,

$$s+s+(s+6)=21.$$

Combining like terms, we have

$$3s+6=21.$$

Subtracting 6 from both sides, then dividing by 3, we have $s=5$ cm. The three sides of the triangle are then YZ = 5 cm, XZ = 5 cm, and XY = 5+6 = 11 cm.

However, the combined lengths of the two shortest sides of a triangle must be greater than the length of the longest side. Since $5+5<11$, this triangle is impossible! *Review the Triangle Inequality in the Shapes chapter of Guide 3A.*

<u>Side YZ is long</u>:

We let s represent the length of short side XZ. Then, $s+6$ represents the length of long sides XY and YZ.

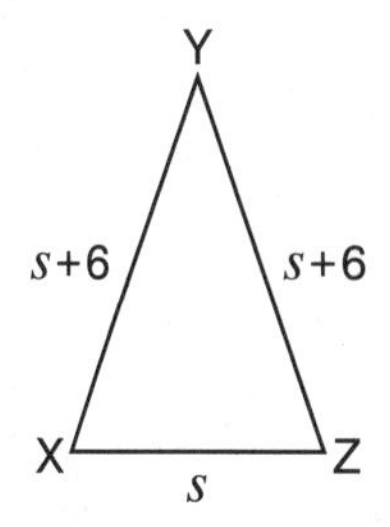

The perimeter of the triangle is 21 cm. Therefore,

$$s+(s+6)+(s+6)=21.$$

Combining like terms, we have

$$3s+12=21.$$

Subtracting 12 from both sides, then dividing by 3, we have $s=3$ cm. The three sides of the triangle are then XZ = 3 cm, XY = 3+6 = 9 cm, and YZ = 3+6 = 9 cm.

These three side lengths can be used to make a triangle. Therefore, the length of side YZ is **9 cm**.

176. When Grogg cuts the rectangle along the dashed line, the resulting shapes have the same combined width as the original rectangle, plus two extra sides of length h.

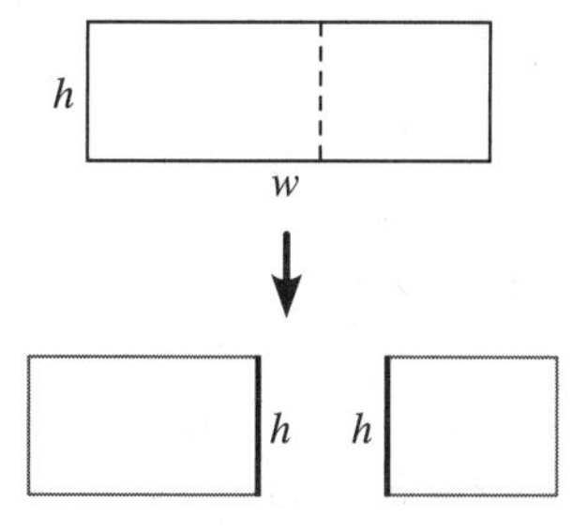

So, the only change in perimeter comes from the two extra sides of length h. The combined perimeter of the two new rectangles is $68-54=14$ inches more than the perimeter of the original rectangle, so $2h=14$. Therefore, $h=7$.

The perimeter of the original rectangle is given by $2w+2h$. Replacing h with 7 gives $2w+2(7)=2w+14$. We are told that this perimeter is 54, so we have the equation

$$2w+14=54.$$

Subtracting 14 from both sides gives $2w=40$. Dividing both sides by 2, we have $w=\mathbf{20}$ inches.

177. We draw the rectangular prism, labeling the missing height h.

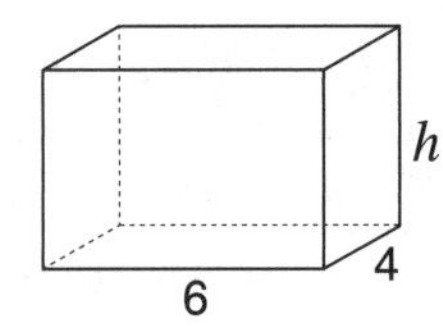

To compute the surface area of a shape, we add the areas of its faces. A 4-by-6-by-h-ft rectangular prism has two faces with area $4 \cdot 6 = 24$, two faces with area $4h$, and two faces with area $6h$.

Since the surface area of the prism is 138 sq ft, we have the equation

$$24+24+4h+4h+6h+6h=138.$$

Combining like terms, this simplifies to

$$20h+48=138.$$

Subtracting 48 from both sides gives $20h=90$. Dividing both sides by 20, we have $h=\frac{90}{20}=\frac{9}{2}=\mathbf{4\frac{1}{2}}$ **ft.**

178. We let t be the age of the youngest sibling, Toby. Since Suzie is twice as old as Toby, her age is $2t$. Rosie is 6 years older than Suzie, so Rosie's age is $2t+6$.

The sum of all three ages is 31. So,

$$t+(2t)+(2t+6)=31.$$

Combining like terms, this simplifies to

$$5t+6=31.$$

Subtracting 6 from both sides gives $5t=25$. Dividing both sides by 5, we have $t=5$.

Since t represents Toby's age, Toby is 5 years old. Suzie is then $2(5)=10$ years old, and Rosie is $2(5)+6=16$ years old.

So, **Rosie is 16, Suzie is 10, and Toby is 5**.

179. The first number Lizzie says is n, and the next number she says is $n+7$. So, Lizzie is skip-counting by 7. Therefore, the next number she says, $2n$, is 7 more than $n+7$. This gives the equation

$$2n=(n+7)+7.$$

Simplifying the right side, the equation becomes

$$2n=n+14.$$

Subtracting n from both sides gives $n=14$.

So, Lizzie is skip-counting by 7, starting with 14. The first three numbers she says are 14, 21, and 28. The next number she says will be $28+7=\mathbf{35}$.

180. We let c represent the number of cards each girl started with.

After Anita gives 18 cards to Beth, Anita is left with $c-18$ cards. Beth *gains* 18 cards from the exchange, so Beth has $c+18$ cards.

We are told that Beth now has twice as many cards as Anita. So,

(# of cards Beth has) = 2 • (# of cards Anita has).

We write an equation:

$$(c+18)=2\cdot(c-18).$$

Distributing the 2 on the right side gives:

$$c+18=2c-36.$$

We can eliminate c on the left side of the equation by subtracting c from both sides.

$$\begin{array}{r} c+18=2c-36 \\ \underline{-c \qquad -c\quad} \\ 18=c-36 \end{array}$$

Then, adding 36 to both sides of $18=c-36$ gives $c=54$.

Since c represents the number of cards Anita and Beth each started with, Anita had **54 cards** before giving some away to Beth.

We check our work: If Anita and Beth each start with 54 cards, and Anita gives 18 to Beth, then Anita has $54-18=36$ cards, and Beth has $54+18=72$ cards. $72=2\cdot 36$. ✓

181. We let d represent the number of duo-dobbles in the exhibit. There are seven more tri-dobbles than duo-dobbles, so there are $d+7$ tri-dobbles.

Each duo-dobble has 2 horns. So, the d duo-dobbles have $2d$ horns all together.

Each tri-dobble has 3 horns. So, the $d+7$ tri-dobbles have $3(d+7)$ horns all together.

Since there are 76 horns in the entire dobble exhibit, we write an equation:

$$2d+3(d+7)=76.$$

Distributing the 3, we have

$$2d+3d+21=76.$$

Combining like terms, this simplifies to

$$5d+21=76.$$

Subtracting 21 from both sides gives $5d=55$. Then, dividing both sides by 5, we have $d=11$.

So, there are $d=11$ duo-dobbles, and $d+7=11+7=18$ tri-dobbles. Therefore, the total number of dobbles in the exhibit is $11+18=\mathbf{29}$.

We check our work:
11 duo-dobbles have $11(2)=22$ horns.
18 tri-dobbles have $18(3)=54$ horns.
All together, this is $22+54=76$ horns. ✓

— or —

We let t represent the number of tri-dobbles in the exhibit. There are seven more tri-dobbles than duo-dobbles, so there are $t-7$ duo-dobbles.

Each tri-dobble has 3 horns. So, the t tri-dobbles have $3t$ horns all together.

Each duo-dobble has 2 horns. So, the $t-7$ duo-dobbles have $2(t-7)$ horns all together.

Since there are 76 horns in the entire dobble exhibit, we write an equation:

$$3t+2(t-7)=76.$$

Distributing the 2, we have

$$3t+2t-14=76.$$

Combining like terms, this simplifies to

$$5t-14=76.$$

Adding 14 to both sides gives $5t=90$. Then, dividing both sides by 5, we have $t=18$.

So, there are $t=18$ tri-dobbles, and $t-7=18-7=11$ duo-dobbles. Therefore, the total number of dobbles in the exhibit is $18+11=$ **29**.